Expedition Mars

Springer
London
Berlin
Heidelberg
New York
Hong Kong
Milan
Paris
Tokyo

Martin J. L. Turner

Expedition Mars

Springer

Published in association with
Praxis Publishing
Chichester, UK

PRAXIS

Dr. Martin J. L. Turner
Principal Research Fellow
Department of Physics and Astronomy
University of Leicester
Leicester
UK

SPRINGER–PRAXIS BOOKS IN ASTRONOMY AND SPACE SCIENCES
SUBJECT *ADVISORY EDITOR*: John Mason B.Sc., M.Sc., Ph.D.

ISBN 1-85233-735-4 Springer-Verlag London Berlin Heidelberg New York Hong Kong
Milan Paris Tokyo

British Library Cataloguing-in-Publication Data
Turner, Martin J. L., 1942–
   Expedition Mars. – (Springer–Praxis books in astronomy and
   space sciences)
   1. Space flight to Mars 2. Rocket engines 3. Mars (Planet) –
   Exploration 4. Mars (Planet) – Exploration – Equipment and
   supplies
   I. Title
   629.4′553

   ISBN 1-85233-735-4

Library of Congress Cataloging-in-Publication Data
Turner, Martin, J. L., 1942–
   Expedition Mars / Martin J. L. Turner.
      p. cm. – (Springer–Praxis books in astronomy and space sciences)
   Includes bibliographical references and index.
   ISBN 1-85233-735-4 (alk. paper)
   1. Space flight to Mars. 2. Mars–Exploration. 3. Manned space flight. I. Title. II.
Series.

   TL799.M3%87 2003
   629.45′53–dc22

                                                     2003058511

Project Copy Editor: R.A. Marriott
Cover design: Jim Wilkie
Typesetting: BookEns Ltd, Royston, Herts., UK

Printed in the United States of America on acid-free paper

Contents

Preface

The human exploration of Mars is felt by many to be long overdue. When the first humans walked on the surface of the Moon, the exploration of Mars was expected to follow inevitably. But this did not happen, and the histories tell us the reasons why. Nevertheless, the desire to expand human presence in space, and to mount expeditions to those planets, moons and asteroids whose surfaces admit of such activity, is endemic, especially amongst the young. Those of us who have grown old and cynical in the space business may have suppressed it, but for many, the initial, youthful interest in space was sparked by just such an idea. It seems a pity to disappoint yet another generation by again allowing this dream to fade away. The arguments for and against human presence on Mars have been endlessly rehearsed. Do we need to explore Mars? Are humans really necessary to do so? Can we afford the high cost of a Mars expedition? Do we need human presence in space at all? It is not always easy to give a positive answer to these questions. The *Columbia* tragedy underlines the risky nature of human spaceflight. The price of the International Space Station and the high cost of Space Shuttle flights are always in the background when protagonists of human exploration attempt to argue their case. The outstanding success of robotic Mars probes, in revealing more and more detailed information about the planet, argues against the need for humans on that planet. Nevertheless, there remains a strong feeling that a human expedition to Mars will happen some day, and that it will happen because we *are* human, with a natural urge to explore and a need to minimise our ignorance of the Universe around us.

Since the termination of the Apollo programme in 1972 there have been moments when the prospects looked good for a human Mars expedition. During the Apollo era, NASA evolved a plan to execute a short-stay mission in 1982; but President Nixon's administration had other preoccupations. In 1987, Sally Ride – the first US female astronaut – chaired a commission that proposed a Mars expedition; but Reagan and his administration were unimpressed. In 1989, President Bush (senior) proposed a programme to land humans on Mars by 2019; but for Congress, the cost was too high. Today, NASA still does not have a firm plan to mount a human expedition to Mars, or even back to the Moon, but it has carried out a number of studies, culminating in the 'Mars Reference Mission' of 1997, and its 'Addendum', published in 1998. While there have been a number of Russian studies for human flight to Mars, there is, as yet, no avowed aim to implement them. The ESA Aurora programme has, as its ultimate aim, the landing of humans on Mars in 2030, but so far has only limited funding. In more practical terms, is not too wide a stretch of the imagination to see the ISS as a precursor to human planetary missions.

These are, however, straws in the wind, and there will be no human expedition to Mars until one or more of the world's space agencies makes a definite commitment. Nevertheless, we should not despair. A human expedition to Mars is neither too costly nor too technically challenging. An amount less than the annual cost of the Space Shuttle programme, spent over twenty years, would pay for it, and most of the technical challenges can be met with existing technology.

This book is about those technical challenges: how to get to Mars and back safely, with a crew of human explorers. In a sense, a human expedition to Mars might seem to be just an extension of the Apollo programme, with longer distances involved. However, as will be made evident, there are qualitative as well as quantitative issues raised by the longer journey to Mars, which challenge both our technology, and the capabilities of human beings, far beyond what was necessary for Apollo. Thirty years on, with much more experience of human spaceflight, we are more than capable of meeting such challenges. In essence there are two main issues: the ability of human beings to sustain very long journeys into space, and the limited capability of chemical rocket engines to power such journeys. The pooled knowledge of the Russian and US human spaceflight programmes, which is now available post-Cold War, and new knowledge emerging from the ISS, will be vital in preparing humans for long journeys. The new and emerging technologies of electric and nuclear propulsion will play a key role in enabling those long journeys.

There are, of course, many books about human exploration of the planets. My excuse for writing this one is that it examines and considers, in a relatively objective way, the physics of the expedition, the engineering challenges, and the implications of the results. While there are many possible ways of imagining such an expedition, there are certain immutable facts that have to be taken into account. An example of this is the stay time on Mars. If minimum energy transfers are utilised, then the explorers will have to wait on Mars for more than 400 days until the planets move into the correct alignment for a safe return journey. This has a huge impact on the quantity of supplies which need to be taken, and on the amount of rocket fuel required. I have attempted to expose all of these boundary conditions and to evaluate them against the possible technical solutions. The boundary conditions are, of course, well known to those designing Mars missions in the agencies, and I hope that by explaining them here in a straightforward way, the decisions of the planners to propose this or that technology will be understood. However, it makes little sense to discuss such matters without numbers, and I have tried to calculate (approximately) the relevant parameters, where appropriate, and to explain how these figures were determined. In some chapters, therefore, there are a number of equations – although they are included only for those who wish to follow the argument in detail, and to check my calculations. The reader will lose very little by skipping these equations and taking the numbers on trust.

Chemical rockets – familiar since the dawn of the Space Age – faultlessly powered the Apollo expeditions; Apollo 13 was compromised by a failure in its power system, not its propulsion system. For the longer journey to Mars, chemical rockets are not optimum – at least for departure from Earth orbit. Better devices are now available in the form of electric thrusters, which are much more fuel-efficient; but these also have their disadvantages – mainly low thrust – and it is increasingly being thought that nuclear thrusters may be the best way of sending the expedition on its way to Mars.

I know that many people will react to such a proposal with disbelief or horror – and perhaps both. Disbelief because it seems like futuristic technology, the stuff of science fiction; and horror at the thought of polluting space with nuclear waste, or indeed the Earth's atmosphere, should there be an accident. The facts speak for themselves, and the reader can check them in this book. The nuclear thruster not only has a vastly better fuel efficiency than the familiar chemical thruster, but it is in a moderately well-developed state, due to the NERVA programme in the United States during the 1960s, and very similar developments in Russia. An unfired nuclear rocket is not at all radioactive, nor is it significantly more dangerous than a chemical rocket, and it would only be fired-up in space, and disposed of safely in a deep-space orbit that will not interact with Earth for a million years or more. Many, of course, will fundamentally object to its use; this book contains only the facts.

Nuclear fission again arises where electrical power for the expedition is concerned. It is very difficult to generate enough power for the explorers' requirements on Mars by using solar panels, as the Sun is too faint, and the efficiency of conversion is too low. Nuclear fission is the appropriate choice to provide abundant power for the life support, the manufacture of rocket fuel, and the protection of the explorers from extremes of cold. Small nuclear reactors are very efficient and lightweight for their power, and are appropriate technology for a Mars expedition. The facts about their use and safety are given in this book.

It is a sad fact that while in 1969 we had a launcher – the Saturn V – capable of lifting 118 tonnes into low Earth orbit, we can today lift only about 23 tonnes. The Mars expedition requires a heavy launcher, and the necessary capability ranges from 80 tonnes to 200 tonnes, depending on the particular expedition scenario under consideration. Two hundred tonnes would have been a relatively modest extension of the Saturn V capability, had it been developed immediately, at the time. Now, even 80 tonnes seems a difficult task. But it is by no means a hopeless task; in practical terms the component parts are clearly available. The engines, for example, are off-the-shelf, and the fuel tanks need only be extended versions of the Shuttle main tank. The stumbling block is one of demand. If there were commercial demand for a heavy launcher, then one would be developed. As it is, there is no current demand for a 200-tonne launcher other than for a relatively ambitious Mars expedition, or lunar expedition, neither of which is approved. On the other hand, an 80-tonne launcher could find commercial application as communication, strategic and military satellites become more capable and heavier. There are long-term plans to evolve the Ariane V vehicle to 37 tonnes in low Earth orbit, and similar forces are in place tending to increase the capability of launchers in the United States. It may well be that a combination of commercial pressure and the needs of a Mars expedition will drive the development of at least an 80-tonne launcher. This is essential for human expeditions, whether to the Moon or to Mars.

The vast majority of space exploration and research, accomplished over the past forty years, has made use of unmanned satellites and probes. Only the fortunate few have had the opportunity to work with astronauts, some with the Apollo astronauts. In more than thirty years working as a space scientist, I have, in common with almost all my colleagues, exclusively used unmanned space observatories and probes. For most space scientists this is an acceptable *status quo*; they have no incentive to support proposals for human space exploration. In the first place it is more difficult to pursue a scientific career using infrequent and high-profile human missions, which often have a limited range of topics that can be studied. More tellingly, the money diverted to human space exploration will most probably be taken from the unmanned programme, resulting in fewer opportunities to carry out one's chosen research. This is a very real problem for those who would like to see a human expedition to Mars, or indeed to the Moon. The opponents of human space exploration are often space scientists, who see these very real risks to their own disciplines. The difficulty is that this continuation, with only unmanned exploration, ultimately leads us to a dead end. It may serve us for a decade-or-so more, but we cannot forever avoid the issue of the human exploration of space. In my view, it is time to lift our noses from the grindstone and take a long-term view. It is time to stop playing about on the fringes of space, and take the necessary step that will lead humanity beyond the confines of the Moon's orbit. A human expedition to Mars is indeed long overdue.

Martin J.L. Turner
Leicester University
August 2003

Acknowledgements

A book of this nature always owes much to many people. In this case I am grateful to my colleagues in the Department of Physics and Astronomy, Leicester University – in particular, Dr Mark Sims, Beagle 2 payload manager – for a number of useful discussions. I am grateful to the Austrian Space Agency for inviting me to participate in its summer school, 'The ISS and beyond to the Moon and Mars', in which I was able to discuss important points with other teachers – world experts on manned spaceflight. I am also grateful to the students, both at the summer school and in my department, who were guinea pigs for some of the ideas presented here, and who showed me where I had taken the wrong path.

I am grateful to NASA for permission to use many of the illustrations in this book, and likewise to the Jet Propulsion Laboratory, managed by the California Institute of Technology; the particular credits for all figures are noted in the captions. Concerning rocket engines, I am grateful to ESA and Arianespace. I am also grateful to the University of Illinois Press for permission to reproduce three illustrations from Werner von Braun's book *The Mars Project*. A number of artists' impressions, made available by NASA, are included here, and I acknowledge their work. I am also grateful to Dr Devon Burr and the AGU, as well as the US Geological Survey. As always, Mark Wade's *Encyclopedia Astronautica* was an invaluable source of detailed information about rocket vehicles and engines. Some of the tabular material in Chapter 6 is taken from *Nuclear Thermal Propulsion*, prepared by students at the University of Texas at Austin, and distributed by the Texas Space Grant Consortium.

All those interested in the human exploration of Mars owe a special debt of gratitude to the team that produced the *Reference Mission of the NASA Mars Exploration Study Team*, edited by Stephen J. Hoffman and David I. Kaplan, Lyndon B. Johnson Space Center, Houston; and its 'Addendum', edited by Bret G. Drake, also of the Lyndon B. Johnson Space Center. I have drawn extensively from these valuable documents.

Finally, I would like to thank Clive Horwood (Chairman of Praxis) and Springer–Praxis, for presenting me with the opportunity to write this book; and my editor, Bob Marriott, for his painstaking care with the manuscript.

List of illustrations, colour plates and tables

Colour plates (between pages 162 and 163)

Tables

1

The return to space

Come my friends
'Tis not too late to seek a newer world

Alfred, Lord Tennyson

So the aging Ulysses cried, as he contemplated his declining years. He had led the most adventurous life of any of the Greek heroes, and in Tennyson's imagination he refused to sit by the hearth, 'To rust unburnish'd, not to shine in use!' We might well think the same, as we sit by the hearth, in the opening years of the twenty-first century, and contemplate an aging and uninspiring space programme. What promise there was! What dreams were fulfilled! What risks were taken in those days! From the first glimmer of hope that space might be reachable – the successful launch of the first A4 rocket, at Peenemunde in 1942 – to the first footprint on the Moon in 1969, was just twenty-seven years. We are fast approaching the fortieth anniversary of the first Apollo Moon landing, and we have little more to show for it. Astronauts make journeys to the space station, and orbit the Earth at a few hundred kilometres altitude – closer than Edinburgh is to London; but despite their frequent flights and expensive training facilities, they go nowhere else, just into Earth orbit and back. Brave men and women that they are, they risk their lives with every launch and with every fiery return to Earth; and some have given their lives to keep the dream alive. But the dream is not what it was. It is a pale shadow of what was dreamed when men first crossed the space frontier.

But it does not have to be like this. Humanity is not trapped on this planet, and we have the technology to explore space and the Solar System. We can land men and women on Mars, we can disturb the dust of that planet with our boots, and we can take the first steps that will lead to the stars. This is the subject of this book: the technology of reaching Mars, and returning, with a crew of men and women explorers, who will indeed disturb the dust of Mars, but who will also shake the dust out of our aging space programme, and in truth out of our aging and declining civilisation.

The Mars expedition programme could begin tomorrow, and before 2020 we could see humans walking on Mars. There is no technological barrier. Everything required is either off-the-shelf, or is a straightforward development of existing technology. The preliminary planning has been done, the concept is prepared and published by NASA, and it only wants the political will to make it happen. When, immediately after the first sub-orbital flight of Alan Shepard in May 1961, John F. Kennedy committed his nation to landing a man on the Moon,[1] the technical problems of getting to the Moon and back were vastly greater than the problems facing us today, should we choose to go to Mars. In 1961, America had orbited nothing heavier than a 120-kg TIROS weather satellite; but by 1964 the Saturn 1 was launching payloads of 17 tonnes, and five years later the 49-tonne Apollo spacecraft, with Collins, Armstrong and Aldrin on board, was on its way to the Moon. To send an expedition to Mars we need to develop a modern rocket capable of launching just 80 tonnes into low Earth orbit, and to develop a suitable deep-space engine to take explorers from there to Mars. In 1967 the Saturn V could place 118 tonnes into low Earth orbit, and the modern 80-tonne launcher ought not to be too difficult to develop. The big difference now is that we have no political will to go to Mars – and without this there will be no Mars expedition.

Figure 1.1. A human explorer on the Moon. (Courtesy NASA.)

---

[1]   President J.F. Kennedy made the commitment in his State of the Union address on 25 May 1961. His famous speech, '... We choose to go to the Moon ...', was presented at Rice University on 12 September 1962.

Nevertheless there are many people who would like to see the Mars expedition mounted; and, as previously mentioned, the technical plans have already been produced, and the required developments have been identified. This book is an attempt to set out the technical issues and the methods by which they can be overcome.

1.1   THE LEGACY OF APOLLO

It seems hardly credible now that, nearly forty years ago, men walked on the Moon. But they did, and they went there using technology that we today would consider archaic, if not primitive. It is a lasting testament to human ingenuity that it *did* happen in those days, and it is sad to think that we who follow after have nothing of this nature to look to for inspiration. For four years – 1969, 1970, 1971 and 1972 – the representatives of our race were exploring the Moon; and then it stopped – nothing more for thirty years. It was the greatest adventure for mankind. It cannot be the last time that humans will go beyond the bounds of Earth's gravity, and there must be a time in the future when we will again walk the soil of another heavenly body. For that time to come, for this generation, the legacy of Apollo must be appreciated, and its lessons applied to the achievement of an expedition to Mars, the next step in the conquest of space.

The technical success of Apollo is abundantly clear. The incredibly fast progress made by the United States – from orbiting grapefruit-sized satellites to sending a real spaceship to the Moon – is there in the histories. The legacy of pictures, transcripts, and rock samples from the Moon, is there for anyone who cares to look, and some of the individual explorers are still with us. It was a dramatic and inspiring success – but why was it so successful?

In many ways it was an organisational as well as a technical triumph. The rocket is not a particularly sophisticated device, and has more in common with a steam engine than with, say, a microcomputer. The basic concept has been known for a thousand years, and the liquid-fuelled rockets that allowed the conquest of space had been in existence since Robert Goddard's first device of 1926. The primary difficulty was to make a multi-stage rocket – for nothing else was powerful enough to get into space – and to ensure that every one of thousands of individual components functioned correctly, and in the correct sequence. This was a problem of global organisation, combined with extreme care in every detail. The origins of Apollo, in the German rocket programme in the 1940s, were basically organisational: determining how to assemble the whole complex of components that went into each rocket vehicle, how to test the system, and how to eliminate faulty or badly designed units. The scheme came to the United States with Werner von Braun, and was ultimately transferred to the Apollo programme.

Applying this process to the Apollo project was very expensive. The design and procurement of components, the long and detailed test programmes, the endless paperwork to ensure that the correct units were fitted, and the following of agreed procedures, all involved huge manpower and salary costs. It is the same today: far more money, in a modern space project, goes into the assembly, integration and test activities than into the actual hardware. The raising of such large sums of money requires political commitment. Only governments can marshal such sums, and where government spending enters the equation, politics dominates. This is not the politics of parties and policies, but the everyday politics of persuading enough people to agree to a certain course of action. In the United States during the 1950s, before Apollo, the Army, the Navy and the Air Force each had its own space programme, and there was intense competition. NASA was a new entrant, and only when the President of the United States made the decision to go to the Moon did NASA have a chance to put the programme together. From then onwards it was a concerted effort, and the best elements of the different rocket programmes were selected and fitted into the overall Apollo concept. For example, the Army Juno vehicle, which ultimately became the Saturn (the Army named its rockets after planets) used liquid oxygen and kerosene, directly inherited from the German programme. The upper stages, however, used liquid oxygen and liquid hydrogen, derived from the Air Force programme, and extending back to the first American liquid-fuelled rockets invented by Goddard. In many cases Apollo took the best from a large number of competing concepts. Many people had to be convinced to sink their own ambitions into this greater adventure, and it was a success.

1.2   THE TECHNICAL ACHIEVEMENT OF APOLLO

The Saturn V rocket had its beginnings in the late 1950s at the Redstone Arsenal, where von Braun and his team were working on large launchers. It was conservatism that led to the use of liquid oxygen and kerosene, but it also happens that, for the first stage of a multi-stage vehicle, this is a good choice, because it produces a high thrust, and high thrust was certainly necessary to raise such a monster off the launch pad. Once the go-ahead for the Moon was given, and everyone was agreed, a programme was established, with test flights and ground tests, which was to end with the first flight of the Saturn V in 1967. In the form that was to carry the explorers to the Moon it was a three-stage vehicle with a total launch-pad mass of more than 3,000 tonnes. The first stage had five F-1 engines, used liquid oxygen and kerosene, and burned for 2 min 40 sec with a total thrust of 39 MN (megaNewtons). The second stage had five J-2 engines, used liquid hydrogen and liquid oxygen, and had a total thrust of 5 MN. The vehicle was, of course, much lighter after the first stage had gone and when the rocket was high in the atmosphere and travelling more horizontally, and the second stage burned for 6 min 30 sec. The final stage had a single J-2 engine that was used both to place the Apollo capsule in Earth orbit, and later to send it towards the Moon. The Apollo spacecraft had a single deep-space engine using the storable propellants nitrogen tetroxide and monomethyl hydrazine. It was used for course corrections on the way to the Moon, for capture into lunar orbit, and for the return journey.

Nothing like this had ever been accomplished. The Saturn V had no precedent in size and complexity, and there has been nothing to match it since. The Apollo spacecraft was partially derived from the Mercury and Gemini orbital capsules, and

Figure 1.2. The launch of Apollo 7. (Courtesy NASA.)

(as can be seen in the space museums) the capsule which the crew entered for take-off and landing was very like those earlier versions. But the main spacecraft – with its deep-space engine, the fuel tanks, and the fuel cells to provide power – was something that had to be developed – the first real spaceship. The Lunar Module – the craft which took the astronauts from lunar orbit to the surface and back – was entirely new, and was the first rocket-powered vehicle that could hover and land without any help from wings, parachutes or similar devices. It could also launch from the surface of the Moon, and rendezvous in orbit with the Apollo spacecraft. In addition it had to carry the astronauts, and later the lunar rover, to the surface, and provide a home for them while they were exploring. When seen in a museum, its strangeness reflects its other-worldly purpose.

Despite the huge capability of the Saturn V, extreme care had to be taken to reduce the mass of every component of the Apollo spacecraft and, in particular, the Lunar Module. The further along the chain of manoeuvres leading to a complete mission, the more important was the mass (as will become apparent later). This is because the fuel needed for the last manoeuvres in the mission has to be carried until it is used. Fuel has to be provided to carry this fuel, and there is a multiplicative effect that necessitates a very large amount indeed being required during the first stages; and so the saving of mass on the Lunar Module and the Apollo spacecraft was very important. This drove a number of new developments. One of these developments was the fuel cell that produced electricity from hydrogen and oxygen (solar cells were not used because their mass would have been much higher, and the mission would be quite short), and the other crucial development was low-power, lightweight electronics. The invention of Large Scale Integration, to enable this, has of course spawned the microchip and the microcomputer that has had such an influence on our lives. There were, besides, other smaller developments, essential for Apollo, which are now part of the background technology of our civilisation.

1.3   THE LOST YEARS

Apollo showed that an expedition to another world is possible. But at that time, far greater obstacles had to be overcome, and they were achieved in record time. We have the technical legacy of Apollo as well as its example – and we also have the thirty barren years during which no human beings have travelled beyond the immediate vicinity of the Earth. The Apollo programme stopped just when it was at its most successful. A youthful and vibrant NASA was ready to continue with the establishment of a permanent Moon base, and to go on to explore Mars. But it no longer had the ear of the President. The forces that were against space travel, and particularly against the use of large sums of government money for space travel, conspired to end it. The programme was abruptly and prematurely terminated, the remaining Saturn Vs and Apollo vehicles went to the museums, and human spaceflight was effectively ended. There was the flight of Skylab, launched on the last Saturn to fly; and there was the promise of the Space Shuttle – a cheap re-usable means of putting men and materiel into Earth orbit; but the days of glory were over. The President and Congress were no longer interested in spaceflight. There were more important matters to attend to, such as the war in Vietnam, and as America had won the 'space race' there was no longer any need to compete against the USSR.

However, it was not only the government that had had enough of lunar exploration; the public had quickly become bored with the achievement. Those who felt that it was wrong for humans to violate a heavenly body had their say, as did those who distrusted technology. So too did most of the scientific establishment. This may seem irrational, but it is true. A new discipline – space science – had emerged along with the Apollo programme, but not *from* it. Once it became possible to place satellites in orbit, it became possible to do 'real science' in space and from space. The TIROS weather satellites were the first to look at the Earth and to do something both strategically and commercially useful: predict the weather. Satellites could be used to measure the properties of the upper atmosphere, and to explore the Van Allen belts (radiation trapped in the Earth's magnetic field), which had been discovered by Explorer 1, the first US satellite.

Astronomy, almost moribund, would have a new lease of life now that telescopes could be lifted above the atmosphere. The first X-ray observatories were launched; and they revealed a totally new Universe, interleaved with the old one: the Universe at 1,000,000°, heated by gravity. There was much more. As a kind of insurance against not winning the race to the Moon, and as a way of preparing the ground for the manned missions, probes were shot at the Moon, and later to the planets. These could take close-up pictures of the Moon and later of Mars. Many failed, but as experience grew, more of the probes achieved both their mark (such as the target planet) and their scientific objectives, and it became possible to explore the planets remotely. Paradoxically, it was the technical developments of the Apollo programme that enabled these probes to become more capable and sophisticated. What attracted the new generation of scientists to these probes was that they were doing real science – method–result–conclusion. Instruments made measurements which could be interpreted and published in the learned journals. Careers could also be built on them – and they were. Satellites and probes were much cheaper than manned spacecraft, and they were affordable. Other nations began their own space programmes wholly based on satellites, and a world community of scientists began to grow and to become engaged in space research using satellites and probes. These scientists had no use for human exploration or Moon bases. They wanted to build and operate their own instruments, not to have to explain to an astronaut what to do, and have the results relayed back by him or her. Human exploration is very expensive, and it is difficult to resist the argument that what money there is for the exploration of space should be spent on many robotic space probes and not on one human expedition. In the end, this argument won. NASA did not have enough money to indulge in further human exploration, but it could maintain a programme of robotic exploration. And it is as well that it did so, otherwise we should never have had the close-up images of the planets that were obtained by the Viking, Mariner and Voyager probes, operated by the Jet Propulsion Laboratory. But the lack of support by scientists for human space flight was a significant cause of the barren years, and today it remains a problem for those who want to revive human exploration.

There is no doubt that if NASA had continued to receive the same amount of financial support that it had for the Apollo programme, it would have advanced to a permanent base on the Moon and to a Mars expedition. In the event it did not, and there was no other space agency that could take up the challenge. In the USSR the old spirit disappeared after the USA reached the Moon first, and the dream was forgotten.

There are two historical views one can take, looking back and wondering why none of these things happened. Either Apollo, in 1969, was a chronological anomaly, an anachronism; or we have sadly fallen off in our technological capability, and our ambition as human beings. Either could be argued. In 1961 there was a unique combination of circumstances. With the success of Sputnik and Vostok, the USSR had shocked the United States out of its technological complacency, and was well on the way to a manned lunar programme. At the same time, the various seeds that were to lead to the Saturn V were already flowering with the Juno and Centaur rockets. There was a team of trained and eager scientists and engineers ready to take up the challenge, and there was a President in place who was prepared to make hard decisions – and he had the backing of Congress to release the necessary funding. Once committed, the programme could not be stopped except by failure – or success. And so it happened. But these circumstances did not recur, and so the barren years followed.

Has our technology declined since then? The answer must emphatically be 'no'. In every field, including rocket technology, we are far more capable than we were then. If we compare the awful grainy images of Neil Armstrong's 'small step' with the images sent across the world, by one child to another, using a mobile telephone, we can see this is so. There is no technological reason for our holding back from the exploration of Mars, but we hold back because we do not have the will to carry on. There is no international rivalry, as between equals; there is no strong scientific drive to explore Mars, other than with robotic probes; there is no strong economic motive; and few people care about the dream.

1.4   A NEW CLIMATE

The dream did not die completely after the end of the Apollo programme. The Marshall Space Flight Center had already commissioned studies of variants on the Saturn V to launch the Moon-base and the Mars expedition. The plan for Mars, evolved about the time of Apollo 11, was to send two very heavy nuclear-powered spacecraft to Mars orbit in 1982. It was a 'short-stay' mission, during which eight explorers would descend from the orbiting spacecraft in a Mars Excursion Module, explore for a month, and then return to the ships for a rapid transit – via Venus[2] – back to Earth. It was, of course, too costly, when storm clouds were gathering round the Apollo budget. The aerospace companies and Vice President Spiro Agnew were enthusiastic, but President Richard Nixon and Congress were not, and nothing came of it.

But from time to time the old ideas were taken out of drawers and dusted off. NASA could not spend serious money on preparations, but a trickle of funding existed, at least for internal studies. This is common in space agencies. Engineers and scientists are always looking forward to what might be done, and small amounts of money and manpower, gleaned from departmental budgets, are applied to studies for future projects. The idea of a human Mars expedition remained alive – just. In

---

[2]   The journey on the normal route to and from Mars takes more than eight months, but if more rocket fuel is available then there is a faster return route, crossing Earth's orbit and returning via a slingshot encounter with Venus. Details are provided in Chapter 3.

1987 the first American woman astronaut, Sally Ride, chaired a commission that proposed a Mars mission. It comprised a cargo ship, sent in advance of a piloted ship, and a two-week exploration on the surface followed by the return to Earth. This time, it was President Ronald Reagan who was unimpressed by the cost.

Then, in the first years of the new administration, in a reversal of roles, NASA received a proposal for a Mars expedition from President George Bush (senior). In a speech delivered on 20 July 1989 – the twentieth anniversary of the first Apollo landing – Bush announced a thirty-year plan to land humans on Mars by 2019. This was the best opportunity yet for humans to go to Mars, and NASA hastened to prepare a new plan: the eponymous '90-Day Report'. It included the construction of a 1,000-tonne nuclear-powered spacecraft at the International Space Station. Once built and fuelled, the spacecraft would depart for Mars, and again an Excursion Module would take the explorers down to the surface for a two-week stay followed by a fast return via Venus. When the study was complete, the cost estimate was $450 billion – and it took rather less than 90 days for Bush to forget it and for Congress to reject it.

The idea of a Mars expedition did not disappear, however, and NASA did not lose hope that something could be salvaged from the 90-Day Report. It was therefore reviewed internally, and work began on a revision. At least, by costing a Mars mission, it had identified the expensive items. This again is a normal approach to new space missions. The first plan is almost always too expensive. The team then usually divide up and consider each element of the mission, to determine whether it can be eliminated, or produced in a cheaper way.

At the same time, Robert Zubrin (now the President of the Mars Society) began to realise that a big mission was not the correct approach for a Mars expedition. At that time he was working for an aerospace company charged with part of the 90-Day Report, and he considered that the right thing to do was not to construct, in Earth orbit, a 1,000-tonne spacecraft, and then send it to Mars; but to cut everything to the bone and create a lightweight mission that could be launched directly to Mars, without any construction or rendezvous in Earth orbit. He also realised that rocket fuel can be made on Mars, using the atmosphere, and so all the fuel for the return journey need not be taken. He therefore produced a plan called Mars Direct, involving a crew of three sent directly to Mars. They would be preceded to Mars by an empty Earth return vehicle, with a rocket fuel 'factory' attached. It would then fill up its fuel tanks with liquid methane and liquid oxygen, made on the surface of Mars. The explorers would stay on Mars for the full 400 days required, using standard minimum energy transfers, and would launch directly from the surface of Mars into the Earth return trajectory. Instead of the 1,000-tonne spacecraft, Mars Direct involved just two 50-tonne payloads: the crew transport vehicle and the Earth return vehicle. To improve safety, a third 50-tonne payload, comprising another Earth return vehicle, would be sent with the crew, so that they had a back-up. At a cost of around $30 billion, Zubrin's proposal, with just three 50-tonne spacecraft going to Mars, was very much cheaper.

During the early 1990s, Zubrin's Mars Direct concept began to penetrate NASA, and many of his ideas were incorporated in the latest and most elaborate plan, prepared at the Lyndon B. Johnson Space Center at Houston: 'The Reference Mission of the NASA Mars Exploration Study Team', issued in 1997. A year later a much-modified version emerged as 'Reference Mission 3.0 Addendum'. This is still the best guess of how a Mars Expedition would be accomplished, if it were to begin today. (The details of these plans are discussed in later chapters of this book, but it is worth noting here the estimated cost of $55 billion – about a tenth of the cost of the '90-Day' mission.)

This is where we are today. There is, within NASA, background activity preparing for a human expedition to Mars, although there is no approved plan to mount one. From time to time, important people make statements suggesting that NASA will mount a Mars expedition very soon; but so far, nothing has come of these statements. The climate is, however, much more positive than it has been since the heady days of the Apollo preparations. This may be the decade in which such a programme will be started. In this book we shall examine the technologies that could contribute to sending the explorers to Mars and returning them home, and how such an expedition could be mounted. Before proceeding, however, we shall consider Mars as *terra incognita* – ripe for exploration by human beings. Why should we explore Mars, and why should we do it ourselves instead of relying exclusively on robots?

1.5   HUMANS OR ROBOTS

Space scientists – even planetary scientists – would be happy to continue to use robotic probes to obtain their data. Robotic spacecraft are relatively cheap, and they are very capable indeed. Modern electronics allows a large amount of intelligence to be built into them, and they have certainly returned some remarkable information about Mars. The question the scientists ask is: why not continue like this? More science per dollar can be obtained by using these probes; NASA budgets, and the budgets of the other space agencies, are limited, and not growing very fast; if we divert resources to a human exploration programme, then robotic science will suffer; and humans are no better than machines in gathering knowledge, and it is vastly more expensive to send them to Mars and to maintain them there. This seems to be a devastating argument against a Mars expedition.

It is perfectly true that the information sent back to Earth from Mars probes, both orbiting and landed, is remarkable. The first probes to land and function were the two Viking landers, in 1976, which sent back the first pictures of the surface, taken *from* the surface; while at the same time the Viking orbiters created a photographic atlas of the surface, from space. More recently there have been both visible light orbiters that have returned images of the surface of the planet at very high resolution, and orbiters that have mapped the planet using other wavelengths and radiations, to try to understand the composition of the planet's surface. The experiments landed so far have produced equivocal results on whether life exists in the soil within a few metres of the landing site. The Viking lander experiment produced a positive result, but since then, scientists have produced many alternative explanations of the result that involve martian chemistry rather than martian life.

Figure 1.3. An artist's impression of explorers on Mars. (Courtesy NASA.)

Other landers failed to make it safely to the surface. The Mars rover which landed successfully on Mars Pathfinder covered an area extending several hundred feet from the landing site. It identified river sand and water-worn rocks, but it did not search for evidence of life. So, the question remains open.

The results of these attempts illustrate some of the difficulties with robotic exploration. It is limited in scope for two reasons. The range over which the samples can be collected is rather small – so far, just a few hundred feet – and the experiments to be carried out on the samples, by the lander, are limited. Once the planned series of experiments has been carried out, that is the end of it. There is no human present to say 'What if we do this?' or 'How about following up that indication?' It would be possible to create a large number of exploration robots with built-in intelligence, land them at many different sites, and carry out very sophisticated experiments with machine intelligence guiding them. This could involve a development programme of similar magnitude to the Mars expedition, and in the long term the robots would not be as clever as humans carrying out exploration. Machines are not as versatile as we are, they are nowhere as mobile as we are in rough terrain, and they are not as bright as we are (at least, not yet). Robotic exploration is therefore worthy of development, it is what we have, and it is producing results; but it is not as useful as human exploration, even if carried out on a grand scale.

1.6 MARS

If humans are to go to Mars then they will attempt to understand a number of questions. The crucial questions relate to life on Mars. Is there, or has there ever been, life on Mars; and could human beings occupy a permanent colony on Mars? These questions relate to fundamental issues for the future of humanity. If life exists or existed on Mars, then it is a common phenomenon in the Universe. It is as simple as that. Earth is not unique. If humans could exist permanently on Mars, then the way is open for human exploration of the Solar System, and eventually of the planetary systems of nearby stars. If the answer to both these questions is 'no', then the Earth may be unique, and also, it is all that remains for humanity, in the long term. Besides the answers to these big questions, there is a myriad of scientific explorations and experiments in martian geology, petrology, vulcanology, meteorology, upper atmosphere and solar wind observations, and in many other disciplines. These have a direct relevance for Earth, because we can test theories of atmospheric circulation, tectonics, wind erosion, stratification, atmospheric exchange with polar caps, and many other theories, in a totally different planetary environment. This is impossible on Earth because we have only one Earth, and the fundamental properties, such as gravity and solar input, have values unique to Earth. It is not possible to 'experiment' with terrestrial weather, for example; but for a new planet, with different gravity, solar input, and atmospheric composition, it will be possible to test theories under new conditions. The result should be a vastly better understanding of our own planet and its mechanisms.

What is this planet like? Why is it so interesting? And how is it different from Earth? No representative of our race has yet set foot on it, and there will be an

Figure 1.4. The planet Mars, showing Valles Marineris. (Courtesy NASA.)

explosion of knowledge when that happens. In the meantime we have the results of robotic probes, which have mapped Mars, landed on its surface, and probed its composition and magnetic fields, as well as its atmosphere. Hubble Space Telescope images are to some extent redundant now because of the high quality close-ups taken from the orbiting satellites. It was not so in the past, of course. Mars is unique in being both close enough to be examined as a planetary disk from Earth, and having a clear atmosphere, so that the surface can be seen. The early observers, using the eye and the telescope, created drawings of features on the surface; and the major features which they saw can now be seen, in detail, in the images and maps made from the orbiting satellites. In many cases the old interpretation was erroneous, although the features themselves are real. The most famous features are, of course, the canals of Schiaparelli. He observed linear features that looked like channels on the surface. The Italian word *canali* was afterwards translated rather badly into English as *canals,* and with the impetus of the wealthy and enthusiastic American, Percival Lowell, they became 'evidence' of civilisation. But they do not exist. They were the result of the human eye–brain connection that synthesised several barely resolved smudges – perhaps craters – into linear features. Alas, the martians of Lowell's imagination do not exist; but Mars remains a very interesting planet.

1.7 MOUNTAINS, VALLEYS AND PLAINS

If we were to see the Earth stripped of the oceans and vegetation, it would not look like Mars. The Earth's crust is formed from a number of tectonic plates that move slowly, driven by convection currents in the molten core. In some regions two plates are moving towards one another; one slides beneath the other, and fold mountains, volcanoes and earthquakes occur. An example of this is the impact of the African plate on the European plate, creating the Alps, and the seismically active regions in Italy and Turkey. In other locations the plates are separating; low-lying regions are created, with volcanoes lining the separation boundary, exemplified by the Atlantic basin and ridge. On Mars there are no tectonic plates. In one of the theories, water is necessary for tectonic action, with the sea acting as a lubricant for the plates sliding under one another. The surface of Mars was formed by other forces, and this is one reason for its interest. Mars certainly once had a liquid core like the Earth's, and recent results indicate that it may still have one; but there seems to be no plate tectonics. The three forces shaping the major landforms on Mars are volcanism, cratering and vast magma upheavals like that which created the Tharsis bulge and the Valles Marineris.

There is no actual *sea level* on Mars, as there is no sea; but a datum for heights and depths has been established as the level where the mean atmospheric pressure is 6.1 millibars (compared with 1,013 millibars for the Earth's atmospheric pressure at sea level). This pressure – 6.1 millibars – is the triple point pressure for water vapour, and so has physical significance. A much more important reason for choosing an atmospheric pressure-based datum is that, like the sea, the atmosphere conforms to the gravity of Mars, and so an isobaric datum closely represents the value corresponding to the supposed sea level. By taking this datum, and measuring

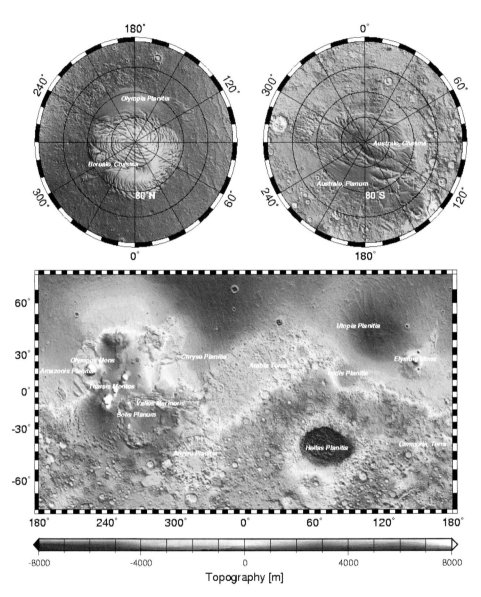

Figure 1.5. The global topography of Mars. The old, heavily cratered highlands in the south give way to newer lightly cratered plains in the north. The dark areas are low-lying regions, while pale regions are highlands. (Courtesy US Geological Survey.)

heights and depths with respect to it, it can immediately be seen that the actual shape of the planet is odd. The northern parts are more or less all below 'sea level', while the southern parts are, in general, above this level. The land mass of Mars would all be in the southern part, as a single continent, and a continuous ocean would lie over the northern half of the planet. It is possible that this odd shape is the result of one or more major impacts on Mars of asteroid-sized bodies. Mars lies close to the asteroid belt, and in the past could well have interacted with stray asteroids. With this hypothesis, the northern part could contain a huge impact basin. The obvious, circular, low-lying area in the south – Hellas Platinas – is an impact basin caused by just a modest-sized asteroid.

Mars is possessed of the largest volcano in the Solar System: Olympus Mons. It covers an area 600 km in diameter and is 27 km high. The lava that formed it was fluid, and so the slope of the sides is very shallow (like the much smaller Hawaiian volcanoes), and there is a huge caldera on the summit. In a nearby region is the Tharsis bulge and the Tharsis volcanoes. The bulge is a large upraised area 9 km above the surrounding territory, and the three volcanoes which sit on top of it reach the same altitude as the top of Olympus Mons. This whole region has been pushed up by magma from the interior of the planet. The size of these volcanoes is perhaps due to the absence of plate movement, so that successive eruptions always happened at the same place. The Hawaiian volcanoes occur on a plate boundary, and the cones are carried away from the magma vent by the plate movement, being replaced by new cones. The volcanoes in the Tharsis region are relatively young. Olympus Mons is the youngest, but none of them have erupted for millions of years.

Volcanism is evident all over the planet from the vast northern lava plains to the oldest volcanoes that surround the Hellas basin in the south; these latter were triggered by the aftermath of the impact that produced the Hellas basin. While we have no lava samples, the ages of volcanoes and other lava features can be roughly

Figure 1.6. Olympus Mons. On the upper side of this image, clouds lap over the huge scarp. (Courtesy NASA.)

estimated by the number of impact craters, with newer regions having fewer craters. Until we have real samples, the age is only roughly determinable by assuming that the cratering rates are similar on Mars and the Moon, and by using the dating of the rock samples brought back from the Moon by the Apollo astronauts. The northern lava plains are relatively young, having few craters, while the southern regions are old, with very heavy cratering. It was this region that was imaged on the first successful Mars fly-by, leading to the erroneous picture of a Mars very similar to the Moon.

There are impact craters all over Mars, with the density varying depending on the amount of subsequent lava flow. The craters themselves differ from those on the Moon because of the presence of the atmosphere and sub-surface water. The former alters the trajectory of the ejecta, compared with the airless Moon, and the latter lubricates the lava flows from the impact, so that they travel much further than on the Moon. Mars shows evidence of very large impacts – the Hellas basin being an example. These deep basins are the result of the impact of asteroid-sized objects, and their study could help us to understand both the probability and the consequences of such impacts on the Earth. The Hellas basin is 1,600 km across and 5 km deep, and during the martian winter, water ice forms in it as a thin layer of frost.

The valleys of Mars were not created by water erosion. Like the Great Rift Valley in Africa, they were caused by crustal movement. The most prominent is the Valles Marineris system of canyons (see Figure 1.4), which extends over a distance of 4,000 km, with sections up to 120 km wide and 7–10 km deep. The western end abuts the Tharsis uplift region, and features here suggest that the withdrawal of magma from underneath the canyon area during the formation of the Tharsis bulge has caused the

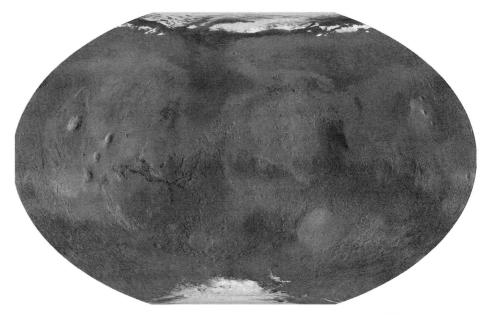

Figure 1.7. The whole surface of Mars. (Courtesy US Geological Survey.)

surface to subside, to produce the complex series of deep valleys. Further to the east, the central region comprises three long parallel valleys that reveal huge landslips along their edges. These valleys eventually join to form one large depression which, at its western end, breaks up into interconnected valleys and chaotic terrain. It is here that some of the famous channels emerge – evidence for running water in the past. Mists seen in the bottom of the valleys are mostly ice particles formed as the atmosphere cools at night; and during some seasons, winds seem to blow dark coloured dust along the valleys. The valley bottoms are very rough, and layering of material can sometimes be seen. As in many places on Mars, the formation of such features can be ascribed either to water – as in stratified rocks on Earth – or to dust-laden winds. The geological history of the Valles Marineris began with the subsidence of several separate regions, as magma was withdrawn. Later, new faults caused further subsidence, resulting in the linkage of the depressions, and the huge landslides seen on the edges of the canyons. It is possible that the first depressions formed were at one time filled with water, which was responsible for the layering in some of them. The later joining of the depressions allowed the water to flow out, and may have been responsible for the channels seen in some areas.

The northern plains were produced by the welling up and flowing out of very fluid lava over huge distances. They are relatively recent, and show little cratering. They lie mostly below the datum level, and would form the ocean bed, were Mars to have one. Closer to the north pole are large dune regions, where the terrain appears much like sand deserts on Earth. These dunes are covered with the polar extension in winter; they do not seem to move.

The rocks and dust on Mars have absorbed oxygen from the atmosphere, and so are heavily oxidised. This produces the red colour of Mars, because of oxidised iron in the rocks, and the dust carried up into the atmosphere is also of a reddish colour. Because there is no reducing agent left in contact with the dust and surface rocks, they are oxidising. This means that in the presence of suitable chemical they will react. Any chemically-based searches for life are complicated by this factor, and it may also be that martian dust is an irritant to human skin or lungs.

1.8   POLES AND ICE CAPS

A martian day is only 3% longer than Earth's day, but its year is equivalent to 687 Earth days. Its orbit – the most eccentric in the Solar System except for Pluto – is much more eccentric than Earth's orbit, and so the heat of the Sun varies significantly during the martian year. On Earth, solar input varies much less because its orbit is nearly circular. The polar axis of Mars is inclined, like the Earth's, but at an angle of 25°. Mars has seasons, just as the Earth does, but the variation of the solar input round its orbit makes the seasons of the northern and southern hemispheres asymmetric. Summer in the southern hemisphere is warmer than summer in the northern hemisphere. This is because Mars is at its closest to the Sun when it is summer in the southern hemisphere, and at its furthest from the Sun when it is winter there. The planet also travels faster in its orbit when closer to the Sun, and

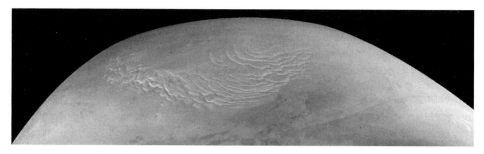

Figure 1.8. The north polar cap of Mars. The layers are water ice and dust deposited year by year. (Courtesy NASA.)

so the southern summer is somewhat shorter than the northern summer. The southern winter is correspondingly both colder and longer, because the planet is further from the Sun and also moving more slowly.

Both polar caps contain water ice and carbon dioxide ice. The permanent (summer) north polar cap consists of mostly water ice, mixed with dust. The temperature does not normally rise sufficiently to cause water to sublime, and at the low atmospheric pressure prevalent on Mars, ice generally sublimates directly to water vapour. The ice cap increases dramatically during the northern winter, extending down to about 60° latitude. This increase is mostly due to carbon dioxide ice that freezes onto the surface as winter advances, and it enters into permanent shadow.

The south polar cap experiences greater extremes of both summer and winter. In its long, cold, winter, it grows extensively, as both water and carbon dioxide freeze on its shadowed surface. In its shorter, warm summer – in martian terms – most of the carbon dioxide ice cap evaporates. The effect of this is rather startling. The global atmospheric pressure falls very considerably towards the end of the southern winter, as a quarter to a third of the atmosphere freezes onto the ice cap; and it rises to a maximum in the middle of the southern summer, as the ice cap evaporates. The range measured by the Viking lander was from 7 millibars at the end of winter, to nearly 9 millibars at the height of summer. This seems to be a very small change – until it is scaled to the situation on Earth. Taking normal sea-level atmospheric pressure – 1,013 millibars – as equivalent to the southern summer pressure on Mars, then the winter pressure would be 790 millibars. This is the global pressure change over the entire planet. We are accustomed to changes of less than 100 millibars at most; and these are in local storm centres, not over the whole planet. It is this remarkable global pressure change that fuels the enormous dust storms that can envelop the entire surface during the southern summer. Smaller dust storms occur in the tropics, and involve winds up to gale force – 110 km/h (70 mph). Global storms seem to start near the Hellas basin in the south, and involve winds up to 400 km/h (250 mph). They can envelop the whole surface of the planet, for a month or two, with the dust reaching to altitudes of 50 km. The atmosphere always contains a certain amount of dust, even between storms, and this alters the colour of the sky. Global dust storms do not take place every martian year, and some years are free of

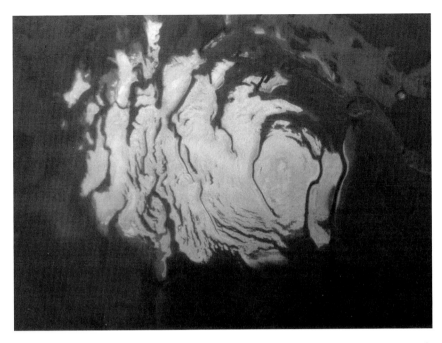

Figure 1.9. The south polar cap in summer. Carbon dioxide ice covers any water ice that may be present. (Courtesy NASA.)

such huge storms. Local cyclonic storms develop near the poles as winter approaches, just as on Earth, with similar spiral cloud patterns. The clouds generally consist of water ice, but at high altitudes it may become cold enough for carbon dioxide to contribute. Tropical clouds also appear at mid-day as the surface atmosphere warms and rises, and tropical mid-day temperatures at the surface can sometimes rise to $20°$ C in small local areas.

The northern ice cap has a smaller effect on the atmosphere than does the southern ice cap, and the topography of the ice caps also differs. The northern cap is low lying, and part of the overall depressed northern hemisphere of Mars, while the southern ice cap is located on high ground. (Here there is a parallel with Earth. The Arctic ice cap rests on the sea, while the Antarctic rests on mountains.) The higher-altitude southern cap is colder in summer, and so the carbon dioxide does not all sublimate; whereas the warmer northern cap loses all of its carbon dioxide in the summer, exposing the water ice cap. The water ice at the southern pole is difficult to see, and estimates of its depth are uncertain. There may be several kilometres of ice mixed with dust at the northern pole.

The major dust storms occur during the southern summer, and so dust can be trapped in the freezing northern cap. This results in a heavily layered appearance of the permanent cap, depending on the prevalence of dust while the ice is freezing. Knowledge of the nature of the polar caps and their influence on martin weather is growing rapidly as results are received from Mars Global Surveyor.

1.9   WATER ON MARS

Water is of vital importance for the Mars expedition, for two reasons. The presence of liquid water is regarded as essential for life to exist; and availability of useful quantities of water, in any form, would be important both for the expedition and for any permanent base.

Mars has no oceans, or standing bodies of water. The famous 'canals' are absent, and would not have worked in any case, because the temperature is now too low for liquid water to be stable on the martian surface. The polar caps have, until now, been difficult to analyse, but as indicated above, the latest results present pretty clear evidence for a large amount of water ice in the north polar region, and possibly a similar quantity at the south pole. Due to the low temperature there is very little water in the atmosphere, which cannot contain more because it is saturated. This is analogous to the cold air near the poles on Earth. Clouds seen in the atmosphere are indeed like terrestrial clouds, and are formed from ice crystals.

Mars was formed further out from the Sun than the Earth, and it is therefore probable that in its primaeval state it had a higher percentage of water than the Earth. On both planets, once the crust had formed and cooled, water was expelled from the interior, through volcanoes. On Earth it condensed to form the seas and oceans; but what happened to it on Mars? Calculations indicate that there would have been enough water present to form an ocean of substantial depth over the whole surface of the planet; but it is clearly not there now.

There is plenty of evidence that liquid water once existed on Mars, and we know that it exists in vapour form in the atmosphere, and in the form of ice at the poles. For water to remain liquid, the atmospheric pressure has to exceed about 6 millibars, and the temperature has to exceed $0°$ C. If the pressure is lower than 6 millibars then only ice or water vapour can exist, and the one will change into the other, depending on the temperature. The atmospheric pressure on Mars hovers around the critical pressure for liquid water to exist. By and large the temperature rarely if ever exceeds $0°$ C, although there is evidence of short-term temperature rises above this level in small and isolated equatorial regions at noon. This means that, all over Mars, water has to exist either as ice or as water vapour, and not as a liquid. If the pressure rises, then liquid water can exist at temperatures higher than $0°$ C. For example, at a pressure of 30 millibars, water would boil at about $20°$ C, and we know that at 1,013 millibars (for example, on Earth) water boils at $100°$ C. So, the absence of liquid water on Mars is more the result of the low temperature than the low pressure. A modest rise in pressure, accompanied by temperatures just above freezing point, would allow the formation of ponds, lakes and rivers.

While some of the primaeval water has certainly evaporated and been lost into space, because of the low gravity of Mars there must still be a substantial amount remaining. The current low temperature, combined with the pressure in excess of 6 millibars, means that most of the present water cannot evaporate and be lost. In the past, of course, different conditions may have prevailed; but for substantial water loss to have occurred, much higher temperatures, combined with low atmospheric pressure, would have been required. In general, while the atmospheric pressure could have been lower, it would be as a result of more of the carbon dioxide freezing onto the poles, which would happen only during very cold periods, when evaporation of water would be even less likely. If the temperature had risen substantially, then this would involve all the carbon dioxide being evaporated from the poles, thus raising the atmospheric pressure. While this is not a strong argument for the retention of most of Mars' water, it is certainly indicative.

Having established that a substantial amount of water should still exist on Mars, it is time to turn to the evidence. The surface of the planet is certainly very dry at present – as dry as the driest desert on Earth – but it cannot always have been like this. The most encouraging result from the Viking orbiters was the observation of many surface features that could only have been formed by water. These features are channels and valley systems that look very much like the dried-up beds of rivers and rain gullies on Earth. They are not precisely identical, and at one time it was thought that they could have been formed by wind effects or by the flow of lava. Recent observations have definitely established that the channels and gullies run downhill, and many have tributary systems that could not have been formed by lava flows.

There are three types of feature. The earliest are small steep-sided valleys with tributary systems, which occur in the old, heavily cratered areas of the martian surface. They look like river systems, but they seem not to have been formed by

Figure 1.10. Steep-sided narrow valleys formed by ancient water flow. The ground falls away to the top right of this image. Aprons of debris are seen at the mouths of the steep-sided valleys. (Courtesy NASA/JPL.)

rainfall; rather, they have the appearance of being formed by the emergence of ground-water. This 'spring sapping' is seen on Earth, and the shapes of the valley heads are identical. An original depression – perhaps a crater or a canyon – cuts through the surface layers and exposes a layer rich in water. The water emerges, and washes away the foot of the slope, which leads to the collapse of the wall above and to the exposure of more of the water-bearing layer. The process is repeated, and the interface at which the water is being released retreats – rather in the same way as a waterfall cuts into the rock – to form a channel. As more and more channels are formed, they combine downstream to form the typical river system with tributaries. However, the rivers were only temporary, as the water quickly evaporated or soaked into the terrain. These systems extend for only 100 km or so. They were not formed by rain, but a rise in temperature was required to set them off. This could have been volcanic, but because of their great age these systems may have arisen during a warmer and damper period in early martian history, when the carbon dioxide pressure was higher, driving the greenhouse effect to raise the temperature for a while. As mentioned above, a small rise in pressure could allow liquid water to exist on the surface at temperatures a few degrees above freezing. There would not have been a water cycle to return the water to the hills, as on Earth, and so once the water-bearing layers had been exhausted, the valleys would have dried up.

The next and larger water feature occurs along the boundary of the old cratered terrain and the flatter northern plains. These are much larger than those mentioned above, and they tend to have a sinuous portion like a terrestrial river, as well as the upstream tributary system. The channels are much wider than the older run-off channels, but also show evidence of spring sapping at the heads of the tributaries. The channel bottoms show a braided appearance like the channels formed by flash floods in terrestrial deserts, or, much smaller and closer to home, by channels in the

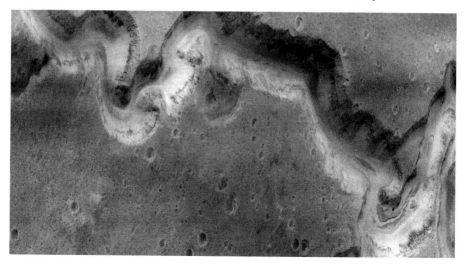

Figure 1.11. A dried-up sinuous water channel flows from the highlands to the plains. (Courtesy NASA/JPL).

beach where sea-water emerges from the sand and runs down as the tide recedes. While the sudden release of water is a probable cause, there is some evidence that glaciers played a part in producing the characteristic shapes of the channels. This implies some seasonal effect by which glaciers advanced and retreated, released debris, and maintained the channels over a significant period.

While the origin of the water release for the small and medium-sized features is obscure, but might have been a global temperature change, there are, on Mars, huge water channels that could only have arisen due to the catastrophic release of water caused by volcanic action. They can be tens of kilometres wide, and stretch for 500 km across Mars. Very recent observations have indicated that their heads are associated with volcanic features, and that successive release of water from deep in the planet, or from volcanic heating of a widespread water-bearing layer, has produced a series of major floods that carved out the channels. In particular, 'tear-drop'-shaped islands in the middle of some channels show characteristics of successive flooding in the channel. These islands are flat-topped, and have their blunt ends pointing upstream. At the blunt end there is some permanent obstruction to the flow; a dyke, or in many cases a crater. The flood was diverted round the obstacle, forming a characteristic tapered deposit of silt. The layering in this silt deposit shows that the floods occurred repeatedly. The Viking photographs, however, were not conclusive, because it was difficult to determine the local slope of the land, and downhill slopes were not obvious. The recent work by Mars Global Surveyor, using a radar altimeter than can measure heights to a few metres, has definitely established that these features were formed by water flowing downhill.

Figure 1.12. A dry water channel many kilometres wide. The water was released by volcanic heating of subterranean ice, and here flowed from right to left to form the characteristic tear-drop-shaped islands. The layering shows successive flooding. (Courtesy Burr *et al.*, *Geophysical Research Letters*, 2003.)

These channels, in their varying sizes and locations, provide unequivocal evidence for the past existence of water floods and streams on the surface of Mars, although perhaps none have their origin in rain; they are all associated with sub-surface water. There is also now indirect evidence for dried-up lakes and ponds, from the remote identification of minerals that could only have formed in standing water. These are not the limestones and chalks of our planet (which were formed by minute animals in the water), but are the result of chemical action in water, such as might arise in salt lakes or hot springs on Earth. This will take some time to interpret properly, but if it is true then the conditions for life might have existed on Mars in the past.

The picture of water on Mars is then very different from that on Earth. Water flowed on the surface in the past, but this seems to have been episodic, driven by temporary atmospheric temperature rises and volcanic action. There is no water on the surface now. The period of channel formation may be over, or the interval between such episodes may be so long that none has happened in recent history. But we may be sure that there is still a lot of sub-surface water present in the form of permafrost, which may be many kilometres in thickness, and located at various depths from the surface. The recent measurements, using neutrons, show the presence of water only metres below the surface over most of the planet. This is frozen, however, and it is not yet possible to determine whether or not the near-surface permafrost is rich in water or rather dry. This is important, of course, for the twin aims of human exploration. If water abounds, then a search for evidence of past life can be carried out near any suitable landing site. At the same time, if substantial quantities of water can be melted out of the local permafrost layer, then both hydrogen for rocket fuel and oxygen for life can be extracted from it, and it can also be used for drinking and for washing. For a future permanent base, water and sunlight are all that are necessary to attempt agriculture, under suitable transparent domes. The carbon dioxide would then be converted to oxygen by the plants.

1.10 MARS: THE PLANET TO EXPLORE

We have seen that Mars is closest to Earth in its composition and nature. But it is not Earth, of course, and life on the martian surface would be very hard. Typical equatorial temperatures range from -80° C to -30° C. At the south pole, the summer temperature is only -110° C. The pressure is far too low for humans to operate without pressure suits, and there is only a trace of oxygen in the atmosphere. The major constituent, carbon dioxide, is not actively poisonous, but even a small excess of it in breathing air results in hyperventilation and eventually death. The atmosphere is relatively thin, and so does not provide much protection from solar radiation, and the absence of a global magnetic field (which the Earth possesses) exposes the surface to bombardment by cosmic rays. As will be shown later, all of these problems can be overcome, and it can be made safe for explorers to spend several years on Mars.

Mars is the obvious place for humans to go in order to continue the exploration of the Solar System. It is relatively nearby, and we have discovered much about its surface and its conditions. It has an atmosphere and weather; it has a wide variety of landforms; it has seasons, and, in the past, it had different climatic conditions; it has water; and it may have at one time supported life. We could learn so much, not only about Mars itself, and possibly the ubiquity of life, but also about the history and future of our own planet, if we could only go there. It is important for the big scientific questions, but it is also important as a goal for space exploration, and, frankly, also as an adventure. Who would not be delighted to stand looking down into one of the vast chasms of Valles Marineris and watch the mist forming far below; or watch the Sun rise from behind Olympus Mons; or trek across one of the huge dry water channels; to say nothing of the eight-month voyage through space? It is right and proper that we should undertake the expedition to Mars.

1.11 NEW TECHNOLOGY

The US or the USSR could have mounted a human expedition to Mars in the 1980s, as the Apollo or Soyuz technology would have been adequate. But they lacked the will and commitment. Now, in the first decade of the twenty-first century, our space technology is undergoing a revolution in almost all areas, and we are much more capable than we were during the 1980s. There are three main developments that will make a Mars expedition much easier now than it would have been then. The first of these is the development of new materials, particularly composite materials and lithium–aluminium alloys that will help to reduce the weight of many of the Mars expedition structures and systems. The revolution in electronics and microcomputers that began with Apollo has now developed far beyond what could have been imagined then. In the areas of control, communications, machine intelligence, and sensors, huge strides have been made. Finally, of these three main developments, rocket propulsion has made great leaps forward, partly as a result of the opening up of Russian technology after the Cold War. Electric propulsion – once a dream – is now a commercial reality, and is used on many communications satellites and increasingly on deep space probes. That old and politically incorrect device, the nuclear rocket, is now close to becoming reality, with all that it implies for easier access to the Solar System. The human frame is not new technology, of course, but from the Russian Mir programme and the International Space Station we have also learned much more about how it reacts to being in space.

1.12 NEW MATERIALS

Weight is the continual obsession with rocket and spacecraft engineers. Konstantin Tsiolkovsky's famous rocket equation states that the velocity that a rocket or spacecraft can achieve depends on its *mass ratio* – crudely expressed, the ratio of the mass of the payload to the mass of the fully fuelled vehicle carrying it. The bigger the mass ratio, the easier it is for the vehicle to achieve the high speeds needed to gain orbit from a planet's surface, or to depart on an interplanetary journey. A large mass ratio means a large rocket with a small payload, and so the mass of the payload is critical.

The lightweight human being has yet to be invented, and there is also not much that can be done to reduce the mass of supplies for the journey; but there is much that can be done to reduce the mass of the interplanetary vehicle, the rocket's fuel tanks, and other equipment, by using new advanced structural materials. Whereas most of the Saturn V rocket and the Apollo capsule was built using aluminium, today we would use carbon-fibre-reinforced plastic for most of the structures and for the fuel tanks. By aligning the carbon fibres along the lines of stress in a particular structure, it can be made very strong and light – far lighter, for a given strength, than aluminium. Where metal is essential – as in the liquid oxygen tanks, for example – aluminium–lithium alloy can be used. This Russian development produces a material of much lower density than aluminium, but of equivalent strength. These new materials will reduce the mass of many structures, both in rockets and in the space vehicles and landers. They are already used in modern spacecraft, partly for commercial reasons. Mass saved in the structure of a communications satellite can be used for more communications equipment and a higher profit.

Another property of composite materials – their flexibility – has a particular application to the Mars expedition. In recent years significant progress has been made in developing an inflatable habitat for use by explorers on Mars, or even for their journey through space. An inflatable cabin has many advantages. In particular it reduces the necessary diameter of the launch vehicle fairing, and so of the vehicle itself. A rigid habitat for the Mars expedition might be 5 m in diameter in order to provide sufficient space for the crew during their long flight. And an inflatable cabin would be lighter, because many of the beams and other structural elements required for a rigid cabin could be eliminated, and it could also be launched, and landed on Mars, in a smaller vehicle.

1.13 ELECTRONICS

Modern electronics are not only lightweight and very modest in power consumption, but are immensely capable. This means that machine intelligence and computing power can be available wherever it will help during the Mars expedition. The first practical rocket vehicle – the German A4 – was guided by a single gyroscope connected by an electro-mechanical linkage to the carbon vanes that deflected the thrust of the rocket engine. The guidance of the Saturn V was rather more sophisticated, but basically the same. Now there can be as much computing power as required in each stage of a launcher. Millions of calculations per second can be carried out, the rocket's position, speed and acceleration can be constantly compared with the ideal trajectory, and corrections can be instantaneously delivered to the thrust vector control system. The same principle can be applied to all the stages of the Mars journey, and to any automatic rovers that are used by the explorers on Mars, to extend their reach. Of course. it is not foolproof, or proof against fools. It is humans who develop and load the computer programs that carry out all this work:

humanum est errare. There have been some spectacular failures because of errors in the programming of launcher computers, or even in converting from metric to imperial units. Human error will always be with us; but in general the application of machine intelligence will ensure that the Mars expedition will be both safer and more capable.

1.14 PROPULSION

Nowhere in the Mars expedition will technical progress be more important than in propulsion. Despite efforts in mass reduction, the success of the expedition depends on getting the crew with as much materiel as possible to Mars. This will not be easy, and crucial to the whole process will be the performance of the rocket engine that drives the interplanetary spacecraft. It could be a conventional chemical rocket engine – late descendent of the first Chinese gunpowder rocket – but the expedition would be much safer and more successful if some improvement over the performance of the chemical rocket could be achieved. There are two modern developments that have significantly better performance than the chemical rocket. These are electric propulsion and nuclear propulsion. One is already in use in space, and the other will probably come into use within the next ten years. (Later chapters in this book deal with the detailed construction and performance of these in-space propulsion devices, and so here we need only a brief summary.)

The other factor in Tsiolkovsky's rocket equation is the exhaust velocity of the rocket. This is the speed with which the exhaust gases leave the rocket engine nozzle. The higher this is, the larger the payload that can be carried for a given amount of fuel. The best chemical rocket engine – the main engine of the Space Shuttle – burns liquid hydrogen and liquid oxygen, and has an exhaust velocity of 4.55 km/s. It is very difficult to improve on this, because oxygen and hydrogen, for many reasons, have the maximum performance amongst available chemical propellants. However, the use of electricity to provide the energy in a rocket engine, rather than chemical reactions, can produce a large increase in exhaust velocity. Different types of electric thruster produce exhaust velocities ranging from 10 km/s to 60 km/s. This would provide a very large increase in payload capability or, alternatively, a faster journey. These devices are now in regular use on communications satellites. They have the disadvantage of low thrust, but for station-keeping this is not a problem, and the reduction in mass of the required propellant allows much more capable satellites to be flown. They also require an electrical power supply. Until now this has been provided by solar panels, which, in any case, have to be used to power the rest of the satellite. Electric thrusters are also used for interplanetary probes. The first one was Deep Space 1 – a probe that visited an asteroid. Here the superior performance of the electric thruster allowed the journey to be made with only 30 kg of xenon propellant. Electric thrusters have been base-lined for many new planetary missions, including ESA's mission to Mercury. Project Prometheus – a NASA initiative to develop and send probes, powered by nuclear electric propulsion, to the outer Solar System – will probably begin this year (2003). In the outer Solar System, sunlight is not strong enough to provide the necessary electric power using solar panels, and so nuclear sources are instead required. Nuclear- or solar-powered electric thrusters may well have a role to play in the Mars expedition, although significantly larger thrusters than have been used so far would be needed for a human expedition.

While the early developments leading to the Saturn V rocket were proceeding, the protagonists of human space exploration were concluding that the immense energy locked up in the nucleus would be advantageous for further exploration of the Solar System. The concept of a nuclear rocket evolved while people were working on the Saturn V. It was a natural outcome of rocket research and the nuclear developments of the time. 'Nuclear' was not then the dirty word that it has now become. Many people thought that nuclear power would provide almost free energy for everybody and revolutionise our civilisation. In those days, to propose a nuclear rocket was a perfectly natural thing to do; and there was even an experimental nuclear-powered aircraft that flew around the United States accompanied by a squad of marines to isolate the site should it crash.

The nuclear rocket is a very attractive technology for a Mars expedition. It is not essential, but along with other new developments it would make the whole thing safer and more capable. The principle is very simple. It is, in fact, just a gas-cooled nuclear reactor in which the cooling gas becomes the rocket exhaust. For those of us whose impression of nuclear reactors comes from nuclear power stations, the idea that one could be made small enough to fit in a rocket seems ludicrous. In fact, nuclear rocket engines are quite small. We are misled by two factors: the deliberate obscurity about the amount of uranium needed to make a reactor, promulgated for security reasons during the Cold War; and the different requirements of a power station and a rocket engine. The former has to run for twenty years or more to be economic, and is located close to civilisation. This requires a large amount of fissile material and moderator, and a great deal of shielding. A nuclear rocket is used only in space, and operates for a total of an hour or so. Much less fissile material is required, and much less shielding. The experimental engines developed in the 1960s were no bigger than the ordinary rocket engines of the time. Indeed, they *were* developed both in the US and the USSR; and they were tested in the Nevada desert and 'somewhere' in the Soviet Union. It was only the Test Ban Treaty that stopped the programmes, because they released a small amount of radioactive material into the atmosphere.

The attractiveness of the nuclear engine for the Mars expedition is that it combines the high thrust of a chemical rocket with a moderately high exhaust velocity – about twice that of the Space Shuttle main engine. This means that a large spacecraft can be accelerated to a high velocity with modest fuel requirements. While in the heady days of the 1960s a launcher with a nuclear upper stage was contemplated, this would not be acceptable. The nuclear engine would only be used to send the interplanetary spacecraft on its way to Mars. It would be launched in an inert condition and would only become fissile once in space, and after use it would be put into a safe orbit round the Sun. An inert nuclear fission core is not radioactive. It is made of enriched uranium, which is no more radioactive than granite road-stone, and it is only when the reactor is operated that it becomes radioactive. A launcher accident would not be any more hazardous because it carried an inert nuclear rocket engine.

Note that this is different from the *radioactive thermal generators* (RTGs) that are used to provide electrical power on deep space probes such as Cassini and the Voyagers to the outer planets. These contain plutonium – which is very radioactive – and it is the means by which they produce power. Extreme precautions are taken in the design of the container so that it will not burst, even in the worst disaster scenario. Nothing like this can happen with an inert nuclear rocket engine, as the uranium it contains is not harmful. (More on these issues will be found in a later chapter.)

The nuclear rocket engine is not essential to the Mars expedition, but it would make it much easier. There is a revival of interest in nuclear engines – solely for deep space use – and many new designs exist based on the developments of the 1960s. Non-fissile tests are already being carried out on the heat transfer from the fissile material to the exhaust gas. A nuclear rocket programme will possibly be approved within the next few years – an appropriate time-scale for the Mars expedition.

1.15 TECHNOLOGICAL READINESS

The above brief summary of new technology that could be appropriate for the Mars expedition indicates just how far we have advanced, in many ways, since the expedition was first proposed in the 1960s. The capability to carry it out existed then, and our capability now is even greater. At the beginning of the twenty-first century we have many preoccupations, and the technology of space flight is obscured in the public imagination beneath the vast cloud of personal technology that everyone can use. The rather pedestrian activities with the Space Shuttle and the International Space Station, and the media interest in disasters, mistakes, and incompetence in the space agencies, all combine to lower our respect for the achievements and the quality of today's space programmes. But the quality is still there, and the achievements are still amazing. All over the world, space agencies are developing new technology and pursuing advanced programmes. Budgets are tight, as governments find new things on which to spend their money, and space is no longer seen as the jewel in the crown of a nation's technical development. It is just one of many demands on limited national research budgets. Nevertheless, the world has advanced steadily in space capability from the first days of Sputnik and Apollo.

The only area in which we fall short of those far-off days is in launcher capability. Away from the Space Shuttle, the almost exclusive preoccupation with satellites has allowed modern launchers to develop in the direction of cheap launches for payloads of about 20 tonnes. The Shuttle is man-rated and can carry about 24 tonnes of payload, but no-one would describe it as cheap. For the Mars expedition we require a launcher that can place 80 tonnes in low Earth orbit – just four times our present capability. In the 1960s the Saturn V could place 118 tonnes in low Earth orbit using rather crude technology. The message is clear. If it could be achieved then, it can be achieved now, and much more easily. The rest of the expedition is no more than normal business. We could use existing rocket engines, as there is no absolute need to go to electric or nuclear propulsion. The spacecraft construction and the human aspects of the expedition are well understood from International Space Station experience. The expedition to Mars is not held back by lack of appropriate technology, but for other reasons.

1.16 THE OLD ARGUMENT

'Space flight is a waste of money'. This refrain is heard everywhere; and if not explicit, the thought is behind most utterances of the media on space. 'Accidents and disasters are just exemplars of its truth, and the achievements trumpeted by the space agencies and the discoveries of the space scientists are all irrelevancies, of interest only to the few.' 'If ordinary activities in space, such as the Space Shuttle and unmanned probes to the planets, are a waste of money, then an expedition to Mars is an even greater waste.' When presented with such statements it is difficult to respond with a single argument – and there is no single argument that justifies the mounting of a Mars expedition. It is a combination of factors, and, indeed, a knowledge of what it means to be human, that leads to the inevitable conclusion that now is the time to go to Mars.

Technically we are ready: there are no technical obstacles that remain to be overcome; the time is now. But if we *can* do it, *should* we do it? There are many reasons why we should. We should satisfy our craving to explore the unknown; we should not wilfully remain ignorant when we have the capability to pursue knowledge; we should not waste the wealth and talent of our civilisation purely on extending the human lifespan and on frivolous pursuits – improving our cooking, making movies, taking ever more expensive foreign holidays, driving better cars, and so on; we should not leave our young people uninspired by great achievements; we should not fail to leave our mark on history. There are many other reasons. The counter-arguments all relate to the 'waste of money' theme. It would be better spent on something else; the tangible benefits to humanity would be small; science is dangerous; only the rich nations can participate; what about famine, war, pestilence? The list is long.

However strong the negative arguments, we cannot avoid the question itself. The technology is ready; the expedition can be mounted; the choice now is whether to go, or not to go. If we choose to stay on Earth, then things will go on much as before; steady progress in life expectancy, at least in the rich countries; more and better personal technology; improvements in agriculture, and perhaps progress against the horsemen of the apocalypse – famine, pestilence, war and death. If the early years of the twenty-first century are anything to go by, a line from the old rhyme comes to mind: 'At present, the other side's winning.' Nevertheless, the argument to stay is honest, at least in intention. Time would show if it had indeed been the right decision. If we do decide to go to Mars, then things on Earth will proceed more or less the same: money not spent on the Mars expedition would not help much with the major terrestrial problems, even if it were to be diverted to that aim; human nature makes such a diversion improbable.

If we do decide to go to Mars, then for twenty years the most talented people we have will work together in an international team, with a single purpose. Humanity will be making a serious step forward, in which everyone on Earth can share, in one way or another. Vast leaps in technology will emerge, as each challenging detail is overcome. Young people in school and college will feel inspired to follow each stage of the mission, and more of them will study science and technology – a sorely-felt need at the moment, and vital for our future. The knowledge that the explorers will bring back with them will have a profound effect on our understanding of life in the Universe. They may find that life also emerged on Mars, or that it did not. It will also change our ideas of the past and future of our planet. Mars is both an image of how our planet once was, with its cratered terrain and absence of water erosion, and an image of the possible future, with a thin atmosphere and no surface water. Understanding the present condition of Mars will help us foretell the future of our planet. Finally, the achievement of the explorers will set a new baseline for human achievement and technological development that will impact on all our lives.

We cannot escape the decision: to stay, or to go.

2

Werner von Braun and the martian dream

The journey to Mars has stimulated the imagination of many, from the seventeenth century onwards. Johannes Kepler, after his great work in determining the orbit of Mars, produced an imaginative story about a journey there. In some cases the technology is identified – swans, or angels – and in others the journey is accomplished by other supernatural or unspecified forces. From the early twentieth century we have the fiction of Edgar Rice Burroughs, the creator of Tarzan of the Apes. Inspired by Percival Lowell's 'canals', Burroughs published a number of martian stories in which his traveller was transported by supernatural power to Mars. In 1937, when the idea of rocket-propelled journeys was already current, C.S. Lewis chose to transport his hero, Ransom, to Mars in a vehicle using an unspecified propulsive force. Lewis despised what he called 'enginry' as a means of transport to the planets. There are, of course, a vast number of twentieth-century stories that use rocket propulsion to make the journey. For nearly all authors, the motivation is focused on Mars itself, and what happens when the travellers arrive there. But there is very little interest in the technology of getting there – and that is the subject of this book.

During the 1930s, the members of the rocket societies in Germany, and the British Interplanetary Society, studied the technical demands of a Mars expedition; examples of some of these papers can be found in the *Journal of the British Interplanetary Society* for those years. The Second World War instigated both the first practical rocket development, and the temporary diversion of attention from the exploration of space to the extermination of fellow human beings. When the dust had settled, one rocket engineer, Werner von Braun – extracted from the German V2 programme with several hundred colleagues, and employed in the United States – prepared a detailed article for *Weltraumfahrt*, a German rocket magazine. Called 'Das Marsprojekt', this article was a careful exposition of the physics and dynamics of the journey, together with a description of the rockets that would be used. Von Braun was impatient with what he called the 'backyard inventor' with his shiny rocket, so popular in Mars stories of the time, and wanted to set down how the expedition to Mars could really be accomplished. He proposed no fancy new technology, but good solid rocket power – the technology that he himself had developed before and during the still recent war. It was 1952, and von Braun had been in America for six years. Two years earlier he had been transferred from Fort Bliss, Texas – where he and the other German experts had been virtual prisoners – to the Redstone Arsenal near Huntsville, Alabama. Here he could begin to develop the series of rockets that would eventually lead to the Saturn V. He was again thinking seriously about space exploration.

Immediately after the Second World War, the primary interest of the military was, of course, missiles, and the many experiments carried out with captured V2 rockets – called A4 by the Germans, before it became a weapon – were devoted to the training and educating of both military and other personnel in the rocket business. This was why the US Army had accepted von Braun's surrender, and had seized so much of the V2 equipment. The Cold War was in being, as was the atom bomb, and the

Figure 2.1. Werner von Braun and the Saturn 1B. (Courtesy NASA.)

potential of ballistic missiles was on the military mind. At the same time, von Braun had always wanted to develop rockets for space exploration. In Nazi Germany he had not been able to express such wishes freely; but in the United States he gradually realised that he could do so.

2.1 THE HISTORICAL CONTEXT OF DAS MARSPROJEKT

The period from 1946 to 1960 was the seed-time of the Space Age; the secrets of the first practical liquid-fuelled rocket were by then available both to the United States and to the USSR. In both nations, the seeds were sown that would lead to the first satellite, the first man in space, and ultimately the first men on the Moon. The arms race was in full swing, and the space race had yet to begin, although on both sides of the Iron Curtain the preliminary rumblings were evident.

It is difficult to overestimate the importance of the inter-war German development of the first practical liquid-fuelled rocket, the A4 (or V2). The father of space travel, the Russian, Konstantin Tsiolkovsky, had, as early as 1903, identified liquid fuel as the necessary technology for the conquest of space, in his paper 'Exploring Space with Reactive Devices'. It was only through the superior exhaust velocity, which could be attained using liquid fuels, that sufficiently high vehicle velocities could be achieved for escape into space. The engineering knowledge, and the special alloys required to make a practical device, were not available in 1903, although by 1926 Robert Goddard was able to fly the first liquid-fuelled rocket, in Aubern, Maryland. It was powered by liquid oxygen and gasoline, in very much the same way as the lower stages of today's launchers. Goddard had independently developed the same ideas as Tsiolkovsky; and once this was known, they exchanged friendly correspondence. Goddard went on to invent many of the practical devices necessary to make liquid-fuelled rockets, and took out numerous US patents. However, government interest in rockets was not strong in the United States during the inter-war years; and even at the outbreak of the Second World War, when Goddard offered his discoveries to the government, they were not particularly interested.

But the situation was very different in Germany, where amateur rocket societies such as the Verein für Raumschiffarht flourished, and attracted the best and brightest young engineers and scientists. In those times, following the Treaty of Versailles the German military was forbidden to manufacture great guns. Recognising the military potential of rockets, which did not break the provisions of the treaty, a young artillery captain approached the VfR with a small grant to build a rocket. In 1932, Von Braun's rocket failed in the demonstration flight; but he so impressed Captain Walter Dornburger that he was hired to lead the technical side of the military rocket programme. Hitler disestablished the VfR in 1933, and by 1934 a team of eighty engineers was working on developments that were to lead to the A4 – later to be used as a weapon called the V2. The V2 weapon was terrible, and the methods of producing it were inexcusable; but the rocket itself was the key to the Space Age.

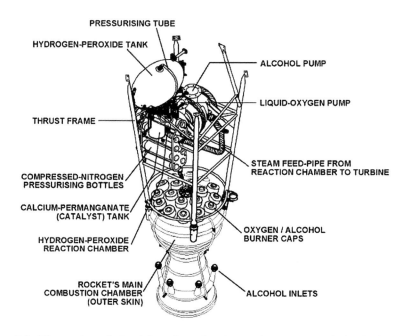

Figure 2.2. The engine of the A4 rocket, showing turbo-pumps and the hydrogen peroxide tank. Hydrogen peroxide was decomposed to provide gas to drive the turbo-pumps, liquid oxygen and alcohol were the propellants, and alcohol was used to cool the nozzle and combustion chamber. This was the first practical high-thrust liquid-fuelled rocket engine.

Everyone knows the theory of the rocket, and quite a few people could cobble one together – and many of us have tried! But they mostly explode on the launch pad, or, if they do lift off, their trajectories range from the ridiculous to the downright dangerous. The A4 rocket, however, was the first to be self-guiding and stabilised. Any rocket, with the thrust acting on the bottom of a long column, is inherently unstable, and one of the many advancements achieved by von Braun and his team was automatic stabilisation, using carbon vanes in the nozzle to direct the exhaust stream. They were controlled by a gyro to keep the rocket pointing in the right direction. The vehicle also had turbo-pumps to deliver the fuel to the combustion chamber at high pressure. In fact, most of the sub-systems found in a modern rocket have their origin in the A4. So it is no surprise that when the Peenemunde team was split up, and part transferred to the USSR and part to the United States, the Space Age really began.

It was the 1950s. Rock and roll had still to appear, and Elvis Presley would make his first trial recording in 1953, the year that von Braun's book was published in America. There was growing self-confidence in the United States, and the love of 'new' things, already endemic in American culture, was growing again. The time was ripe for new ideas, and exploration of the planets began to seem possible. Von Braun was released from the strictures of the Nazi régime, but was still working for the

Figure 2.3. The first launch from Cape Canaveral: an A4 rocket with a Corporal upper
stage. The assemblage was called 'Bumper'. (Courtesy NASA.)

military in the United States. Missiles were on the agenda, but he was thinking about
other things: the dream that had led him into rocketry as a youth; the dream that
now began to appear as if it could become reality; the dream of putting men into
space. In his spare time he began to work out the details for the human exploration
of Mars.

At the Redstone Arsenal, work continued on successors to the A4 rockets
captured from Peenemunde; and in particular, the Redstone missile was being
developed. It used liquid oxygen and alcohol as propellants and eventually was used
for nuclear warhead tests; it was subsequently developed into the Jupiter-C vehicle
that launched the first US satellite. Tooling and designs from the Jupiter were later
to be used for the Saturn V vehicle development. So, while von Braun was thinking
about the mission to Mars, he was also developing rocket technology that would
eventually lead to the Saturn V.

2.2   THE MARS PROBLEM

The correct method for reaching the planets had been determined by Hermann
Oberth and Hohmann during the 1920s. It was necessary to escape from Earth into
an elliptical orbit, round the Sun, which would take the vehicle out to the orbit of
Mars. Von Braun knew all this, and it irked him to see the kind of 'Buck Rogers'
approach, of blasting off in a direct flight to Mars, seen in so many films and novels

of the time. Science fiction afficionados will know that the late 1940s and the 1950s were the golden age of the genre, and so everyone was aware of the possibility of human spaceflight. Stories involving colonisation of the Solar System, and the different life forms to be found on the other planets, were part of the background. Venus had yet to be shown to be inhospitable, and for Mars there was the lingering idea of the canals, and of an old, possibly extinct, civilisation. But in most cases the technological descriptions of space flight were wildly inaccurate. Von Braun knew exactly how a rocket-powered journey to Mars should be undertaken. With his knowledge – perhaps unique in the western hemisphere – his slide-rule, and his tables of logarithms, sines and cosines, he could calculate the mission precisely. He would determine the proper trajectories, he would calculate the amount of fuel required, and he would select the appropriate rocket technology for the vehicles. He would set down the complete Mars expedition, and how it could be carried out. He would create Das Marsprojekt.

2.3   THE MARTIAN DREAM

In the early twentieth century, the journey to Mars had triggered the imagination of many, including pioneers such as Tsiolkovsky, Oberth, and later von Braun. Mars came from the nineteenth century, trailing the idea, put forward by the American, Percival Lowell, that the *canali* of Schiaparelli were real canals, built by martians to husband the dwindling water resources of their planet. This theory was, of course, hotly disputed, and by the time that the first space pioneers were writing, it had been fairly well discounted. The martians and their canals were effectively debunked, both by better observations, and by Alfred Russell Wallace (who independently developed

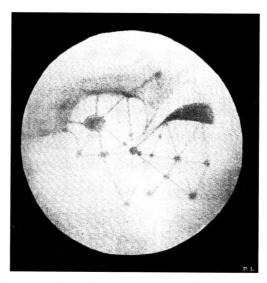

Figure 2.4. Canals on Mars. (From *Mars and its Canals*, by Percival Lowell.)

the theory of evolution). He argued that Mars was too cold, that the polar caps were frozen carbon dioxide, and that water would evaporate from canals more or less instantly. Nevertheless, Mars became the most interesting planet for humans to explore – and it remains so today.

When the pioneers envisaged spaceflight, Mars was in their minds. Lowell had published his book *Mars and its Canals* in 1906. Three years earlier, Tsiolkovsky published *Exploring Space with Reactive Devices,* in which he discussed both liquid-fuelled rockets and soft landings on planets, and also predicted the founding of settlements. All the early rocket enthusiasts considered the exploration of space and the planets. In the USSR, in addition to Tsiolkovsky, Freidrich Tsander, during the 1920s, wrote about a journey to Mars, and in 1929 Yuri Kondratyuk discussed many of the techniques for a real Mars mission, including atmospheric braking. In Germany there was the work of Herman Oberth. He was the driving force behind German rocket enthusiasm, and was the main inspiration for von Braun and his colleagues. Von Braun joined Oberth as a new member of the VfR. German space enthusiasts had been enthralled by Oberth's book *Die Rakete zu den Planetenräumen,* published in 1923, and von Braun was no exception. For a while, Oberth actually supervised the activities of von Braun and two student colleagues, who were attempting to fly liquid-fuelled rockets from a disused ammunition dump near Berlin. When von Braun, having been supported by a military scholarship at University, began work (as a civilian employee of the military) that would lead to the A4, exploration of the planets was still discussed by the team members, out of hours. In fact, in 1944 von Braun was arrested by the Gestapo and the SS for spending too much time talking about space exploration, to the supposed detriment of the Fatherland's defence programme, and was released only on Dornburger's direct appeal to Hitler. Nevertheless, the idea of a manned mission to Mars was a constant background to practical rocket developments.

Looking back, it is possible to see that the major technical advances required for space exploration did not at all arise from the need to explore space. The exploration

Figure 2.5. Walter Dornburger and Werner von Braun. (Courtesy NASA.)

of space was, however, a need strongly felt by many of the protagonists, as indeed it is felt by many people today. The only way to engage the engineering and financial support necessary to develop rockets was to exploit more pressing national interests. Without the German war effort, there would have been no A4; if Khruschev in the USSR, and Kennedy in the USA, had not seen superiority in space as an essential element of the Cold War, then there would have been no Saturn V and no Apollo. There would be no footprints on the Moon. The engineers and scientists wanted to explore space, and make rockets that worked; the people with the money, governments and the military, had different priorities. In such situations, a purist attitude cannot be adopted. The space race was good for science, for our knowledge of the Solar System, and ultimately for humanity; but it happened because of the Cold War. None of the scientists and engineers involved was going to point out the impure motivation of their governments.

The dream was to set foot on Mars, but the Moon was an easier target, and eventually the Moon was set as a national goal, both for the United States and for the USSR. It was a genuine race, with a very similar level of government backing. Finally, however, the USSR, having achieved so many 'firsts' in space, did not succeed in winning this particular race. Many people believed that after the Moon, NASA would exploit its success and go on to explore Mars, but in the event, this did not happen. With the USSR effectively out of the space race, there was no need for the United States to exert itself further, and the Apollo programme was terminated by the US government. The follow-on to Mars was shoved back in the drawer, to await a more propitious time.

Figure 2.6. Werner von Braun, President Kennedy, and a model of the Saturn launcher. (Courtesy NASA.)

But when von Braun put together Das Marsprojekt, all of this was in the future, and the dream was still intact. Von Braun and others hoped eventually to achieve human spaceflight, but in the meantime there were large government resources available to design, build and fly bigger and better rockets. We should not criticise these men for accepting military money, as they were effectively instructed to work on military vehicles; and in the 1940s they would not receive funding to build rockets by insisting on planetary exploration as the aim. The positive aspect was that while, in wartime Germany, von Braun was arrested for talking about space exploration, he was not going to be arrested for writing about Mars exploration, or indeed anything else, in post-war America. He was free to express his hopes and dreams. He was also at a happy stage in his life. He had recently married, he was soon to be appointed Director of the embryonic NASA Marshall Space Flight Center; and his adopted country was pleased with him.

It is probable that most of the ideas that von Braun used to facilitate the Mars journey had been discussed many times, from the 1920s onwards, by enthusiasts. Many of the above-mentioned pioneers were instrumental in specifying the problems to be overcome, and in suggesting how they might be solved. Von Braun's book was the first appearance of all of these ideas put together to form a complete mission plan.

2.4   THE MARS PROJECT

Most people are familiar with the concept of escape velocity – the speed necessary to escape from Earth's gravity; but it is not enough set off from Earth at the escape velocity. Mars is moving round the Sun, as is the Earth, and because the journey takes a finite time, the planets will move during the journey. The rocket carrying the explorers has also to travel in an orbit round the Sun – an elliptical orbit this time – that links the orbits of Earth and Mars. Then there is the business of capture by the gravity of Mars when the spacecraft arrives, and how to land on Mars. And for the return journey, all of this has to be carried out in reverse. Von Braun knew how to calculate and determine all of these factors, and he set out to explain them, not for the uninitiated (due to the complex mathematics), but for the interested amateur who would be keen enough to follow the mathematical and physical arguments. He chose the German space magazine *Weltraumfahrt* (*Spaceflight*), to set out his ideas. He was assured of an informed readership, as Germans were still interested in rockets and spaceflight, and he had sufficient space, in the long article, to deal with everything in detail. And it is probable that the editors were keen to have a German, writing for their magazine, who was already making a name for himself in post-war rocket science (von Braun did not take up US citizenship until 1955). In the event, in the United States there was so much interest that he was able to publish an English translation the next year, 1953, as a book, *The Mars Project*, published by the University of Illinois Press.

Like many Mars enthusiasts today, von Braun clearly saw the difficulty of predicating a Mars expedition on some exotic future technology. To do this would be to make the proposal more like science fiction; and, of course, the start of the project would be delayed until the technology had been developed. He hoped to see men on Mars in his lifetime, and the best way to achieve this was to use the technology current at the time. This was the technology with which he was thoroughly familiar, having to a large extent developed it, and he knew it could be used to achieve a practical Mars expedition. He proposed to use simple liquid-fuelled chemical rocket engines, very similar to those used on the V2 and to those he was developing for the Redstone missile.

Von Braun was not only a rocket engineer, as he had, in addition, vast experience of the organisational and logistical problems associated with a major rocket project. He had been technical director of the Peenemunde centre, and no less a person than Oberth called him 'the ingenious organiser and engineer who brought the ideas of space flight to practical realisation'. So, he brought to the Mars project his unique knowledge of rockets, his interest in spaceflight, and his proven ability as a leader and organiser.

The first myth that he wished to dispel was that of the lonely inventor and his backyard rocket. While he was a student and an engineering apprentice, he had been one of a small group trying to develop rockets. Even with Oberth as their mentor, the students achieved only modest progress, succeeding in sending a tiny liquid oxygen and gasoline rocket to about 300 m altitude. But in 1934 he became the technical leader of the eighty-strong team put together by Dornburger. In that year the A2 rocket was successfully fired, and by 1936 the A4 was in design. The first successful flight took place on 3 October 1942. Von Braun knew that small teams could not achieve success in rocketry and that a large team of specialists was required to solve all the problems; and he applied this logic to his Mars project – an expedition which would take seventy people in ten interplanetary spacecraft. He argued that many different specialists were required, and that the possibility of accidents had to be taken into account, so that one ship of the flotilla could aid another in distress. He even quoted the mission of Columbus, with his three ships, as an example of a successful expedition that would almost certainly not have returned if it had consisted of only one ship. This was partly the result of his conviction that only big teams could achieve real success; but it also reflected the times, when men were required to do so much that is, today, carried out by machines. So, this pioneer of the super-modern in technology still looked to humans to carry out almost all the activities necessary for the expedition. As an example of the flavour of von Braun's past experience, his Peenemunde team had been supported by large numbers of trained women 'calculators', who, using slide-rules, analysed all the data from engine and flight tests.

This large expedition required that the spaceships, all the fuel and supplies, and the explorers themselves, should be assembled in Earth orbit prior to departure for Mars. To do this required what he called the three-stage ferry vessels, to carry all this tonnage into Earth orbit. His expedition vehicles consisted of these ferry vessels, the spaceships themselves, and the charmingly named 'landing boats' to transfer the expedition from Mars orbit to the surface of Mars. He envisaged the ten spaceships departing from Earth orbit together, and some ship visiting, using 'space boats', would take place during the long voyage to Mars, perhaps to help a vessel in trouble.

The flotilla would arrive at Mars and go into orbit round the planet, and the landing boats would then transfer the expedition to the surface. Once the exploration tasks were completed, the voyagers would return to the orbiting spaceships, leave Mars orbit to return to Earth orbit, and descend to Earth's surface.

Each stage was carefully considered, and the necessary calculations carried out. But this was no flight of fancy. It was a solid design exercise by perhaps the leading expert of his day.

2.5   THE SPACESHIPS

The celestial mechanics that control the actual journey to Mars are explained later in this book, but here we need a summary in order to understand von Braun's concept of the Mars Project. His flotilla was to be assembled in Earth orbit, and he chose an orbit with a two-hour period, at about 1,730 km above the surface. To remain in such an orbit a spacecraft has to be accelerated to a speed of more than 7 km/s, and at the same time it has to be raised, against the force of gravity, to an altitude of 1,730 km. This is the Earth launch problem, and the vast quantities of energy released in a Space Shuttle launch, for example, demonstrate this energetically difficult procedure. Once in Earth orbit, comparatively smaller increases of speed are sufficient to send the flotilla on its way to Mars. The spaceships never enter a planetary atmosphere, neither do they experience the force of gravity in a landing or take-off, so they can be very flimsy compared with launchers or landers. This was why von Braun wanted to assemble them in space. Much less massive structures would be required to hold everything together, and this would save weight to be carried up by the ferry vessels. Everyone knows that in space there is no need for aerodynamic shapes or streamlining, and von Braun was keen to dispense with the shiny, pointed rocket ship. These vessels could, in fact, be a conglomeration of fuel tanks, cargo containers and crew compartments, attached to the rocket engines by a lightweight structure. The only acceleration they would experience would be the departure manoeuvre from Earth orbit – to be duplicated in reverse on return – and a similar pair of manoeuvres at Mars.

The departure manoeuvre of the fully assembled spaceships is a simple acceleration from the circular orbit into a hyperbolic (open-ended) orbit. To achieve this a precise velocity increment is required, calculated so that when the spacecraft has travelled far enough away from Earth to be effectively free of its gravity, it will have enough velocity left to enter the so-called Hohmann transfer orbit to Mars. This is done by firing the rocket engines once, for a stipulated length of time. The firing has to take place at the correct position in the orbit round the Earth, so that the spaceship is travelling in the right direction, once it leaves Earth's influence. Von Braun calculated both the velocity increment and the position in the orbit necessary to send the expedition on its way to Mars. The velocity increment was 3.31 km/s, in addition to the 7.07 km/s required to stay in the parking orbit – a total velocity, relative to the Earth, of 10.38 km/s. The position of the start of the departure manoeuvre was calculated to be 242° before the point in the orbit where it is crossed

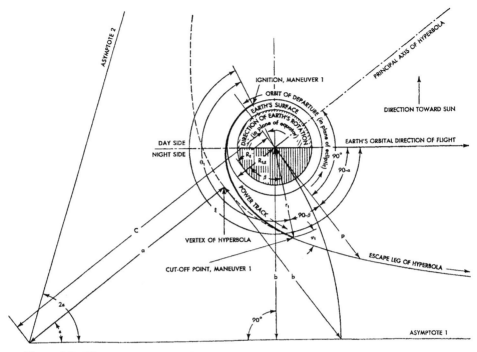

Figure 2.7. Werner von Braun's drawing showing the manoeuvre for Earth departure. (From *The Mars Project*, courtesy University of Illinois Press.)

by the extension of a line passing through the Earth from the Sun. This latter position is the real departure point for Mars, and at this time the spaceship is travelling in a course perpendicular to the Earth–Sun line. A moment's thought will show that it is not travelling towards Mars at this point, but more or less at right angles to the line joining Earth and Mars. (This is one of the vagaries of celestial mechanics, and will be explained later.) It is also one of the facts of the case that von Braun wished to explain to the public. All these calculations were undertaken using calculus and geometry, with paper and pencil, plus the tables of functions such as logarithms, sines and cosines. There were no electronic calculators, and no computers; only a slide-rule.

To calculate the amount of fuel that the rockets would burn, von Braun needed to know how much mass had to be shifted to Mars. There were the seventy men; but how much food, water and oxygen would they require? This depended on how long they would be away. He computed that the journey to Mars, and the return journey, would each take 260 days, and supplies would therefore be required for 520 days in space. But how long would they remain on Mars? This is fixed, because the planets have to be in the correct position relative to one another for the return journey to be successful, for Earth to be in the right place when the spaceships return. This is 449 days. The supplies then have to last for a total of 969 days (2 years 8 months). The consumables for the mission are based on 1.235 kg of oxygen per day, 1.2 kg of food per day, and 2 kg of drinking water, for each expedition member. In addition there is the equipment and many other items, producing a payload mass of 26 tonnes for each of the seven crewed ships. The empty ships themselves weigh 25 tonnes, and the three cargo ships contain the landing boats and 195 tonnes of additional materiel. The difference in payload mass for the cargo vessels is due to the fact that they will remain in Mars orbit when the crewed vessels return to Earth, so they do not contain fuel for departure from Mars.

The amount of fuel required for the rocket engines also depends on the engines themselves, and which propellant combination they use. Here, von Braun chose to be conservative. He had most experience with liquid oxygen and gasoline or alcohol as propellants; but although these are effective and powerful, they have a disadvantage in that liquid oxygen can be stored only under refrigeration. This would not suffice for a long voyage in space, where electrical power would be very limited and refrigeration equipment would be very heavy. Instead, he chose not to use the traditional 'burning' process of a fuel with oxygen, but to use the different chemical reaction of nitric acid and hydrazine. He was effectively looking for oxygen in the form of a room-temperature liquid. Nitric acid is just such a room-temperature liquid, and contains three oxygen atoms for each nitrogen atom; by weight it contains 70% oxygen. Hydrazine is a chemical relative of ammonia, with a higher molecular weight. It is a liquid at room temperature, and, moreover, has the useful property that when mixed with nitric acid it spontaneously ignites. This combination of propellants is therefore very useful for in-space propulsion, because the propellants can be stored in sealed tanks (as can petrol), and the rocket engine

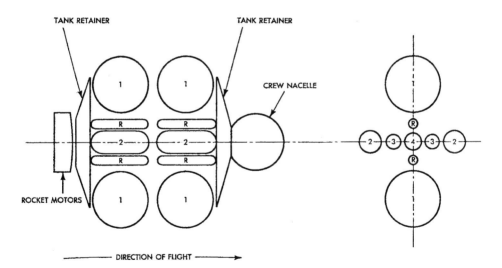

Figure 2.8. Werner von Braun's drawing of a spaceship, showing the fuel tanks for the Earth departure manoeuvre (1), Mars capture manoeuvre (2), Mars departure manoeuvre (3), and Earth capture manoeuvre (4). (From *The Mars Project*, courtesy University of Illinois Press.)

can be automatically fired up by simply opening the valves to allow the two propellants to mix in the combustion chamber, where they will spontaneously ignite. So useful is this property that this combination, in a modern form, is used for all manned in-space manoeuvres even today – for example, in the Space Shuttle orbital manoeuvring system. It is comforting to know your rocket will start when you want to come home. The method by which von Braun and his VfR colleagues started their early liquid oxygen and gasoline rockets was to toss a flaming petrol-soaked rag into the mouth of the nozzle, after the propellant valves had been opened.

Nitric acid and hydrazine is not the most efficient rocket fuel combination. It has a relatively low exhaust velocity, so that more fuel has to be burned per tonne of payload to achieve a given vehicle velocity. Nitric acid and hydrazine are both very toxic; there were several fatal accidents in the USSR space programme involving these propellants. They therefore need careful handling. Nevertheless, their advantages are so great that they are base-lined, in their modern form, for many future space missions. Von Braun liked the idea that they could be stored in inflatable tanks. He suggested that nylon could be used to create a reinforced flexible fuel tank that would be much lighter than a metal tank, and could be brought up from the Earth, deflated. He also proposed the same approach for the cargo and crew modules; nylon was, after all, the archetypal modern material in the 1940s. In fact, very similar ideas are proposed today, using carbon fibre instead of nylon. All in all, many of von Braun's ideas were very forward looking, and cannot be faulted even today, half a century after he first proposed them. Given the low efficiency of the chosen propellants, von Braun calculated that he would require 3,663.5 tonnes for the crewed vessels that would have to return to Earth, and 3,306 tonnes for the cargo vessels that would stay in Mars orbit.

2.6   THE LANDING BOATS

The landing boats were needed to transfer the expedition to and from the surface of Mars. At the time, the atmosphere on Mars was thought to have a surface pressure of about 80 millibars – 8% of that on Earth. Given that the gravity of Mars is only 38% of that on Earth, winged vehicles were perfectly feasible. However, we now know that the pressure is about 7 millibars – a tenth of what von Braun thought – and the use of wings would be more difficult with such a low pressure. His landing boats were winged, and were designed to fly down to the surface, rather as the Space Shuttle does today. He calculated everything, as usual, estimating the size of the wings required, and how much they would heat up in traversing the atmosphere. The wings would have a span of 153 m and would heat up to about 650° C, and the boat would land at 196 km/hour – similar to a modern jet airliner. But he was worried about the landing. Such a high landing speed requires a long runway, or at least a very long cleared area; and running into rocks at 196 km/hour would be very unpleasant. His solution was to land one boat, using skids, on the polar cap, where the ice would be smooth. The polar cap is a very inconvenient place for the departure because it would be virtually impossible to meet up with the orbiting spaceships, using only rocket power. The winged landing boat could fly to the required position, on the way down, but the wings – which weighed 30 tonnes – would have to be discarded for the return journey and the flight made by rocket power alone. It would have to take place from somewhere near the equator of Mars, because the arriving and departing spaceships could only enter and leave an orbit not far from the equator. The alignment of the planetary orbits absolutely requires such a manoevre. The plan was to land one boat on the polar cap, after which the crew could then prepare an airstrip for the succeeding boats to land on, nearer the equator. Von Braun ignored the huge difficulty of a journey of thousands of kilometres across totally unknown country. Homer might have nodded; but even today there remain the difficulties of landing on unprepared sites. The Apollo 11 lunar lander had to dodge unsuitable terrain, and several Mars probes have been lost on landing, presumably due to topographical anomalies – or, as we say, rocks.

For the return journey from the surface of Mars, the landing boats needed to be stripped down and their wings removed. They would have landed horizontally on the prepared runway, but needed to be raised to a vertical position for launch. This is one of the areas in which the large expedition was an advantage, as amongst seventy men there would be plenty of muscle power to unbolt the 15-tonne wings and to raise the rocket into a vertical position. Presumably they would use sheer legs and blocks and tackle to do this, very much in the tradition of the navy during the nineteenth century. Von Braun drew up various schemes, recorded in his personal notebook, for the erection of the rockets. Once the six landing boats had been erected, the crew would enter and lift off, to rendezvous with the orbiting return ships. They would leave behind the wings and the boat that had been landed on the ice cap, and would bring their samples, and anything else found on Mars, back to Earth.

It is curious to reflect on why von Braun did not use the ablative re-entry techniques – which were beginning to be developed for ballistic missile warheads – to slow the spacecraft, followed by a vertical, rocket-cushioned, soft landing. He was certainly familiar with the problem of re-entry, because a significant number of V2 warheads exploded prematurely due to re-entry heating. The wings, despite their huge area, would have been very useful for manoeuvres such as polar diversion; nevertheless, they were a huge complication for the mission, as they required two separate landings at locations thousands of kilometres apart. Vertical landing allows the landing site to be changed a little while descending, and only a small clear area is needed. The then current erroneous idea of the thickness of the martian atmosphere nodoubt played a part in the selection of wings, and it might also be wondered whether the re-entry techniques were top secret in those days, applying, as they did, to nuclear warheads. Certainly, within ten years, the Mercury programme was using ablatively cooled re-entry capsules to bring back the pilots, and in practice it is the only method by which astronauts are returned to Earth – even now, forty years later. As will be seen later, ablative cooling and atmospheric braking play a large part in the latest Mars expedition concepts. Von Braun also used wings to return his ferry vessels to Earth; but there was complete silence regarding ablatively cooled re-entry.

2.7   THE FERRY VESSELS

Having calculated the amount of propellant necessary – more than 3,000 tonnes for each spaceship – a powerful ferry vessel was required to place such a large mass into Earth orbit. Von Braun conceived a three-stage launcher, with a winged orbiter that could fly back to the base after delivering its cargo to orbit. This winged orbiter had a length of 15 m, a wingspan of 52 m, and a diameter of 9.8 m. Its dry weight – without fuel – was 27 tonnes, and it could carry a payload of 39 tonnes. It is interesting to compare this with the Space Shuttle orbiter, which is 37 m long, has a wingspan of 24 m, weighs about 80 tonnes empty, and has a payload capability of 25 tonnes. Given that not much was known about hypersonic flight at high altitude, and that no multi-stage rocket had yet been constructed, there is a remarkable similarity between the ferry third stage and the Space Shuttle, designed thirty years later. The complete three-stage rocket was 60 m high 'without fins' – a 'Buck Rogers' touch – was 20 m in diameter, and weighed 6,400 tonnes. On the launch pad, the Space Shuttle is 56 m high and 8.7 m in diameter; and weighs 2,000 tonnes. The Saturn V, designed by von Braun ten years after the publication of 'Das Marsprojekt', had a height of 102 m, a diameter of 10.1 m, and launch-pad mass of 3,000 tonnes. The payload to Earth orbit was 118 tonnes, compared with the ferry orbiter all-up weight to Earth orbit of 78 tonnes. The most remarkable difference, however, is in the launch pad mass. The Saturn V is slimmer and taller than the ferry vessel and has a larger payload capacity, but it uses less than half the fuel. This is because von Braun decided to retain the use of nitric acid and hydrazine, for the ferry vessel, rather than liquid oxygen and liquid hydrogen. In contrast to the in-space manoeuvres, it is perfectly sensible to use cryogenic liquids such as oxygen and hydrogen for a launch from Earth. The liquid can be loaded into the vehicle just before launch, and kept topped-up, to replace that evaporated, right up to the time of launch. Kerosene, as used in the huge F-1 engines of the Saturn V first stage, or liquid hydrogen, as used in the upper stages, are much more efficient than the propellants chosen by von Braun, and very probably cheaper too. He was, of course, familiar with liquid oxygen, as he had used it since the 1930s and for the A4 – and his conservatism later became clear. Here, I think, he was simply choosing the most conservative approach as part of his attempt to convince everyone that it could be done at that time, in 1952.

By comparing the ferry vessel with the Shuttle and with von Braun's Saturn V, we see many similarities. It is clear that the ideas – which would eventually lead to the actual 'ferry vessels', the Shuttle and the Saturn V – were very much in von Braun's mind when writing 'Das Marsprojekt' in 1949–50. He was also very advanced in planning the reuse of the stages. The ferry orbiter, of course, returned to the launch pad, but the other stages were to be landed by parachute, so that they could be reused. Ideas of liquid fly-back boosters are today regarded as advanced; but von Braun was using the idea fifty years ago, before any multi-stage rocket had been built. As always, he calculated the necessary rocket deceleration to de-orbit the boosters, and the size, number and weight of the parachutes.

With characteristic Teutonic thoroughness von Braun calculated all the numbers relating to his mission, from the oxygen and water needed to support a man for a day, to the tonnes of propellant required and the temperatures of wing surfaces. He also calculated – again with paper and pencil – the precise trajectories of the ferry vehicles, the landing boats and the spaceships; and he carefully included the simplifying assumptions that he had used, and the possible corrections required for a real mission. It was a remarkable *tour de force*; and it might be asked whether modern-day engineers have the numerical and mathematical skills to carry out such calculations, with only tables and a slide rule. He not only displayed his versatility, but also, for the first time, set out a properly evaluated Mars expedition, showing exactly how it could be achieved with the then available technology. He even calculated the number of ferry launches and the cost of the fuel. He required a total of 950 ferry flights to put the expedition into Earth orbit, using 5.3 million tonnes of fuel for the launchers. This must have given him pause for thought, but he bravely calculated the cost – $500 million – and compared the quantity with that carried in aircraft for the Berlin Airlift. It was ten times as much. By comparison, each Space Shuttle launch has a marginal cost of $63 million – a total of $60 billion for the 950 flights. The Saturn V would have cost $430 million per launch, which was much more expensive than the Space Shuttle.

Everything that has followed owes something to von Braun's vision – not only for interplanetary flight and expeditions, but also for the technology that was to become standard as the Space Age developed. When he carried out this work, the Redstone missile was the epitome of *racketshiffart*. Sputnik was still five years away, and Apollo 11 was twenty years in the future. Yet the ideas come fresh off the page as if these things had already happened. It could be said that von Braun was gifted with remarkable foresight, or alternatively that we have not advanced very far since his day. In a sense, both are valid. The true situation is that von Braun hoped – how he hoped – that he would see a Mars expedition in his lifetime. In this he was disappointed, as we are disappointed; but he did see men walking on the Moon – and, what is more, they reached there and back by using his rocket, the Saturn V. This was, perhaps, enough for one man's lifetime.

2.8   MARS AFTER DAS MARSPROJEKT

By the time von Braun's book was published in America, he and his team had become part of the Army Ballistic Missile Agency at Redstone Arsenal, and the beginning of the Space Age was rapidly approaching. There was almost Byzantine in-fighting between the various sponsors of rocket research, and the Army, the Navy and the Air Force all had their programmes. The Army had the heritage of von Braun's team and the V2, while the Navy and the Air force were developing indigenous technology. They all wanted weapons – ballistic missiles – and almost no-one cared about space exploration. At the same time, the USSR was steadily preparing to launch the first satellite, with advanced plans for manned spaceflight already in being. Having set out how to undertake a Mars expedition, von Braun concentrated on the first step: a suitable satellite launcher. He knew that the manned launcher was a long way off, and that his paymasters were mostly interested in defending the United States rather than putting people in space. With the capabilities of rockets improving, the possibility of carrying out space science and upper atmosphere science began to emerge. As the end of the decade approached, the International Geophysical Year prompted both the US and the USSR to promise to orbit a satellite, although few, outside the Russian team, took this promise seriously. As is well known. the Soviets wiped the floor with the US, launching Sputnik and Sputnik 2 – the latter with a dog on board – before the US could put anything into space. The real Space Age had begun. The western world was clinging on by its fingernails, and its one success was due to von Braun; at least he had become a US citizen in 1955.

The story is worth the retelling. In 1954 a group including von Braun had met to examine the possibility of launching a satellite; and they agreed that a Redstone rocket, with a modified upper stage, could achive it. A proposal was therefore produced, and the government approved it. But the Redstone was German technology. It had been developed, from the old A4, by Germans in the United States, and, like the V2, it even used liquid oxygen and alcohol as fuel; and it was also a military vehicle. Since 1946, the Navy had been developing indigenous rocket technology in the Viking sounding rocket, and by August 1955 had lobbied for and secured the right to launch the first US satellite with its Vanguard launcher, based on the Viking. The Redstone project was subsequently rejected. In September 1956, before the Vanguard contract had been awarded to industry, the Jupiter C – a three-stage launcher based on the Redstone, with two solid rocket clusters as upper stages – had put a payload into a ballistic orbit peaking at 1,096 km and returning to Earth. Von Braun was ordered to ballast his upper stages with sand to avoid inadvertently launching the first Earth satellite. Vanguard, in the event, failed, and did not succeed in launching a satellite until March 1958. The day after Sputnik was put into orbit, von Braun promised that he could launch a satellite within 60 days, using the Jupiter C. On 8 November 1957 he was ordered to do so, and on 1 February 1958 he succeeded, thus taking the United States into the Space Age.

All of this was changing the situation for a Mars expedition. It was all very well to plan Mars expeditions while the space frontier had yet to be crossed, and one imaginative idea was as good as another. But now there were real spacecraft and real launchers; and as reality overtook imagination, the range of possibilities narrowed. The Mars expedition required payloads of 50 tonnes in Earth orbit, but the actual launchers were launching satellites of about 1 kg in weight. If reaching Earth orbit was so difficult, could it ever be hoped to send anything to Mars? The Moon looked more inviting.

Von Braun had a day job, of course, as technical director of the Army Ballistic Missile Agency, which was in intense competition with both the Navy and the Air Force. The Navy had been effectively put out of the game by the generally recognised failure of the Vanguard Project, while the Army went on to launch satellites of up to 20 kg, compared with the 1-kg satellite launched by Vanguard. The ABMA wanted to push ahead with the derivative of the Jupiter C launcher, called Juno, while the Air Force was pushing for the use of the indigenous technology that it had developed for ballistic missiles – in particular, the Atlas rocket with the first

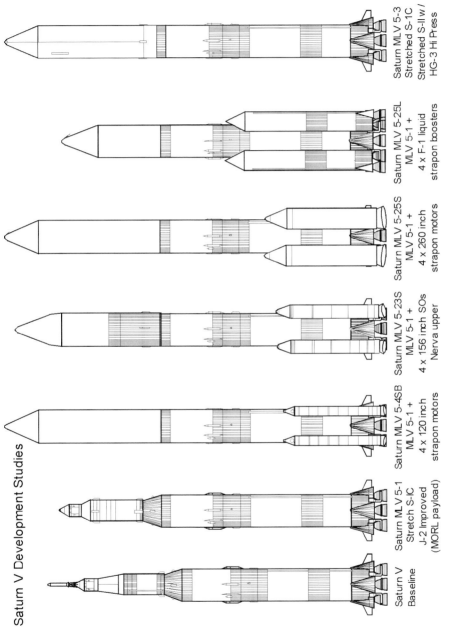

Figure 2.9. Saturn V variants proposed for Mars expeditions. (Courtesy Mark Wade.)

liquid hydrogen–liquid oxygen upper stage, called Centaur. These had been developed with no view to space exploration, but simply to put the largest payloads possible into intercontinental ballistic orbits, with the aim of extermination. In fact, it happened that the engines developed for the Air Force would prove to be essential for further manned exploration of space. The Juno went on to launch the early Mercury astronauts into sub-orbital flight, but the old engine technology – a direct heritage from the V2 – was past its sell-by date.

The next visionary step in human exploration was generated by the ABMA, instigated by von Braun. This was nothing less than the establishment of a military base on the Moon: Project Horizon. This project, conceived during the early 1950s, used advanced versions of the Juno, which would later be called Saturn. The plan was to establish the base by 1965, and typically involved large numbers of Saturn flights – not of the Saturn V, but of the much smaller Saturn 1. However, it was not awarded funds, although many of the ideas resurfaced once Kennedy's decision had been made to go to the Moon. Von Braun's team wanted to develop a large space launcher, as did the Air Force. There then ensued a protracted argument over the use of liquid hydrogen as a fuel. Hydrogen is the most efficient rocket fuel because it produces the fastest exhaust stream; but it is very difficult to use. Von Braun was very much against its use because of its difficulties, and so his proposed large vehicles all used conventional propellants. By 1960 the proposal for the large vehicle had lost the support of the Air Force, and had become the province of the newly created NASA. Eventually von Braun was convinced of the need to use hydrogen, as it allowed the same payload with one fewer stages on the launcher. Once project Apollo was approved, he went on to develop the Saturn V, using all-indigenous engines, with the upper stages using liquid hydrogen and liquid oxygen. The difficulties were overcome.

At the same time, the Mercury and Gemini flights were confirming the United States' place in human spaceflight, and the Apollo programme – which would enable twelve men to walk on the Moon – was developing. But the Mars expedition was not forgotten, and it appeared in several projections prepared in the 1950s as the ultimate goal of human space exploration. There were even versions of the Saturn V proposed which could be used to transport the expedition. These studies continued from 1965 through to 1967, but with the demise of the Apollo programme, the development of the Saturn V ceased, and there is now no available vehicle that matches its capabilities. The largest lift capability available today is 24 tonnes into low Earth orbit, whereas the Saturn V, as used for Apollo, could place 118 tonnes into low Earth orbit. As will be seen later, the Mars expedition for the twenty-first century requires a new heavy launcher to be developed, to re-create the technological capability of the 1960s.

2.9   MARS IN REALITY

Manned space exploration was effectively stopped by the Nixon government in 1972, and the remaining Apollo spacecraft and Saturn Vs were placed in museums. The

Space Shuttle programme then began, but it had no capability to take humans anywhere but to near-Earth space. The space race was still going on, and in particular the Russians were pushing on with large robotic payloads to the Moon and planets. This new capability of exploring by proxy, as it were, was much cheaper, and could salvage national prestige. Once the first satellites were launched, missions to Venus and Mars, using small robotic probes, began to appear feasible. After all, we knew so very little about the planets that even a few close-up TV pictures would advance our knowledge tremendously. And then there emerged yet another player in the complex game that was being played between the various US organisations involved with space: the Jet Propulsion Laboratory, of the California Institute of Technology. Founded in 1936 as a student project, by 1940 it had acquired Army research contracts, and by 1944 had its present name and was charged with developing tactical ballistic missiles. It was a major force in the post-war rocket business, and provided the Sergeant rockets for the upper stages of the Jupiter C that launched the first US satellite; the satellite that was launched – Explorer 1 – was also built by JPL. The Director, William Pickering, was ready with a proposal to send versions of Explorer to the Moon, almost immediately after the first Sputnik. However, this came to nothing; but it was the beginning of JPL's planetary exploration programme that continues to this day. JPL became involved in Pioneer and Ranger Moon-shots; Mars was also on the agenda; and in 1964, two Mariner probes were sent to Mars using the Atlas–Agena launcher. The first did not succeed because of a failure of the launch fairing to separate properly; but the second *did* succeed, and after a 228-day cruise it flew past Mars at a distance of 9,000 km. Rather disturbingly, it showed a cratered surface, rather like the Moon's – but this was later discovered to be atypical. Since then, many probes have been sent to Mars, not all of them successful. The first spacecraft to orbit Mars was Mariner 9, launched in 1971, and the first soft lander was the Russian Mars 3, although it transmitted data for only 20 seconds after arriving. Both Mariner 9 and Mars 3 arrived at Mars at the same time, and Mariner 9 mapped the planet. The successful soft-landers were the Viking probes, 1 and 2, that arrived in 1975, and which sent back pictures from the surface and conducted the famous searches for life – which are still regarded as inconclusive. After discharging the landers, the Viking orbiters remained in orbit and completed photographic mapping of the whole planet, and for the first time we had a proper idea of the surface of Mars and of its various land-forms.

The dream of human exploration of Mars had been temporarily subsumed into human exploration of the Moon; and this was of course an amazing achievement and scientifically very successful. The concentration on this led to planetary exploration being conducted using robotic probes; and this is where we stand today. There is no firm plan for the human exploration of Mars, and there is feverish concentration on exploration of the planet using probes. The latest images are remarkable in their definition, and exploration using different wavebands and radiations has shown the probable presence of buried water over much of the planet. But probes are no substitute for human exploration, and the dream has yet to be fulfilled.

2.10 THE MARSPROJEKT JOURNEY

Werner von Braun and Willy Ley – a famous German science writer who had been responsible for introducing the 18-year-old von Braun to Herman Oberth – later refined the Marsprojekt by reducing the crew to twelve men and eliminating the polar landing. Instead they assumed that a place with flat sandy plains could be found where the whole expedition could land. In their jointly authored book, they included a description of the voyage.

The crew are taken up to the orbit where the spaceships are waiting, and enter their ships. Strapped down in their 'contour couches', they await the moment – calculated, of course, by von Braun – to start the engines. The propellants are first trickled into the combustion chambers, and then the main flow starts and the engines roar. The thrust, which is quite small, lasts for 15 minutes, and then the spaceships are sent on their slow voyage to Mars. The 'whine of the gyros' then dies down. (Gyros play a large part in early space fiction, and they always seem to whine.) Day and night, of course, have no meaning, as the ships are always in sunlight, and a system of watches replaces the normal sequence of light and darkness. The crew seem to spend a great deal of time on maintenance of the equipment, the life-support (here called air conditioning) and the electrical power supply. A nuclear-powered turbine generator system is used (exactly as proposed for nuclear power supply today), and the chief electrician has to maintain it. In an emergency, one of the other ships can be called on for help.

As time goes on, small differences in trajectory lead to the drifting apart of the ship, and closing-up manoeuvres have to be initiated to bring them back within a distance such that the space boats can move between ships in an emergency. Crew sanity is helped by the daily radio show – broadcast from Earth via a high-power transmitter – which includes news, a lecture, and music. Radio is also used for tracking, and to exchange navigational information with Earth. Seventy-three days out, the Earth is seen to transit the Sun's disc – a true expression of how far they have travelled. To help deal with crew boredom, some of the crew-members exchange ships, for a change of scene. Much of what we would consider automatic is here the responsibility of the crew: the manual chemical analysis of the atmosphere in the spaceships, routine maintenance of the equipment, navigation using sextants to measure the relative positions of the planets, and, of course, calculations using slide-rules and tables. The nitric acid and hydrazine must not be allowed to freeze, and so the temperatures are monitored by the crew, and appropriate measures taken, either by changing the attitude of the ship so that sunlight falls on the cold tanks, or by introducing electrical heating. Should the atmosphere inside the ship become poisoned – possibly by release of mercury from a broken pressure gauge – then it will need to be filtered, or possibly released into space and replaced by a new oxygen–helium mix from the store. It is a long voyage, and it is as well that there is so much work for the crew, just to keep everything working on the spaceships.

After eight months in space, the ships approach Mars. In fact, they have slowed down so much, as they drift away from the Sun, that Mars is approaching them from behind, as their orbit intersects that of Mars. The navigators float in their

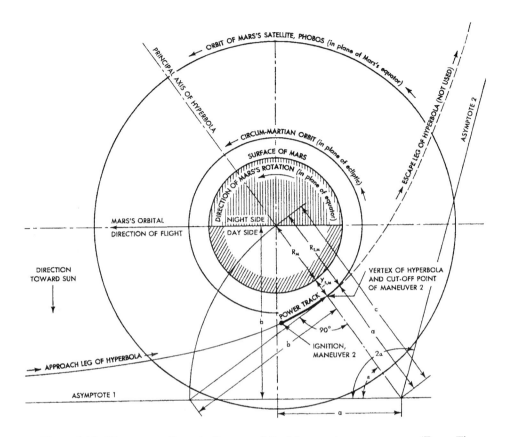

Figure 2.10. Werner von Braun's diagram of the Mars capture manoeuvre. (From *The Mars Project*, courtesy University of Illinois Press.)

astrodomes, taking fixes on as many planets as possible, more or less continuously, so that they can establish the relative velocities of the spaceships and Mars. After furious calculation it is established that a velocity change of 63 feet/s is required to ensure that the spaceships and Mars pass within 5,470 miles of each other, as the free space orbit peaks just inside the orbit of Mars. The ships are tilted so that the thrust of the engines is in the correct direction, and the engines give a short burst. The velocity is again checked; it agrees with the predictions; and they now only have to wait for Mars to capture them with its gravity. During this interval the huge, empty fuel tanks, along with empty food containers, broken tools and other unwanted itemsm are discarded to lighten the ships as much as possible. They continue to drift out to the rendezvous point, where they are drawn into the gravitational field of Mars, and enter a hyperbolic orbit round the planet. The spaceships are then rotated until the engines point exactly forward along the trajectory. Mars grows, and fills much of the sky. The spaceships go faster and faster as they fall towards the planet in a graceful sweep – the hyperbolic orbit. They will pass 620 miles above the surface, and would then rise again to fly off into interplanetary space if they were to stay in the hyperbolic orbit. This does not happen, however. When the speed is 3.2 miles per second, and they are near the position of closest approach, the engines fire to slow the spaceships. The firing instant, and the duration of the burn, are controlled by a chronometer-timer 'like an alarm clock'. For nine minutes the crew feel the thrust at about one-third gravity; and after nine months in weightless conditions, this is quite stressful. The engines then shut down, and the ships are safely in orbit round Mars.

As they orbit the planet they can see the surface in detail for the first time, and can check the proposed landing sites to determine whether the terrain is as they expected. While they make all necessary preparations for the descent to Mars, and while the planet continues to rotate under them, the final landing site is confirmed. As a final check they send 'sounding missiles' to the surface, and these follow the same kind of trajectory as will the landing boats, except that they parachute to the surface. The sounding missiles send information back to the spaceships about the atmospheric conditions at the site, and the precise altitude can be measured using a radio beacon. While the calculations are checked and double-checked to ensure that they can reach the proper place on the surface, the landing boats are separated from the mother ships. The 'interplanetary radio station' is then removed from the landing boat bay, and assembled onto the mother ship. Three crew-members will stay in orbit as 'ship keepers', to relay information back to Earth and to keep the return ship fully maintained while the explorers are on Mars – a 'somewhat thankless and certainly monotonous assignment'. They are under orders to return from Mars after the 400 days are up – whatever happens to the surface explorers.

In this modified expedition, the landing team has been reduced to nine men, and only one landing boat is required for the expedition. The crew-members enter the boat, and strap themselves in. The correct moment for departure is selected, and the engines fire to de-orbit the boat, which then follows an elliptical path almost half-way round Mars, descending all the time. A 'whispering sound' grows in intensity as the atmosphere is entered, and soon the boat, with its enormous wings, becomes manoeuvrable. The pilot will first direct the boat down into the atmosphere (negative lift) to ensure that the enormous glider does not bounce off into space again. Later he will alter the angle of attack to generate positive lift to slow down the vehicle. At a height of about 24 miles it will have slowed to sound speed and become like an ordinary aeroplane; and as the landing site approaches, the glider will manoeuvre into the correct glide path. A smoke bomb will be dropped to determine the wind direction; then, after final adjustment to the path, the glider will deploy its undercarriage and land. To help slow it down without sinking into the sands of Mars, it will have wide tyres and skids under the enormous wings.

A safe landing accomplished, the crew don their pressure suits and, one by one, step out of the airlock and stand on the wing. They see a desert landscape 'like the American south-west', under a dark blue cloudless sky. There are major tasks before the explorers, including the erection of the forward section of the landing boat that will be used for the return to the mother ships, and the inflation of the 'tent' that will be their home for 400 days. But we will leave them at the summit of their achievement: the landing on Mars.

2.11 RETROSPECT

Seen from our twenty-first-century perspective, there are as many 'retro' aspects as there are parallels. The false measurement of the atmospheric pressure changes the whole aspect of the landing. The graceful glider with its huge wings is, alas, not possible, and it has been replaced by the dumpy, reversed-cone entry capsule. With the wings goes the horizontal, aircraft-type landing, as the lander will descend vertically. The manual approach to nearly everything seems very old-fashioned. The equipment is a mixture of the archaic – dials, and mercury-filled gauges – and the modern – the nuclear-powered turbo-generator, as proposed for a manned mission today. Radio technology is primitive. At the time, vacuum tubes were the norm, and so sets were large, power-hungry, and unreliable; and thus the chief radio technician was also chief electrician.

The biggest change is the computer. The Mars Project was to be achieved using techniques that were not much different from those used for navigation during the eighteenth and nineteenth centuries: navigation with sextants, calculation by hand, checking and double-checking, and primitive clockwork timers – the latter, on the Mars Project, being used to control the engine burns. But the crew of a new Mars Project would have no such requirements, as computing power is now so enormous that navigation and astronautics would all be non-manual. The dials and mercury gauges would be replaced with digital instruments directly read and monitored by computers, the landing pilot would need to act only in an emergency, and the landing would now be completely automatic.

The loss of the huge wings severely reduces the mass that can be landed on Mars, and it is doubtful whether the crew of twelve, with all their stores and two caterpillar tractors, could today be landed in a single vehicle.

The Mars Project may have its 'retro' aspects, but it was nevertheless a *tour de force* of its time, and the only things that we would carry out differently today are those that derive either from advances in our knowledge about Mars or from developments in technology that von Braun could not have foreseen. We all owe a great debt to him and his colleagues – first for enabling the first step in human space exploration – the Apollo project – and finally for setting out so clearly the means for the next step – the exploration of Mars.

3

Getting there

For any journey it is necessary to know the starting and finishing points and the route to be taken, whether it be a visit to a friend or a journey to Mars. Celestial journeys are, of course, very different from terrestrial journeys, and many of the unspoken assumptions regarding terrestrial journeys do not apply to their celestial counterparts. We assume that home and the friend's house remain fixed on the Earth's surface during our journey there and back, and that we can take any path between them; but neither of these assumptions is true for a journey between planets. The planets are in continuous motion, and the possible paths between them are strictly defined by Kepler's laws of planetary orbits and by Newton's laws of motion and his law of universal gravitation. Only certain paths can be taken, and due allowance has to be made for the motion of the planets during the journey to Mars, during the landing, during exploration, and during the return journey. Moreover, the technical requirements on the spacecraft depend quite critically on the nature of the journey and the total time away from Earth. To establish the fundamentals, we need to understand orbits, and the way in which the planets and spacecraft move.

3.1 ORBITS: INTERPLANETARY MOTORWAYS

When the plague closed the universities in 1666, the 24-year-old Isaac Newton had to leave Trinity College, Cambridge, and return to his family home in the safety of rural Lincolnshire. Here, in the intellectual isolation of Woolsthorpe Manor, he continued to think about the motion of material bodies; and somewhere, perhaps in the orchard, he made the intuitive leap from the apple falling to the ground to the Moon drifting through the heavens. He realised that the force that pulled the apple from the tree to the floor of the orchard was the same force that held the Moon in its sphere. Heavenly and earthly bodies behaved in the same way, and were moved by the same forces. This intuitive leap became Newton's law of universal gravitation.

In formulating this law, Newton was completing a chain of scientific discovery and philosophical enquiry that extended back over many centuries. The motion of the heavenly bodies has been a preoccupation of mankind from the dawn of

Figure 3.1. Woolsthorpe Manor – Isaac Newton's family home in Lincolnshire.

recorded history; and the ideas of the ancient world, codified by Ptolemy – a Greek astronomer working in Alexandria in the first century AD – held sway over western civilisation from the time of Aristotle until the fifteenth century. The established truth, supported by the Church, was that the Earth was at the centre of the Universe and that the planets, including the Sun and Moon, circled the Earth. Each planet had its own crystal sphere, and each sphere rotated about the Earth at a different speed. Beyond the outermost planetary sphere were the fixed stars on their own crystal sphere. The rotation of the crystal spheres was accompanied by celestial music – the music of the spheres – and everything in the heavens was perfect and in perfect harmony. So persuasive is Ptolemy's picture of the Universe, and so useful, that modern astronomy still uses it for practical purposes. We still talk about the motion of the Sun and planets against the fixed stars. This is, after all, what we actually observe from the rotating and moving Earth.

The chain of ideas, careful astronomical measurements, mathematical analysis, intellectual inspiration, perspiration, and sometimes martyrdom, can be only briefly summarised here. Around 300 BC, the minor philosopher Aristarchus of Samos proposed that the Sun should be at the centre of the Universe; but his idea was submerged beneath those of the philosophical heavyweight, Aristotle, who taught that the Earth was at the centre. It was not until Copernicus revived the idea of a heliocentric Universe that it again became a living theory in his *De revolutionibus orbium coelestium*, dedicated to Pope Paul III, and published in Nuremberg in 1543, more or less while he was on his deathbed. The political climate was harsh, and Copernicus did not dare publish his book while he was still open to attack by the establishment. Later, Galileo, convinced by his telescopic observations, took up the idea in his *Siderius Nuncius*, published in 1610. He was persecuted by the Church over many years, but his influence and cunning were sufficient to avoid martyrdom – unlike his predecessor, Giordano Bruno, who in 1600 was burnt at the stake for promoting the Copernican Universe.

The final chapter of this tale commences in the same year, 1600, when Johannes Kepler arrived at the observatory of Tycho Brahe in Prague. Brahe was the observational astronomer *par excellence*. He had obtained from Erasmus Habermel a sextant that allowed him to measure the positions of stars and planets to a

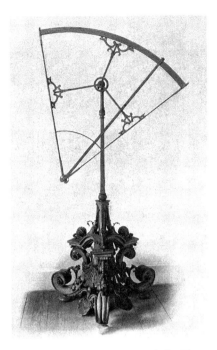

Figure 3.2. Tycho Brahe's sextant, made by Erasmus Habermel.

precision of 2 arcminutes (1/30 of a degree) – before the invention of the telescope. Settled in his observatory at the Royal Summer Palace near Prague, Brahe worked on his series of very precise measurements of the apparent motion of the planets against the background of the fixed stars.

3.2   THE WANDERING PLANETS

In our light-polluted city skies, the planets are not so obvious as they were to previous generations; but with dark skies, the bright planets were very much a feature of the night sky, and their peculiar motion against the background of the stars was well known from the earliest times. Indeed, the word 'planet' is derived from the Greek word for 'wanderer'. The name 'wanderer' arises because the planets move against the background of the fixed stars, from constellation to constellation. Generally, this motion is in the same direction, but occasionally a planet appears to backtrack on its path, moving, for a time, in the opposite direction, before returning to its original motion. This was known from ancient times, and many explanations were put forward, right up to the time that the Copernican system was adopted.

It was important to be able to predict the movements of the planets, for astrological and navigational reasons. From the mistaken point of view that the Earth was fixed at the centre of the Universe, it was obviously difficult to explain their peculiar behaviour with a simple theory. The best attempt, as adopted by Brahe

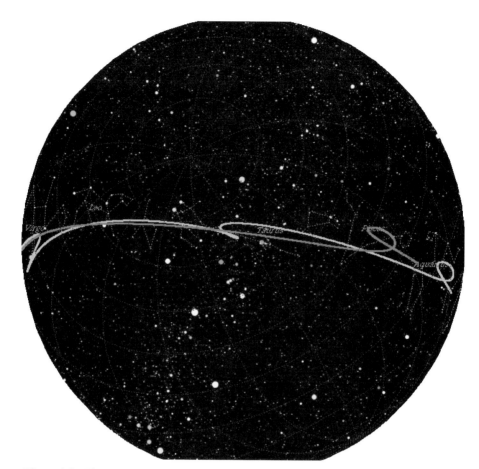

Figure 3.3. The 'wandering' motions of Mars and Venus, seen against the night sky.

– who was unconvinced of the Copernican view – was that planets moved along curved paths, called *epicycles*. An epicycle is the path of a fixed point on the rim of a wheel as the wheel rolls along a surface, and such a point describes a loop from one contact with the surface to the next. The idea was that the planet, for whatever reason, followed an epicyclic path as the 'wheel' rolled round the inside of the planet's particular crystal sphere. This worked quite well in predicting the motion of the planets, with more and more complicated epicycles added to explain observed discrepancies. Tycho Brahe was hoping to eradicate the remaining discrepancies, using his precise observations, when he died quite suddenly in 1601, a year after Kepler joined him. Kepler made sure that whatever else went to Brahe's heirs, he (Kepler) would keep the records of the observations, made painstakingly over many years, by his deceased employer.

Kepler quickly realised that the epicyclic theory could not explain the motions of the planets, now that Brahe's accurate measurements were available. They simply did not fit. Adopting the Copernican view, he tried to explain the measurements

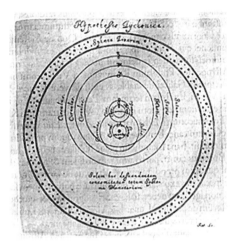

Figure 3.4. Tycho Brahe's interpretation of the epicycles.

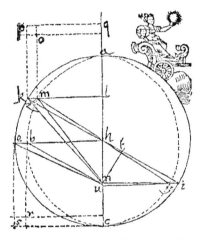

Figure 3.5. Johannes Kepler's geometry applied to the Earth's orbit round the Sun.

using circular paths for the planets about the Sun, including the Earth. But these did not fit either. It was only when he realised that the motion of the Earth around the Sun varied throughout the year that he understood the paths of the planets to be neither circles nor epicycles but, in fact, ellipses. Unlike a circle, an ellipse has two centres, or foci, separated by a distance that depends on the elongation of the ellipse. When the foci are close together, the ellipse is almost circular; and when they are far apart, it becomes very elongated. The Earth's path is not very elongated – the ratio between the smallest and greatest distance from the Sun is only 0.98 – but Kepler's analysis of Brahe's measurements was clear: the path was not a circle. Once he understood this, Kepler was able to assign the correct motion to the Earth, and, using the new orbit for the Earth as a baseline, he was able to analyse the motions of the other planets. He found a simple geometrical theory that precisely fitted all of them.

There were three simple rules in Kepler's theory. The first statement was that the paths of the planets were ellipses, with the Sun at one focus. The second rule was that the radius vector – the line joining a planet to the Sun – swept out equal areas in equal times. This was a geometric statement of the fact that the planets moved faster in their orbits when they were closer to the Sun, and that the increase in speed was exactly proportional to the change of distance. The final rule stated that the square of the period – the time taken to complete one orbit – was proportional to the cube of the length of the semi-major axis of the ellipse; that is, the cube of the average distance of the planet from the Sun. (Here 'square' and 'cube' refer to the period multiplied by itself, and the distance multiplied by itself, twice. In modern notation we would write P^2 and L^3.) Note how geometrical this is; calculus had not yet been invented, and mathematics was still dominated by geometry. Note, too, that the laws not only state how the planets move, but provide precise mathematical rules for their motion. Anyone could apply these rules and, starting from current observations, could predict where the planets would be in the sky for any time in the future. In 1609, Kepler published his first two laws in *De Motibus Stellae Martis ex observationibus G.V. Tychonis Brahe* (*On the motion of the Planet Mars from observations by Tycho Brahe*), and the third law was published ten years later.

Knowing Kepler's laws, it is now easy to explain why Mars sometimes turns back on its path amongst the stars. It is a simple consequence of the fact that Earth, closer to the Sun, moves faster than Mars moves in its orbit. Earth periodically overtakes Mars, and then Mars, seen from Earth, appears to move backwards, until the

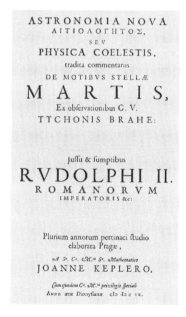

Figure 3.6. The title page of Kepler's book on Mars, dedicated to the Holy Roman Emperor Rudolph II.

planets, in their paths, have separated from one another by a sufficient distance. To be more precise: it is the Copernican system that explains why Mars appears to move backwards sometimes, and it was Kepler who explained when and by how much.

Kepler's laws explain exactly how the planets move, but despite several near misses, he was unable to determine *why* they moved in this way. It was left to Newton, in his orchard at Woolsthorpe, to determine and formulate the physics of planetary motion, the physics of orbits, and the physics of interplanetary travel. Newton's law of gravity allows us to calculate the motion of any planet round the Sun; but what is more important for us, is that it also allows us to calculate the path of a spacecraft in orbit around a planet, or on a journey between planets.

Newton realised that any material body – be it a planet, a moon, a comet, or simply a lump of rock or a falling apple – was subject to the same laws of motion. It must continue to be at rest, or in a state of uniform motion in a straight line, unless acted upon by a force. The force he had in mind was the force of gravity, which he had 'discovered'. The force of gravity, he said, attracts every material body to every other material body. The reason everything is not stuck to everything else is that moving bodies have *momentum*. It is easy to have an intuitive appreciation of momentum. The Rugby full-back, approaching at a run, has momentum, and it can be experienced if an attempt is made to stop him. Newton gave momentum a precise definition: the amount of momentum a body has is just its mass multiplied by its velocity; a heavy body moving slowly and a lighter body moving quickly can have the same momentum. It is momentum that resists the attractive force of gravity, and stops the Moon falling on to the Earth and the Earth falling in to the Sun. If anything has sufficient forward momentum, then gravity cannot cause it to fall directly downwards, and it must move in a curved path. In most cases this will not be a collision course. This does not agree with everyday experience, only because the Earth is so big compared with everything else on it. For us, the Earth's gravity is much the strongest force around, and so dominates what happens; and the Earth is

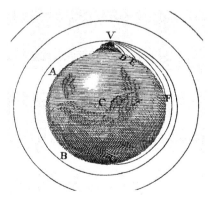

Figure 3.7. Newton's illustration of the connection between falling bodies on Earth, and orbits. It is also the first illustration of how to place a satellite in orbit round the Earth (although Newton had no such intention).

so big that we, living on its surface, find it difficult to miss. In space, the distances between objects are very large compared with their size, and so collisions are rare.

The other rule which Newton understood, but which had puzzled Kepler, was that the force holding the planets in their orbits decreases precisely as the *square* of their distance from the Sun. For example, doubling the distance from the Sun, quarters the force; and tripling the distance reduces the force by a factor of nine. For Mars, Newton's law of gravity is expressed in the following simple equation,[1] which involves the mass of the Sun and of Mars, and the square of the distance between them:

$$F = G \frac{M_{Mars} M_{Sun}}{R^2_{Mars}}$$

The scaling factor, which determines how big the force will be between Mars and the Sun, is G – Newton's universal constant of gravitation. It is universal in that it applies throughout the Universe and to the attractive force between any pair of material bodies: planets, stars, rocks, spacecraft, or apples. The attractive force between Mars and the Sun uniquely depends on their masses multiplied together, divided by the distance between them, multiplied by itself.

Since an orbit is essentially a balancing act between gravity and momentum, it follows that the speeds of the planets in their orbits decrease with increasing distance from the Sun. Mars moves more slowly than the Earth, as it is further from the Sun, while Venus moves more quickly. This is not quite the same as the period – the time taken to orbit the Sun once – which increases with distance from the Sun, both because of the slower velocity, and because the orbit is bigger. In fact, the martian year, or orbital period, is 687 days. Kepler's law covers the combined effect of decreasing gravity, *and* increasing orbital circumference, by relating the *square* of the period proportionally to the *cube* of the mean distance from the Sun. The circumference of the orbit is proportional to the distance, and the square of the velocity is decreased proportionally to the distance; hence the cube of the distance appears against the square of the period. This is much easier to understand using the following simple equation, in which P represents the periods and R represents the distances:

$$\frac{P^2_{Mars}}{P^2_{Earth}} = \frac{R^3_{Mars}}{R^3_{Earth}}$$

Kepler's law can be simply tested by using data for the orbits of Mars and Earth around the Sun. The period of the Earth's orbit – the year – is 365 days, while the period for Mars is 687 days. The mean distances from the Sun are 150 million km for Earth and 228 million km for Mars. According to Kepler's law, the ratio of the squares of the two periods should be equal to the ratio of the cubes of the two distances. The reader can easily demonstrate this, if wished, by substituting the numbers, given above, in the equation.

---

[1]   If you did not read the Preface, and dislike equations, please note that these mathematical paragraphs can be skipped without losing the thread of the argument.

3.3 PHYSICS AND GEOMETRY; ORBITS AND CONIC SECTIONS

Before examining the practical applications of these laws in interplanetary travel, we should return to the question of the shape of the orbits; the ellipses that Kepler realised were the true orbits of the planets. In keeping with the geometrical frame of mind of the pre-calculus era, we note that an ellipse is member of a whole family of curves called 'conic sections'. This arcane expression comes about from the following experiment. If one slices (sections) a paper cone and examines the cut end, different shapes are seen, depending on the angle of the cut. If the cut is made parallel to the base, then the shape is a *circle*. If the cut is made at a moderate angle to the base, then the shape is an *ellipse*, with the elongation or eccentricity of the ellipse increasing as the angle increases. There are, however, other ways to slice the cone. It can be sliced vertically downwards, at some distance from the point. In this case the shape is an open-ended curve. However tall the cone, the two open ends of the curve will never meet, but will continue to diverge, to eventually follow the angle of the cone. This particular curve is called an *hyperbola*. If the cone is sliced at some distance from the point, and at an angle away from the point, then another curve is produced. The two ends of this curve diverge, but become more and more parallel with distance, and, in fact, become exactly parallel at an infinite distance from the point. This curve is called a *parabola*. A simpler practical demonstration can be carried out using the diverging beam of light from an electric torch. By shining the patch of light on a flat wall, different curves will be shown, from circular – when the beam is pointed straight at the wall – through elliptical, to the parabolae and hyperbolae produced by shining the beam nearly parallel, and parallel, to the wall.

Rather surprisingly, these curves all appear as the shapes of different kinds of orbit, and the family of curves all derive from Newton's laws. Newton, in fact, used calculus to derive the equation, although he did not publish his method for a long time. The equation expressing the orbit of a planet, moon or spacecraft, based on Newton's work, is given below in modern notation. This *orbit equation* is rather complicated, but it is important, and is worth examining in detail.

$$\frac{1}{r} = \frac{GM}{r_0^2 v_0^2} \, (1 \, + \, \varepsilon \cos \theta); \, \varepsilon \, = \, \left[\frac{r_0 v_0^2}{GM} - 1\right]$$

The left-hand equation – for the path of a planet or any other body moving in space – is an expression of Kepler's laws. In this equation, r is the distance from the Sun at any time, and GM/r_0^2 is a constant, representing the attraction of gravity at the part of the orbit closest to the Sun. It is clearly similar to Newton's law, and is in fact derived from it, with G and M representing the gravitational constant and the mass of the Sun respectively. At the distance of closest approach – here called here r_0 – the velocity of the planet is v_0. We know from Kepler's second law that v_0 is the maximum velocity of the planet in its orbit, because it is closest to the Sun here. The elongation of the orbit is represented by ε the *eccentricity*, given by the equation to the right. This is determined by the ratio of the product of the distance squared and the velocity squared, to the mass of the central object – the Sun, in the case of a planetary orbit. *G* is, as before, Newton's universal constant of gravitation, sometimes known as 'Big G'. The structure of the Universe decrees that, in our units, *G* is equal to 6.67 10^{-11} m^3/kg/s^2. Substitution in any equation involving *G* will show why its units are as they are.

The orbit equation is expressed in polar coordinates; that is, in a system in which the changing distance from the central point – in our case, the Sun – and the planet, is represented by *r*; and the position of the planet in its orbit is represented by the angle θ, which varies from 0° to 360° in one complete revolution of the planet about the Sun. The amount by which *r* varies is determined by the eccentricity.

The eccentricity is a measure of the precise degree of elongation of the orbit. It will perhaps be obvious that the eccentricity increases as the velocity at closest approach is raised. If the eccentricity is zero, then the distance from the Sun does not vary and the orbit is a circle; which can be checked by substitution in the orbit

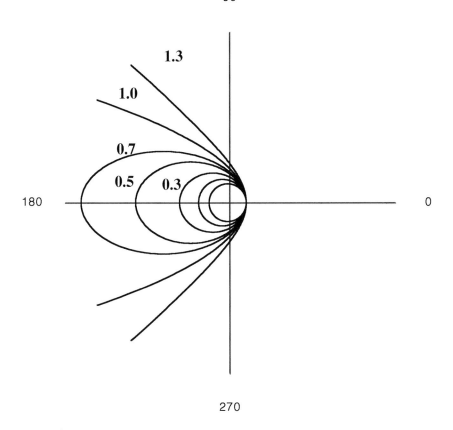

Figure 3.8. Orbits with different eccentricities according to the orbit equation. Eccentricities less than 1 produce ellipses; an eccentricity equal to 1 produces a parabola; and eccentricities greater than 1 produce hyperbolae. The four cardinal points show the progression of the angle θ, as the planet or spacecraft moves along its orbit.

equation. If it is between zero and 1, then the orbit is an ellipse with the elongation of the ellipse dependent on the precise value. If the eccentricity is equal to 1, then by the time the planet has travelled half-way round its orbit, the distance from the Sun will be infinite; in other words, the planet has escaped from the Sun's gravity. The shape of this orbit is a parabola, which has the property that the track becomes parallel to the axis, at infinity. If the eccentricity is greater than 1, then the orbit is divergent and is an hyperbola. So, from Newton's law of gravity are derived: two closed orbit-shapes – circles and ellipses – and two open shapes – parabolae and hyperbolae.

But what determines the eccentricity, and the shape of the orbit? It is the ratio of the momentum of the planet to the gravitational potential, GM/r_0. So, the greater the momentum of the planet, compared with the force of gravity at its closest approach, the more eccentric the orbit. This is a very important principle of orbital mechanics, and is crucial to interplanetary travel. Of course, the planets are bound in their orbits, following elliptical paths; but a spacecraft, for example, could be given an excess of momentum over the gravitational potential, and travel in a different kind of orbit; perhaps a parabola or an hyperbola. Such an orbit might allow it to voyage between the planets. Ultimately, however, velocity is the prime factor. The faster a spacecraft can be made to travel, the greater will be the eccentricity of its orbit. As its initial speed is increased, its orbit will change from a circle to ellipses of increasing elongation, then to a parabola and ultimately to hyperbolae with increasingly open angles. All of these possible orbits can play a part in the journey to Mars.

It is important to realise that these orbits are the *only* paths that a spacecraft can follow. They are strictly defined by the action of gravity, and the spacecraft's momentum at its starting point. In this sense they are like motorways: once embarked upon, the only way is forward until a suitable exit manoeuvre can be carried out. Similarly, spacecraft cannot suddenly change direction, as depicted in the movies; all they can do is change to another orbit, strictly following Newton's and Kepler's laws. Of course we, as children of the Space Age, are familiar with the concept of orbits – at least in as far as it concerns spacecraft and satellites in orbit round the Earth. The first man-made object to voyage in the heavens was launched by Russia (or rather, the USSR) in 1957, and was called Sputnik – 'fellow traveller'. This translation later came to have political connotations, and perhaps 'travelling companion' would have been better. Sputnik was in orbit round the Earth, and was also a true travelling companion to the Earth in its journey round the Sun.

Gravity is universal, and a spacecraft is always under the influence of gravity, wherever it may be. The greatest influence is that of the closest large body. In the case of Sputnik, it was of course the Earth; but the Sun's gravitational pull is not negligible. It does, after all, keep the Earth in its orbit. Then there is the Moon. Its pull is also not negligible, and it causes the ocean tides. It would appear, therefore, that the forces acting on a spacecraft are very complicated. This is in fact true, and for precise calculations of the path of a spacecraft, all these forces – even the gravitational attractions of the other planets in the Solar System – have to be taken into account. This can be accomplished using complicated mathematics, and nowadays computer programmes; but in former times these calculations were carried out by hand and published in the *Nautical Almanac*, for maritime navigation.

Fortunately for us, and for the space pioneers, a reasonable approximation can be calculated with quite simple mathematics, using the concept of 'spheres of influence'. This is no more complex than considering only the largest of those forces acting in any portion of space. For a satellite in Earth orbit, only the Earth's gravity needs to be considered; on the journey to Mars, when the spacecraft is distant both from Earth and Mars, only the Sun's gravity needs to be considered; at Mars, only martian gravity is important – and so on. This works because the force of gravity decreases with the *square* of the distance. Thinking, just for the moment, of a spacecraft leaving the Earth, each time the distance from Earth doubles, the force is quartered, while the distance from the Sun hardly changes at all, and so its constant gravity eventually becomes the dominant force on the spacecraft. The paths to be taken by a spacecraft in its journey to Mars can therefore be broken down into a series of orbits about the different heavenly bodies involved, with smooth transfers from one to the other.

3.4   REACHING ORBIT: ROCKETS AND LAUNCHERS

The first and greatest barrier to be overcome is to launch the spacecraft, or its component parts, into orbit round the Earth. In the closing years of the nineteenth century and the first years of the twentieth century, Konstantin Eduardovich Tsiolkovsky – fittingly, a *Russian* schoolteacher – first determined how to achieve this. Using Newton's laws, he calculated that for a spacecraft to remain in a circular orbit around the Earth it would require a velocity in the region of 7.6 km/s. Anything much less than this would result in an elliptical orbit that intersected the Earth's surface, and the spacecraft would return to Earth. At the other end of the velocity spectrum, he – and, of course, the military – were familiar with the 'orbits' of shells fired, much more slowly, from guns; the faster the shell left the barrel, the higher and further it would travel. A super-gun could project a shell over the horizon, and if sufficiently energetic it would be possible to cause the shell's trajectory to miss the Earth's surface; the falling shell would descend more gradually than the surface of the round world curves away. This is a fairly good analogy, but it does not quite work in practical terms. A shell projected from a gun with sufficient speed can be in orbit; but, alas, Newton's and Kepler's laws ensure that the orbit passes through the gun emplacement – a nasty shock for the gunners when it returns from behind, having travelled round the Earth. To produce a safe orbit the spacecraft has to be lifted up to a sufficient altitude to make sure that it misses obstructions, and in addition be given sufficient *horizontal* velocity: Tsiolkovsky's 7.6 km/s. Since the atmosphere has a retarding effect on any spacecraft, it is best to ensure that the chosen height is beyond the point where this can happen. A typical safe orbit would be at an altitude of about 500 km, at which height, atmospheric drag should be small enough to allow the spacecraft to remain in orbit for five years or more. Although elliptical orbits are acceptable, the lowest safe orbit is circular, because it is beyond the atmosphere at all points. This type of orbit is known as *low Earth orbit* (LEO), and is the typical first step in any space endeavour. The task of the launchers with

Figure 3.9. Konstantin Eduardovich Tsiolkovsky. (Courtesy Kaluga State Museum.)

which we are familiar – Ariane V, Proton, and the Space Shuttle – is generally to put spacecraft and components into LEO; from where other orbits can be reached with comparative ease. The task of a launcher is then to reach about 500 km above the surface of the Earth, and to boost its spacecraft to a horizontal velocity of 7.6 km/s.

With typical vehicle speeds (of railway engines) at the end of the nineteenth century of 100 km per *hour*, Tsiolkovsky was no doubt depressed – initially. The only possibility, at the time, before the invention of the aeroplane, was the rocket; and Tsiolkovsky produced his second discovery: his famous *rocket equation*. This defines the speed which can be reached by a rocket, in terms of the amount of fuel it carries and the velocity of its exhaust. This is perhaps the most important equation for space travel, and is presented here with no apology:

$$V = v_{exhaust} \log_e \frac{M_0}{M}$$

The ratio M_0/M – the *mass ratio* – represents the amount of fuel that the rocket carries in terms of its total mass. M_0 is the total mass of the rocket at the beginning of the firing, M is the mass at the end, and the difference is the amount of fuel exhausted through the rocket nozzle. The more fuel that is burned for a given mass of rocket, the faster it travels; and the greater the exhaust velocity, the greater the final speed for a given amount of fuel burned. To achieve the very high velocity of 7.6 km/s, the rocket needs to burn a large amount of fuel, and to have a very high exhaust velocity. Tsiolkovsky knew enough about chemistry and thermodynamics to determine that the basic gunpowder rocket of the time stood no chance at all of achieving the necessary high velocity, as a typical gunpowder rocket has an exhaust velocity of about 1,500 m/

s. Designing a rocket to carry enough gunpowder to achieve the necessary velocity by 'brute force' is complicated, because more fuel requires bigger fuel containers. The empty or 'dry' mass of the rocket will consequently be increased, and it is therefore very difficult to achieve a mass ratio of more than about 8. Such a mass ratio implies that 87.5% of the mass of the rocket is fuel, and only 12.5% is left for structure, fuel tanks, and the all-important payload – the object to be launched into space. The physics problem, with a mass ratio of only 8, is that it is the *natural logarithm* of 8 that appears in the equation – log_e. The logarithm is a slow function of its argument: log_e 8 is only 2.08, so even with a mass ratio of 8 and an exhaust velocity of 1,500 m/s, the final speed of the rocket would be only about 3 km/s.

But Tsiolkovsky was not daunted, and he instead produced his third great idea: the 'cosmic rocket train' – which we now call a multi-stage rocket – the means by which all payloads are launched into space today. The idea is simple. If the fuel tanks are dropped off as they become empty, the average dead weight of the rocket will be reduced as it climbs. It happens that the best approach is to keep the mass ratio of each stage the same. The first stage is massive and contains most of the fuel, and has as its 'payload' the combined second and third stages. When its fuel is exhausted and the rocket has reached a speed of about 1,600 m/s, the whole first stage drops off, and the second stage – which has the third stage as its 'payload' – takes over. It is much lighter than the first stage, and by now is travelling at about 20° to the horizontal.[2] The second stage then reaches a speed of about 3,000 m/s, and in turn drops off. The final stage then takes over – usually already at the necessary altitude for the required orbit – tilts fully to the horizontal, and burns the remaining fuel to reach the final speed of 7,600 m/s to inject the payload into a low Earth orbit.

It was, of course, to be a long time between Tsiolkovsky's first enunciation of the means for getting into space, and its achievement; but when it came, everything happened just as he had depicted in his *Dreams of Earth and Sky* of 1895, and his *Cosmic Rocket Trains* of 1924. Today, we take for granted the use of multi-stage rocket launchers to place in orbit our communication and exploratory satellites, our astronauts, and the International Space Station. It will be these launchers that will bear the burden of placing the explorers, their vehicles, and their materiel, in low Earth orbit, for an expedition to Mars.

From the point of view of propulsion, the most difficult task is that of travelling from the Earth's surface into a safe Earth orbit. We live on a planet with strong gravity – which for us is a great advantage. It retains the atmosphere which we breath and which protects us from the dangerous ultra-violet light from the Sun and other radiation. But it also greatly hinders our efforts at sending objects into space. It is no accident that the rocket – invented in China more than 1,000 years ago – has become a space vehicle only recently, as the multi-stage rocket, and the high-tensile and high-temperature resistant materials needed to construct it, became available

---

[2]   The three-stage launcher travels in a curved path. It initially rises vertically to clear the dense atmosphere, and then gradually tilts over towards the horizontal, gaining both altitude and velocity. Finally, it moves in a true horizontal path to gain the maximum horizontal velocity.

only during the last century. A bamboo rocket using gunpowder could not have the necessary mass ratio and exhaust velocity to reach orbital speed, however big it might be. On a planet with lower gravity this would be possible – but then the atmosphere might possibly leak away before the inhabitants could develop the necessary technology to build even a gunpowder rocket.

3.5   INTERPLANETARY JOURNEYS: ELLIPTICAL TRANSFER ORBITS

Once a circular orbit around the Earth is achieved, what are the next steps in achieving the transfer to Mars? The orbit equation indicates how this may be achieved. In a circular orbit the spacecraft has a constant velocity. If the rocket engine is fired, to increase the velocity, the spacecraft will enter a new orbit that is elliptical. The initial conditions of the new orbit are the distance from Earth at closest approach – which is the same as the constant distance in the old circular orbit – and the new velocity at closest approach – which is greater than the old circular orbit velocity. The eccentricity ε will no longer be zero, and its exact value will define the degree of elongation of the ellipse. Following Kepler's law, the velocity will be a maximum where the spacecraft approaches closest to the Earth, and a minimum at its greatest distance. In effect, the spacecraft is travelling 'uphill' as it moves away from its closest approach, and its kinetic energy is exchanged for potential energy as it rises. At the apex of its flight it begins to fall back towards the Earth, gaining velocity as its potential energy is converted back into kinetic energy. The new elliptical orbit will be just as stable as the old circular orbit, and the spacecraft will continue to orbit the Earth, rising and falling, unless it is disturbed. The distance of closest approach is called the *perigee* of the orbit, and the furthest point is called the *apogee* (both from the Greek). An important point to note is that the perigee is located at the point where the rocket was first fired to increase the spacecraft's velocity from the old circular orbit velocity, and the apogee is located exactly opposite this point, $180°$ round the ellipse. Since pedantry is rife, it would be as well to note here that the 'gee' in 'perigee' refers to Earth as the centre of the orbit. When the orbit is about the Sun, the correct termination is 'helion', as in 'perihelion'; and for a star the termination is 'astron', as in 'apoastron'. The general term used when the body being orbited is not specified, is 'apsis' – as in *periapsis* and *apoapsis*.

An elliptical orbit such as that described above is often called a *transfer orbit*, because it is the means of moving a spacecraft from a lower orbit to a higher orbit. It is used, for instance, when a communications satellite is to be placed in a *geostationary* orbit; that is, an orbit with a period the same as the rotation period of the Earth. As viewed from Earth, a spacecraft in such an orbit appears to be stationary, because both are moving round at the same rate. For this orbit the spacecraft needs to be at an altitude of 36,000 km where, following Kepler's law, it will have a period of 23 hrs 56 min – the rotation period of the Earth. (The 24-hour day arises because, in one rotation of the Earth, our planet also moves sufficiently in its orbit to make the Sun appear overhead about four minutes later each day.) The communications satellite is first launched into a low circular orbit, in which it may remain for several circuits. Then, when it is at the point in its orbit that is exactly opposite the place where it is desired that the satellite should take up its final station, an auxiliary rocket is fired to provide it with enough velocity to place it into an elliptical orbit with an apogee at 36,000 km, while its perigee is naturally at the same altitude as the low circular orbit. The spacecraft then travels along its new orbit until it reaches the apogee. If no further action were to be taken it would continue round its elliptical orbit and return to the perigee; but if the auxiliary rocket is fired again when it is at apogee, the spacecraft is provided with sufficient additional velocity for it to enter a new circular orbit at the altitude of the apogee, which in this case is 36,000 km. The elliptical orbit is thus the *transfer* orbit between the circular low Earth orbit, and the much more distant, circular, geostationary orbit.

It is important to keep in mind the relative magnitudes of the velocities associated with these different orbits. Kepler's law dictates that the lower the orbit, the faster the spacecraft moves. Therefore, the spacecraft in the circular low Earth orbit is moving faster than it would move in the much higher circular geostationary orbit. However, the situation is different for the elliptical transfer orbit. At perigee, where it is closest to Earth, it is moving *faster* than it moved in the circular orbit having the same altitude. As it rises towards apogee it slows down, because it is working against gravity; and when it reaches apogee it is moving more *slowly* than it would do in a circular orbit at that altitude. This is the reason why an extra boost to velocity is needed here, in order to secure the new circular orbit. So, to transfer from a low circular orbit to a higher orbit it is necessary to initiate *two* velocity increases – one at perigee, to place the spacecraft in an elliptical transfer orbit, and another at apogee, to circularise the orbit at the new altitude. It may be clear, therefore, that this process will be echoed in the business of transferring from Earth to Mars.

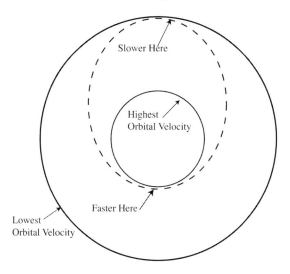

Figure 3.10. An elliptical transfer orbit between two circular orbits. The spacecraft velocity is faster than the circular orbit at perigee, and slower than the circular orbit at apogee

3.6   ESCAPE VELOCITY: PARABOLAE AND HYPERBOLAE

Before moving on to consider how to move between the planets, the use of the two other orbit forms – the parabolic orbit and the hyperbolic orbit – should be considered. As the velocity given to a spacecraft in low Earth orbit is increased, the altitude of the perigee will not change, but the spacecraft will move further and further away from Earth at the apogee of its elliptical orbit; that is, the eccentricity will increase. Of course, as the spacecraft moves further away from the Earth, the force of the Earth's gravity will drop according to Newton's law, with the distance squared, so that the retarding force – which will ultimately bring it back on the return half of its orbit – decreases. It may seem obvious that at some point the gravitational force could become be so weak that the spacecraft would not return. Although gravity decreases with the square of the distance, it never becomes zero; and even in the far reaches of the Solar System the gravity of Earth still has a finite, if small, attractive force, and the spacecraft will feel it. This might seem complicated, but fortunately the physics, and the calculation of the real result, is all contained in the orbit equation. When the eccentricity $\varepsilon = 1$, the farthest distance of the spacecraft from Earth (the apogee) becomes infinite. This is expressed mathematically in the equation when $\varepsilon \cos 180° = -1$, and the right-hand side becomes equal to zero, so that $1/r = 0$. For $1/r$ to be zero, the distance r should be infinity. So if $\varepsilon = 1$, the orbit becomes open, and the spacecraft will move away from Earth forever, but at an ever-decreasing velocity. The velocity only becomes zero at an infinite distance. This is the parabolic orbit; and the velocity required for $\varepsilon = 1$, and to enter a parabolic orbit, is the well-known *escape velocity* – the minimum velocity necessary to escape from the Earth's gravity.

Infinity is not a very useful concept when one only wants to go to Mars. Fortunately, while it is true that the Earth's gravity can be felt even on the surface of Mars, the spacecraft will feel the dominant force of the Sun's gravity long before it arrives there. As the Earth's gravity diminishes it will eventually become much less influential than that of the Sun, which is more or less constant, anywhere in Earth's orbit. The spacecraft will then have to move according to a new set of orbital rules. For our purposes it is sufficient to define 'infinity' as the point where this occurs. This transition will, of course, be gradual, but as mentioned before, it is sufficiently accurate for our purposes to treat the two cases separately.

The importance of the parabolic orbit is that it tells us the minimum velocity that the spacecraft has to achieve in order to escape the Earth. Another interesting deduction can be made from the orbit equation or from Figure 3.8: that the spacecraft will eventually move away from the Earth in a direction at *right angles* to the thrust of the rocket that places it in the parabolic orbit. This is just one of the peculiar rules that make travel in space so different from travel on Earth. It is because gravity continues to control the path of the spacecraft after the engine has ceased thrusting. It will also have been noticed that the thrust of the rocket lies along the tangent to the original circular orbit, the object being to increase the speed *along the path* of the orbit. If escape from the Earth is required, it might seem best to set off in a direction directly away from it. However, this is the worst approach, because the rocket has to fight gravity all the way, and at any instant, most of the thrust is wasted in lifting the as yet unburned fuel against the force of gravity – known as *gravity loss*. On the other hand, by thrusting along the path of the orbit, at right angles to the gravitational force, no lifting against gravity takes place, and all the thrust can be directed into accelerating the spacecraft. So, this oblique approach results in the easiest escape from the Earth's gravitational influence. We have assumed here that the period during which the rocket is firing is short, so that the spacecraft does not move very far along its orbit while the thrusting is taking place; and this is generally the case. If this were not so, then there would still be some small gravity loss due to the increase in altitude as the spacecraft begins to lift out of its circular orbit into the parabolic orbit (we shall return to this in a later chapter).

The parabolic orbit will enable a spacecraft to escape from the Earth; but it will effectively have zero velocity, with respect to the Earth, at the point of escape when it falls under the gravitational influence of the Sun. It will, of course, have a velocity with respect to the Sun that is precisely identical to the Earth's. A spacecraft launched into such an orbit would forever orbit the Sun at a little distance from the Earth, but in the same orbit as the Earth, because both have the same velocity. For many purposes it is important for the spacecraft to instead have some residual velocity when it leaves the Earth's influence, so that it can go somewhere. This is where the hyperbolic orbit assumes its role. By accelerating the spacecraft to *more* than the escape velocity, the eccentricity of the orbit becomes greater than 1; and in the orbit equation, $1/r$ becomes equal to zero much earlier in the orbit. The function $\varepsilon \cos \theta$ becomes equal to -1 – not at $\theta = 180°$, but at some smaller angle, depending on the actual value of the eccentricity. This implies that the spacecraft escapes from the Earth's influence earlier in the orbit, and, more importantly, with some residual velocity. The larger the initial excess velocity, the sooner in its orbit does the spacecraft escape, and the higher will be its residual velocity. As we shall see, this hyperbolic orbit will be the orbit required to set the spacecraft on its journey to Mars.

It is possible to calculate the velocity necessary to be given to the spacecraft at perigee to achieve these different orbits, using only the orbit equation and the expression for the eccentricity. They are repeated here for convenience:

$$\frac{1}{r} = \frac{GM}{r_0^2 v_0^2}(1 + \varepsilon \cos \theta); \qquad \varepsilon = \left(\frac{r_0 v_0^2}{GM} - 1\right)$$

For a circular orbit, $\varepsilon = 0$, and so the required velocity for a circular orbit is that which produces

$$\frac{r_0 v_0^2}{GM} = 1$$

$$v_0 = \sqrt{\frac{GM}{r_0}}$$

The necessary speed can then be calculated for any planet and for any starting point by substituting the appropriate planetary mass and distance. By analogy, for escape along a parabolic orbit:

$$\varepsilon = 1$$

$$v_0 = \sqrt{\frac{2GM}{r_0}}$$

It is interesting that the velocity of escape is only $\sqrt{2}$, or 1.414 times the circular orbit velocity. In other words, the most demanding requirement is to reach orbit. From orbit to escape requires an additional velocity of only 41%.

   The velocity in the above equations is the *initial* velocity, and the equation tells us what shape the orbit will be for a given value. We actually need to be able to calculate the necessary initial velocity for escape along a hyperbolic orbit that will set the spacecraft on its way to Mars. The above equations can be simplified by substituting the eccentricity in the orbit equation. This is a useful form to calculate the *distance* from Earth at any time, for any orbit where the eccentricity is known:

$$\frac{r}{r_0} = \frac{1 + \varepsilon}{1 + \varepsilon \cos \theta}$$

It is also possible to calculate the *velocity* at any point in the orbit, from a similarly derived equation:

$$\frac{v^2}{v_0^2} = 1 - \frac{2r_0}{1 + \varepsilon} \left(\frac{1}{r_0} - \frac{1}{r} \right)$$

   These equations appear to be a little complex, but they apply to all orbits and simplify to Kepler's laws for elliptical orbits, and are very useful. For example, they can be used to calculate the velocity of the spacecraft at the apogee, for any type of orbit. Imagine an ellipse with eccentricity 0.5. Then, at apogee, $\theta = 180°$ and $\cos \theta = -\frac{1}{2}$. This gives, by substitution in the equation above, for the distance at apogee, $r/r_0 = 3$; and so the distance from the Earth at apogee is three times the distance at perigee if the eccentricity is 0.5. The velocity equation can be used to calculate the speed at this point, again by substitution, giving $v^2/v_0^2 = 1/9$, or $v = v_0/3$. This proves the earlier assertion that the velocity at apogee is smaller than at perigee. For an ellipse, we could have derived this from Kepler's second law, which effectively states that the speed is inversely proportional to the distance.

   This example shows how the equation works for elliptical orbits; but we can also use the same equations for open orbits; and in particular, the hyperbolic orbit. As an example, let the eccentricity be 1.5 (> 1); then, of course, the distance at apogee will be infinite, as discussed above. The velocity for any distance less than infinity can be calculated using the equation as before, simply by substituting the value of r. But the most useful thing is to calculate the excess velocity that the spacecraft has when it leaves the Earth's gravitational influence; this could be, for example, the velocity with which a spacecraft sets out for Mars. From the point of view of the physics describing the motion, the point of escape is at an infinite distance from the Earth, although in reality it will be at about ten Earth radii, when the Sun's gravity takes over. We find, substituting $1/r = 1/\infty = 0$, that $v^2/v_0^2 = 1/5$, and so the excess velocity is $v = \sqrt{v_0/5}$, or about 45% of the initial velocity. This is the speed, relative to the Earth, that the spacecraft takes away with it after climbing up out of the potential well of the Earth's gravity.

The physics and mathematics outlined in the preceding few paragraphs may seem complicated, but they are all that we need in order to be able to calculate and determine the orbits and velocities necessary for a voyage to Mars and back.

3.7   THE EARTH–MARS JOURNEY: A COMPLEXITY OF ORBITS

The general scheme will be a series of rocket firings and orbital manoeuvres, taking into account the physics described above. The spacecraft, its fuel and its payload will first be placed in a circular low Earth orbit. A firing of its rocket engines will place it in an hyperbolic escape orbit that leaves it with an excess velocity when it escapes the Earth's gravitational influence. From this point it becomes a small planet orbiting the Sun. Because it has excess velocity with respect to the Earth, its orbit will not be the same as the Earth's orbit, and this excess velocity will put it into an elliptical orbit. If the excess velocity is exactly right, the ellipse will be a transfer orbit that touches the Earth's orbit at perihelion and the orbit of Mars at aphelion – which is, of course, an orbit round the Sun. Then, it is only a matter of entering orbit round Mars, and later descending to the surface. The physics of such transfers from one planetary orbit to another was worked out by Hohmann, and these orbits are called 'Hohmann orbits' in his honour. It will perhaps be obvious that to achieve these orbits – which just graze the inner and outer circular orbits – only the smallest velocity increase is required. They are therefore also known also as 'minimum-energy transfer orbits'. In most cases these will be the only transfer orbits that we shall consider.

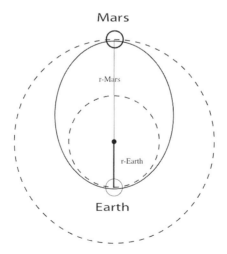

Figure 3.11. The Earth-Mars transfer orbit. The major axis of the transfer orbit is the sum of the radii of the orbits of Earth and Mars round the Sun. This is a Hohmann orbit, or minimum-energy transfer orbit.

3.8   THE MOVING PLANETS: WAITING FOR THE RIGHT TIME

This is the general scheme, but the details are both complex and interesting. One reason for this is the fact that Earth and Mars move in their orbits – a facile statement considering what has gone before, but one of fundamental importance. The most obvious consequence is that in order to rendezvous with Mars, the spacecraft must reach aphelion at the instant that Mars is passing through the point where the orbits touch. If not, then the spacecraft would simply return to Earth's orbit along the remainder of the elliptical path. It will not have escaped the reader's attention that if Earth is then not at the spacecraft's perihelion, but elsewhere in its orbit, the spacecraft will again travel out to Mars' orbit. This is not what a careful astronaut wants, or needs. The timing of all these manoeuvres must be precise, and the planets must be at their appointed places at the appointed times. As might be expected, this complicates the whole business enormously. The fact that Earth and Mars are in motion round the Sun defines very tight limits on the length of any expedition to Mars, and provides one of the major technological challenges for such an expedition.

For the spacecraft to arrive at the orbit of Mars at precisely the right moment, it must leave Earth when both Earth and Mars are at particular positions in their orbits. These positions are defined by the requirement that in the time taken by the spacecraft to reach the orbit of Mars, Mars itself will have travelled round its orbit to the point where the spacecraft has arrived, and they must meet. This translates into the requirement that the spacecraft should be launched on its journey to Mars when Earth and Mars are at specific positions in space, relative to one another. Since Earth travels round its orbit in a little more than half the time it takes Mars to complete one circuit, Earth will overtake Mars once in each martian year of 687 days. So for the journey to Mars to begin, the correct relative positioning of the planets will occur roughly once during that time. It is, in fact, easy to calculate the true interval between such propitious moments, knowing the orbital periods of the two planets. They occur at intervals of about 779 days (2.13 years) – the *synodic* period of Mars. This is the first constraint on the scheduling of a Mars expedition.

But what should the relative positions of Earth and Mars actually be for the journey to begin? To determine this, the length of the spacecraft's journey to Mars needs to be known. This is just half the period of the elliptical transfer orbit. It can be calculated using data on the orbits of Earth and Mars, and Kepler's laws. These are shown, together with the spacecraft orbit, in Figure 3.11. As can be seen, the point of departure from Earth and the point of arrival at Mars are diametrically opposite to each another. The transfer ellipse has to extend from the orbit of Earth to the orbit of Mars, and so its major axis is simply the sum of the radii of the orbits of Earth and Mars. These values, and the period of the Earth's orbit – a terrestrial year – are all that need to be known in order to calculate the period of the transfer ellipse. Astronomers, as is their wont, use the mean distance of the Earth from the Sun as a unit of length called the Astronomical Unit (AU). In these units, the mean distance of Mars from the Sun is 1.524 AU:

$$r_{s/c} = \frac{r_{Earth} + r_{Mars}}{2} = \frac{1.524 + 1}{2} = 1.262 \text{ AU}$$

$$T_{s/c} = T_{Earth} \sqrt{\frac{r^3_{s/c}}{r^3_{Earth}}}; \text{ Kepler's third law}$$

$$T_{s/c} = 365 \sqrt{\frac{1.262^3}{1^3}} = 518 \text{ days}$$

The time for the spacecraft to enter Mars orbit from Earth orbit is *half* the transfer orbit period: 259 days. So, when the spacecraft leaves Earth, Mars has to be 259 days back in its orbit from the point where the spacecraft and Mars are to meet. Earth moves faster round its orbit than does Mars, and the spacecraft, being in an elliptical orbit, moves faster than the Earth. Mars must therefore be ahead of Earth at the moment of departure. Its orbital period is 687 days, and the difference between half its period and the 259 days of the journey to Mars gives the number of days Mars that should be ahead of the departure point: Mars has to be 85 days ahead of Earth at the moment of departure.

For the return to Earth, the transfer path starts at Mars and finishes at the Earth's orbit, diametrically opposite to the departure point. In the time taken to traverse the half ellipse – again, 259 days, the same as for the outward journey – Earth has moved so as to be at the arrival point to meet the spacecraft. Because the Earth travels faster in its orbit than Mars moves, it has to be further back from the departure point at the instant of departure. Earth will travel for 259 days to reach the same point as the spacecraft, so it has to be behind Mars at the departure time: 259–365/2 = 76 days.

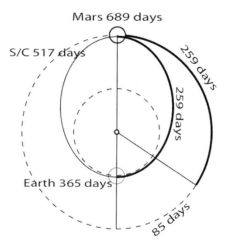

Figure 3.12. The timing of the journey to Mars. When the spacecraft departs, Mars is 85 days ahead of Earth, which ensures that the spacecraft and Mars both arrive at the encounter point at the same time.

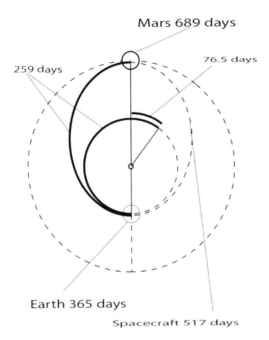

Figure 3.13. The timing for the return to Earth. The Earth moves faster than Mars, and it therefore has to be 76 days behind Mars when the explorers leave.

It will be clear from the above that neither the departure time from Earth nor the departure time from Mars can be selected arbitrarily, and are strictly defined by celestial mechanics. The entire expedition to Mars therefore has a strictly defined duration. The journey there takes 259 days, as does the journey back to Earth; but the expedition has to stay on Mars *until the planets are in the correct relative positions for the return journey*. At the time of *arrival* at Mars the Earth will be 259 days from the departure point – well past Mars and, in fact, 76 days ahead. The expedition must stay with Mars until the Earth is in the correct point for departure, 76 days *behind* Mars. Both Mars and Earth are moving in the same direction, and so it is not sufficient to count only the time needed for the Earth to move: $365 - 2 \times 76 = 213$ days. Mars is also moving, and so the departure point is moving further round the orbit, ahead of the Earth. It is the rate at which Earth is overtaking Mars that is important: $(687-365)/687 = 0.468$ days per day. So, the time taken for the Earth to arrive at the correct position for departure from Mars is $213/0.468 = 455$ days. The expedition must remain on Mars for this length of time in order to find the planets in the correct alignment for the return journey. A quick addition shows that the explorers will be away from Earth for a total of 2 years 8 months – a rather long time.

The length of time the explorers are away of course implies that a large quantity of stores will be required – including water and oxygen as well as the more mundane food, fuel and medical supplies. There are also many other issues such as the exposure of the crew to radiation during the flights and also on the surface of Mars,

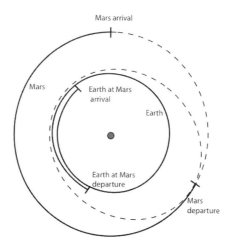

Figure 3.14. The wait on Mars for the planets to position themselves for the return journey. Earth has to execute more than one complete orbit in order to be in the correct position – 76 days behind Mars – for the return journey to begin.

which is not so well protected as Earth, having a very tenuous atmosphere and no magnetic field. All of these will impact greatly on the design of the vehicles and the size and amount of cargo they have to carry. But before examining this we should first examine the necessary velocity changes that the Mars spaceships will have to undergo in order to carry out the transfers outlined above.

3.9   HOW FAST AND HOW FAR

For the purposes of this chapter we shall assume that the spacecraft, crew and cargo are all placed in a low circular orbit at about 500 km above the surface of the Earth, before the expedition departs. The next procedure would be to accelerate the spacecraft to a new, hyperbolic, orbit, such that the residual velocity is sufficient for it to enter the elliptical Earth–Mars transfer orbit when it has 'escaped' from Earth's gravity. For all the manoeuvres associated with the Mars expedition – except, perhaps, for the landing on Mars – the spacecraft are in vacuum, and moving under the combined effects of momentum and gravity. When the rocket engines are fired they increase or decrease the velocity and hence the momentum of the spacecraft. Momentum is conserved – as dictated by Newton's laws – and so at any one time the spacecraft has the algebraic sum of all the previous momentum changes as its instantaneous momentum. The rocket equation takes care of the change in mass during a rocket-powered manoeuvre, so we can always speak of a velocity change rather than a momentum change. For each manoeuvre we calculate the velocity change. This is distinct from the actual velocity of the spacecraft, which depends on what has happened previously. It is velocity change that is important, because this determines how much fuel is required. Space engineers refer to velocity change as *delta-V* – the etymology is obvious. The task in this chapter is to calculate the total velocity change required for the mission – the total delta-V – because it is this that will determine the total amount of fuel required.

3.10 LEAVING EARTH

We know that the transfer orbit has a period of 518 days, and we also know the smallest and largest distance from the Sun: the inner distance is equal to the radius of the Earth's orbit and the outer one to the radius of the martian orbit. Here, and throughout, we assume for the purposes of calculation that these planetary orbits are circular; and although not strictly true, it is sufficiently accurate for a rough calculation. We need to determine the eccentricity of the transfer orbit in order to calculate its initial velocity. This can be derived from the radius equation given earlier:

$$\frac{r}{r_0} = \frac{1 + \varepsilon}{1 + \varepsilon \cos \theta}$$

Setting $\cos \theta = -1$, r becomes the outer radius, and the equation becomes:

$$\frac{r_{Mars}}{r_{Earth}} = \frac{1 + \varepsilon}{1 - \varepsilon}$$

$$\varepsilon = \frac{\left(r_{Mars} \middle/ r_{Earth} - 1 \right)}{\left(r_{Mars} \middle/ r_{Earth} + 1 \right)}$$

Substituting the values gives $\varepsilon = 0.208$, and substituting this in the velocity equation, where r_0 is the radius of the Earth's orbit (149.6 million km), gives:

$$v_0 = \sqrt{\frac{GM_O}{r_0}} (1 + \varepsilon) = 32.74 \text{ km/s}$$

This is a rather frightening velocity – much greater than the 7.6 km/s which is about the current capability of our rockets. Fortunately, nature is kind to us. On escaping from Earth's gravity into the Sun's gravitational field, the spacecraft will have, in addition to any excess velocity from its hyperbolic orbit, the Earth's *orbital velocity*, absolutely free: 29.8 km/s. Of course, if the astronauts perversely decided to escape in the direction opposite to the Earth's motion round the Sun, then 29.8 km/s would be subtracted from the spacecraft's velocity. To set out – in the right direction – on the Earth–Mars transfer orbit, the spacecraft requires an excess velocity of only 2.94 km/s, which is well within the capability of our present-day rockets.

This is the residual velocity, after escape from Earth; so the actual velocity given to the spacecraft in its low Earth orbit, in order to place it in the appropriate hyperbolic orbit, has still to be calculated. This will be greater, because the spacecraft has to climb up, out of the Earth's gravity. The hyperbolic orbit is defined by its eccentricity – somewhat greater than 1 – and by the starting distance from the Earth, which is the same as the radius of the low Earth orbit. With an open orbit, like an hyperbola, the relations between radii cannot be used to calculate the eccentricity. Fortunately, however, there is a relatively simple method of deriving the necessary velocity, using just the other velocities involved, and the definition of the eccentricity. To calculate the value of v_0 – the necessary velocity to place the spacecraft into the correct hyperbolic orbit – we can use energy conservation. The kinetic energy of the spacecraft, when it leaves the hyperbolic orbit and becomes a miniature planet in the Sun's gravitational field, is just the difference between its kinetic energy at injection into the orbit and the kinetic energy required for a parabolic escape. The energy consumed in just escaping from the Earth is simply subtracted from the total kinetic energy, to give the residual energy. In terms of simple equations:

$$\frac{1}{2} mv^2 = \frac{1}{2} mv_0^2 - \frac{1}{2} mv_{esc}^2$$

$$v^2 = v_0^2 - v_{esc}^2$$

In these equations, v is the residual velocity required to enter the transfer orbit. Because we know that $v_{esc}^2 = 2v_{circ}^2$, it is possible to put everything in terms of the initial circular orbit velocity – the current velocity of the expedition spacecraft – before it sets off for Mars:

$$v_0 = \sqrt{v^2 + 2v_{circ}^2}$$

The velocity required to escape from low Earth orbit into Mars transfer orbit is 11.16 km/s. Again, since the spacecraft is already travelling at 7.62 km/s in its circular orbit, it is only the difference that needs to be given: 3.55 km/s. It will then enter the transfer orbit at a speed of 2.94 km/s, having climbed out of Earth's gravity. The eccentricity of the orbit is given by

$$\varepsilon = \frac{v_0^2}{v_{circ}^2} - 1$$

and is equal to 1.149 – greater than 1, as required for an hyperbolic orbit. Using these figures and the radius of the circular orbit, the hyperbolic orbit can be drawn, and is shown in Figure 3.14. Note that this is an hyperbolic orbit with its focus at the centre of the Earth. The orbit it will enter on escape will be an elliptical orbit with the Sun as its focus.

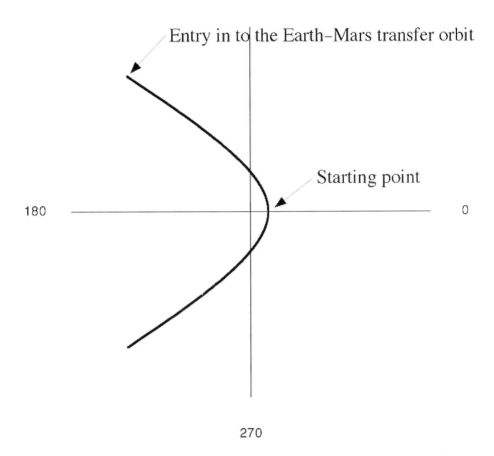

Figure 3.15. The departure hyperbola that carries the spacecraft into the Earth–Mars transfer ellipse. The upper half of the curve, only, is followed from the starting point in low Earth orbit.

3.11 A FASTER JOURNEY

It is legitimate and interesting to ask what would happen if more velocity were to be given to the spacecraft, so that it would leave the influence of Earth's gravity with more than the 2.94 km/s calculated above. Our knowledge of orbits, as outlined above, shows that as the velocity is increased, the eccentricity of the transfer orbit will increase from 0.208 to some other larger value. This increase in eccentricity raises the aphelion of the orbit so that the spacecraft crosses the orbit of Mars much earlier in its orbit; that is, before it has moved 180° round its orbit. It takes little imagination to realise that this would shorten the journey to Mars, at the expense of

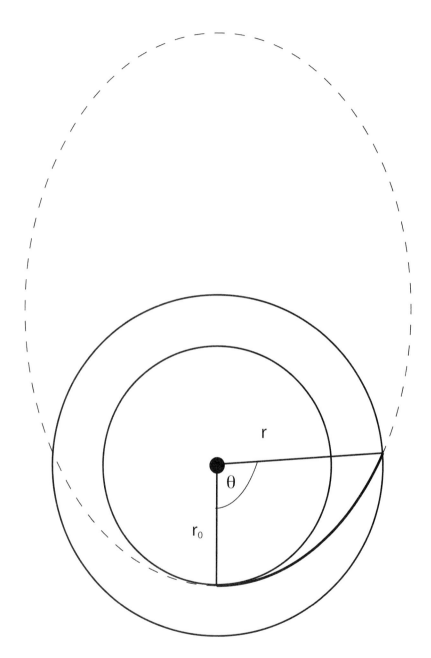

Figure 3.16. The layout of a fast transit orbit to Mars. The higher velocity increases the eccentricity of the ellipse and the aphelion, and the spacecraft therefore crosses the orbit of Mars earlier, when it has travelled round its orbit by $\theta°$ rather than $180°$. This reduces the time necessary to reach Mars.

a higher initial velocity. A shorter journey would be beneficial – especially for the crew, who would spend less time cooped up in the spaceship, and less time exposed to zero g and to excess radiation. As we shall see later, this is an important consideration. The question is: how much shorter could the journey be, and how much additional velocity would be required?

The general idea is presented in Figure 3.16. The distance that the spacecraft has to travel is much less than in the minimum energy Hohmann orbit; and, moreover, because the spacecraft is in the early part of its elliptical orbit its average speed will be greater. This reduces the time of transit. On encounter with Mars it will be travelling at an angle to the direction of travel of the planet, and the calculations of approach velocity will need to take this into account. Nevertheless, the encounter and capture can be accomplished in just the same manner as for the Hohmann orbit.

Calculation of the time of transit is beyond the scope of this book. Whereas it is easy to discover the period of a particular orbit, from Kepler's law, and halve it to calculate the time to aphelion for a Hohmann orbit, it is more complicated to calculate the time to cover the portion of the ellipse between Earth and Mars (shown in Figure 3.16), because the spacecraft's velocity is continually changing. The results of this calculation are presented in Figure 3.17, as a function of the eccentricity of the fast transit ellipse. The right-hand scale shows the necessary velocity, including the 29.8 km/s from the orbital velocity of the Earth, while the left hand scale shows the journey time. It can be seen that a quite modest velocity increase will drop the first 50 days from the journey time, and that it thereafter becomes harder and harder. The best compromise for a shorter journey, without undue velocity increase and hence fuel use, is a journey time of about six months.

It should be emphasised that we assume throughout this book that the orbits of Earth and Mars are circular. This is not the case, however, and Mars' orbit is even more eccentric than Earth's orbit. The net result of the eccentricity of the planetary orbits is that for some conjunctions the journey time will be shorter or longer than the average. The launch slot in 2003 is an example in which significantly less extra velocity is required for a short journey to Mars. This is taken advantage of by, among others, the ESA Mars Express mission, with only 3 km/s excess velocity, which will reach Mars in a period of six months.

As mentioned above, the fast transit is important for the human element of a Mars expedition; but as will be seen below, the fuel penalty may be exorbitant. In what follows, and throughout, we assume the use of the minimum energy, or Hohmann transfer ellipses. We need to establish the minimum resources that the expedition will require.

3.12 MARS ENCOUNTER

For the next eight months, the spacecraft moves along its transfer ellipse towards the orbit of Mars. It is slowing down all the time, as the Sun takes its gravitational toll of the spacecraft's momentum. The rocket engines are turned off after the entry manoeuvre, except for a necessary course correction; otherwise the spacecraft drifts

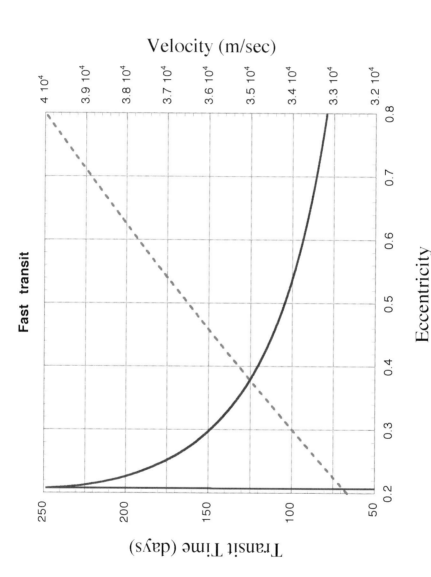

Figure 3.17. A fast transit to Mars. The dotted curve shows the injection velocity into the Earth–Mars transit orbit (right-hand scale), and the plain curve shows the transit time (left-hand scale). The velocity shown is the total velocity, including the 29.8 km/s derived from the Earth's orbital velocity. The eccentricity of the elliptical transfer orbit is shown on the horizontal scale.

silently on its way, moving further from Earth and from the Sun. At the same time, Mars is moving along its orbit towards the meeting place, where the spacecraft and Mars will encounter each other. As the time of Mars encounter draws near, the planet will be seen from the spacecraft, approaching obliquely, from behind. By this time the spacecraft will have slowed down, so that it is moving more slowly than Mars. In fact, at aphelion, when the two orbits touch, the spacecraft will be moving at 21.47 km/s, while Mars will be moving at 24.14 km/s. So, the difference in speed will be 2.67 km/s; and Mars will be overtaking the spacecraft at this relative speed. Here we have used the average speed of Mars in its orbit. Because the orbit of Mars is somewhat eccentric, its speed varies by about 2 km/s either side of the average value, in its journey round the Sun. This has important consequences for a real mission (as we shall see later), but for the moment we shall assume that the encounter takes place with Mars travelling at its average speed.

In the absence of martian gravity, the planet would glide past the spacecraft – unless, of course, the orbits were to be so precisely aligned as to cause Mars to crash into the spacecraft from behind. What actually happens is more complicated, because as the spacecraft and Mars approach closer to each other, martian gravity, obeying the inverse square law, begins to dominate over the Sun's gravity. When this happens, the focus of attention switches from the Sun to Mars – as, indeed, does the focus of the spacecraft's orbit. It is in many ways the reverse of what happens when the spacecraft leaves Earth's orbit. Seen from Mars, the spacecraft would appear to be approaching from ahead of the planet with a speed of 2.67 km/s. It will be attracted by the martian gravity, and will begin to accelerate towards the planet until it reaches its closest approach. In most cases this will occur some distance above the surface of Mars. It will then pass on, increasing its distance from Mars, and slowing down again as martian gravity takes its toll. *But it will not be captured by Mars.* Gravity is a *conservative* force – which means that there is exact accounting for energy and momentum in a gravitational field – and so a spacecraft approaching Mars with a velocity of 2.67 km/s must leave it with the same velocity, no matter what happens in between. The spacecraft will fly off into the cosmos at 2.67 km/s relative to Mars, and enter some other orbit round the Sun. This would be a severe disappointment for the explorers. The spacecraft would have executed an hyperbolic orbit around the planet Mars – and that is all. To enter a closed orbit round Mars, some energy must be extracted from the moving spacecraft so that its velocity relative to Mars at closest approach corresponds to an eccentricity of less than 1. When this happens, the spacecraft will enter an elliptical orbit about Mars, and will not shoot off into interplanetary space.

The hyperbolic Mars approach orbit will have an eccentricity represented by

$$\varepsilon = \frac{v_0^2}{v_{circ}^2} - 1; \text{ or}$$

$$\varepsilon = \frac{v^2 + 2v_{circ}^2}{v_{circ}^2} - 1$$

We know v – the approach velocity – and we can calculate v_{circ} for Mars at an assumed altitude of 500 km. Inserting the mass of Mars into the equation produces a value of 3.31 km/s for the circular velocity, and 1.67 for the eccentricity of the hyperbolic approach orbit. The mass of Mars is about one tenth of the mass of Earth, so the orbital velocities are correspondingly smaller. The velocity of closest approach in the hyperbolic orbit is v_0, which is, from the equation, equal to 5.83 km/s. To enter a circular orbit at 500 km altitude, 2.52 km/s must be removed from the spacecraft's velocity. If a rocket engine is used for this then the spacecraft will need to be reversed so that the nozzle points in the direction of motion. Then, at the moment of closest approach it will be fired to slow the spacecraft down by 2.52 km/s.

Having entered orbit around Mars, with a circular velocity of 3.31 km/s, the next task for the expedition will be to land on the surface. Ignoring, for the moment, the effect of the martian atmosphere, this landing will involve placing the lander into an elliptical orbit with an apoapsis which touches the circular 500-km orbit, and a periapsis which skims the surface of Mars. A quick calculation, using the equations already described, shows that the lander has to slow to 3.2 km/s to enter the landing ellipse, and that it will be moving at the higher speed of 3.66 km/s as it skims the surface. The speed near the surface of Mars is greater than the speed in the circular orbit, because the spacecraft effectively falls towards the surface, gaining energy from Mars' gravity. This speed must be removed for a safe landing – again, perhaps, by using a rocket engine.

To land on Mars, however, slightly less than this amount of braking is required. If the landing site is near the equator, and the landing direction is the same as the rotation of the planet, then 0.242 km/s – the velocity of the surface of the planet due

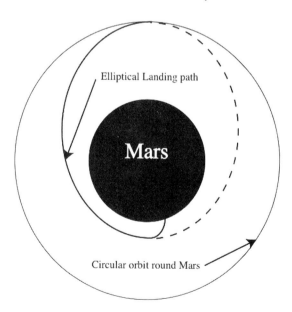

Figure 3.18. The manoeuvre for landing on Mars.

to its spin – need not be removed. It is for this reason that the Mars transfer orbit has its aphelion slightly inside the orbit of Mars. When the spacecraft encounters the planet, the net velocity of approach is in the *opposite* direction to its orbital motion, because Mars is moving faster than the spacecraft. The hyperbolic orbit then passes Mars at closest approach in the same direction as the planetary rotation, which this ensures that 242 m/s of surface velocity is available to cushion the landing.

3.13 THE RETURN JOURNEY

For the return journey, the total velocity change is, to a first approximation, the same as for the outward journey. The spacecraft lifts off from the surface of Mars, and must then follow an ascent ellipse to take it up to the altitude of the circular orbit round Mars. To take advantage of the spin of Mars, it will follow the path of the dashed half ellipse in Figure 3.15. The velocity required to do this, including the circularisation burn, will be approximately the same as the landing velocity change. Once safely in a circular orbit about Mars, the explorers will wait for the correct moment of departure. The rocket engine will be fired to place the spacecraft into an escape hyperbola, with the same shape and parameters as the capture hyperbola, but in reverse. This will put the spacecraft into the transfer orbit to Earth. Mars, in its circular orbit, is moving faster than the velocity needed to enter the transfer ellipse, so the arm of the hyperbola followed by the spacecraft will be the arm driving it backwards with respect to the direction of motion when it enters the transfer ellipse, as shown in Figure 3.19. The velocity of the spacecraft then subtracts from the planetary velocity of Mars to produce the necessary total velocity reduction to enter the ellipse.

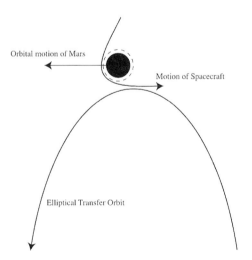

Orbital motion of Mars

Motion of Spacecraft

Elliptical Transfer Orbit

Figure 3.19. The hyperbolic orbit used to place the spacecraft in its Earth return transfer orbit.

The spacecraft then drifts along its return orbit, reaching Earth orbit after 256 days of travel. It is then travelling faster than the Earth in its circular orbit. The spacecraft will then begin to overtake Earth as it approaches perihelion. The transfer orbit is designed to have its perihelion just outside the orbit of Earth. Earth moves more slowly than the spacecraft, because the spacecraft is at perihelion and therefore moving at its highest velocity, and so the relative velocity of approach lies in the same direction as the orbital motion. Again, if nothing is done the spacecraft would pass Earth in a hyperbolic orbit and shoot off into the Solar System. The relative velocities of the spacecraft and Earth are the same as at departure, but in the opposite direction. The spacecraft is reversed, and the engines are used to slow it down to be captured into Earth orbit, which for these calculations is the end of the mission. By placing the closest approach of the spacecraft in its hyperbolic orbit, under Earth's gravity and outside the Earth's orbit, it is again moving in the same direction as the planet's spin, so that when the time comes to land on Earth it is moving in the right direction to gain from this. The gain here is 465 m/s. Earth has a larger circumference than Mars, and the Earth day is about the same length as the martian day.

3.14 THE TOTAL VELOCITY CHANGE FOR THE EXPEDITION

The purpose of these calculations is to determine the total velocity change that must be induced by rocket engines powering the expedition, and hence the amount of fuel that will be required for the whole expedition. Our rocket-powered spacecraft uses the same amount of fuel to accelerate and decelerate. There is no equivalent of braking as with terrestrial vehicles – although, as will be seen later, atmospheric braking can lead to important fuel savings for a Mars expedition. Whether the spacecraft is climbing out of the Earth's gravitational field on its way to Mars, or slowing down in order to land on Mars, fuel has always to be provided and burned. From the previous paragraphs we can see that to move from a 500-km orbit about Earth into Mars transfer orbit requires 3.55 km/s – remembering that the expedition is already moving at 7.62 km/s round the Earth and that the Earth is already moving at 29.8 km/s round the Sun. Capture by Mars requires 2.52 km/s, and landing on the surface requires another 3.42 km/s. The total is therefore 9.49 km/s. The return journey will just double value to 19.0 km/s because, in this example, all the return manoeuvres are the same as the outward manoeuvres, but with their direction reversed.

In this section we have explored the way orbits have to be used to transfer a spacecraft from Earth to Mars, and have calculated the velocity changes and orbit changes necessary to carry out the journey. In a way, the requirement of 19.0 km/s is the central factor for the arguments presented in this book. It is the basic requirement for us to reach Mars, and as we shall see, it is a very challenging one. But how accurate is the figure derived here? It makes some simplifying assumptions – the most important of which is that Mars travels in a circular orbit. The actual marked eccentricity of the orbit of Mars (first noticed, of course, by Kepler) means that our calculations – particularly for Mars capture – are only true on average. For different starting dates, the relative velocity of the spacecraft and Mars when they meet will vary by as much as 17%, and this is equally true of the departure from Mars. This means that rather than the opportunities for an Earth–Mars expedition occurring with equal value every 2.13 years, some recurrences of the correct planetary positions will be more advantageous than others in terms of the required velocity change.[3]

Nevertheless, the requirement remains substantially the same. If it is to be met in full by the use of rocket thrust, then the amount of fuel that has to be carried will be a major obstacle to realising the expedition (as will be discussed later). However, the requirement can be substantially reduced by the use of other means for changing orbits which do not involve the expenditure of rocket fuel, but which instead utilise planetary atmospheres to slow down spacecraft, and the gravitational fields of other planets to change the paths of spacecraft. These are known as *aerobraking* and *aerocapture*, and *gravity assist* manoeuvres

3.15 AEROBRAKING AND AEROCAPTURE

Landing on a planet that has an atmosphere is very different from landing, for instance, on the Moon, which has no atmosphere. The Apollo crew capsule, and every Space Shuttle, uses the Earth's atmosphere to provide most of the necessary braking to bring the spacecraft to rest on the surface. Mars has an atmosphere, and it can be used in the same way to slow down the landing spacecraft and allow it to alight safely. The amount of braking necessary is just the landing velocity calculated above: 3.66 km/s. The small mass of Mars, and its low gravity, result in an atmosphere that is very much less dense than Earth's atmosphere: the surface pressure is around 1/100 of the sea-level pressure on Earth. At the same time, because the martian gravity is weaker, the atmosphere extends further out into space; and so paradoxically, at high altitudes the martian atmosphere is denser than the Earth's atmosphere. It might (correctly) be deduced that the early braking of the spacecraft while it is still at high altitude will be more effective than during the final stages, and that rockets might still be necessary. A typical landing will use the friction of the upper atmosphere to slow the spacecraft while it is still distant from the surface. Once the speed has been reduced, parachutes designed for use in a thin atmosphere and at high velocities will take over, with rockets used only for the final soft landing For the moment it will be assumed that most of the 3.66 km/s can be removed using the martian atmosphere. This can possibly be subtracted from the total of 19 km/s.

The martian atmosphere can, however, be used for more than just the landing. It can also be used to enable the capture of the approaching Mars transfer vehicle by

---

[3]   The recent (August 2003) close encounter between Earth and Mars is a unique example of this. It is the best opportunity for 60,000 years.

martian gravity. This process – *aerocapture* – has only recently been introduced for unmanned missions to Mars and to Venus, and it can certainly be applied to human expeditions if the circumstances are right. However, it requires very precise guidance when approaching the planet's upper atmosphere. When the vehicle approaches the planet on its hyperbolic orbit, only the very smallest rocket burn is used – just sufficient to ensure that the eccentricity of the orbit – which is greater than 1 for an hyperbola – falls below 1, and changes the orbit into a very elongated ellipse. The altitude of the periapsis of the hyperbola, and so of the subsequent ellipse, is chosen to be just low enough to ensure that atmospheric drag slows the spacecraft a little. If the encounter is at too low an altitude then the slowing is too drastic, and the spacecraft plunges into the atmosphere and is burned up. If the altitude is too high, then the atmosphere has little effect. If the altitude is chosen correctly, then at each periapsis passage the spacecraft is slowed a little. Since we know that reducing the speed of a spacecraft at periapsis decreases the altitude of the apoapsis, it is easy to understand that this process will gradually convert the very elongated ellipse into the desired low circular orbit around Mars. Figure 3.20 illustrates a real instance of this process used for the Mars Odyssey robotic mission. The orbit smoothly changes from an ellipse to a circle over hundreds of circuits of the planet (those shown are each separated by about a hundred circuits). This kind of orbital manoeuvre requires great precision in spacecraft guidance. The 'window' of altitude that must be aimed for is only a few kilometres wide; and this window is very small when seen at the end of a journey of more than 100 million km. In the early space programme, computational speeds and navigation were insufficiently precise to allow a reasonable chance of success for planetary missions, although it could of course be used for the return to Earth. Even with modern computers and navigation techniques, it can still go wrong. The loss of the Mars Climate Orbiter mission in 1999, due to a navigational error that plunged the spacecraft too deeply into the atmosphere, is evidence for this. In terms of manned missions, the safety of the crew is paramount, and the system would need to be foolproof. The other aspect of aerocapture for manned missions is the length of time taken to circularise the orbit, as the crew will not want to spend several months in orbit around Mars while this takes place. A more rapid braking process will be needed, and this will place even greater demands on guidance precision. Such a rapid aerocapture, without the need to use an intermediary set of elliptical orbits, was the approach used to return the astronauts from the Moon during the Apollo missions. It may well be applicable to a Mars landing, and will certainly be used for the Earth capture and return to the surface.

3.16 GRAVITY ASSIST

There is another non-propulsive means of changing velocity that is very often used for interplanetary journeys with robotic probes. Known as 'gravity assist', it is a means of increasing the range and velocity of a spacecraft without the expense of direct thrusting with a rocket engine. It is nothing less than the theft of some of the

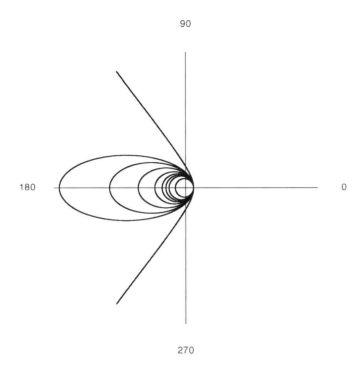

Figure 3.20. Aerocapture ellipses. Orbits are plotted every hundred orbits. Also shown
are the initial hyperbola and the first ellipse, which has an apoapsis of 60,000 km.

momentum of the very planets in their orbits. To see how this theft can take place,
again consider the approach of the spacecraft to Mars, on its hyperbolic orbit. The
intention, of course, is to slow the spacecraft down with its rocket engines so that it
can be captured by Mars and set in a circular or elliptic orbit. But what would
happen if the spacecraft were not slowed down, and were instead allowed to pass the
distance of closest approach with its velocity unchanged? It would rise along a
hyperbolic orbit away from Mars, the mirror of the approach orbit, and then
journey off into space with a velocity, relative to Mars, exactly equal to the velocity
with which it approached: in this case, 2.67 km/s. Initially this appears to be a useless
exercise, because the velocity when moving away from Mars is exactly the same as
the velocity when approaching Mars. However, recall that before the encounter with
Mars the spacecraft was moving at 21.47 km/s in its elliptical transfer orbit with
respect to the Sun, while Mars was moving at 24.14 km/s in the same direction.
Leaving Mars on the outbound leg of the hyperbolic orbit the spacecraft would have
a velocity *relative to Mars* of 2.67 km/s; and to this must be added the velocity of
Mars in its orbit, 24.14 km/s. So the total velocity of the spacecraft with respect to
the Sun is now 26.81 km/s, while its velocity before the encounter was 21.47 km/s: a
clear gain of 5.34 km/s. The spacecraft is now moving away from Mars towards the
outer Solar System, and has undergone a significant change in direction as well as a
gain in velocity. The gravity of a planet or the Moon, or indeed the Sun, can change

the direction of a spacecraft without any loss of kinetic energy. At its most fundamental, the familiar circular orbit is an example of this effect. The spacecraft is continually changing direction, but no energy is lost.

The first effect of the planetary fly-by that we have just considered is a radical change in direction; and it is sometimes used to change the direction of a spacecraft without expenditure of fuel. The second effect is a real gain in velocity, although at first this seems to be a mere mathematical sleight of hand. It is, however, not 'free' energy; rather, it is the result of the minute slowing down of the planet caused by the mutual attraction between the planet and the spacecraft. Isaac Newton first realised that the planet attracts the apple, but that the apple equally attracts the planet. They both move towards one another, but it is the apple – vastly lighter than the planet – that moves the most. In the same way, Mars is slowed down by the encounter with the spacecraft, but this slowing down is minute compared with the equivalent speeding up of the spacecraft.

This example used Mars, as we have already examined in great detail the way Mars and the spacecraft interact; but it applies to any encounter between a spacecraft and a celestial body, and can be used to speed up or to slow down a spacecraft. To speed up, the spacecraft has to pass behind the planet in its orbit, as in the example; and to slow down, the spacecraft has to pass in front of the planet in its orbit. Mars clearly cannot be used in this way if it is to be the destination of our journey, but we can use any of the inner planets, including the Earth and the Moon, to help the spacecraft on its way to Mars, and indeed to slow it down on the return journey. The use of gravity assist requires some quite complicated navigation. Unmanned missions like the Japanese Nozumi have used a combination of Earth and Moon fly-bys to boost the spacecraft to Mars, from a rather inadequate initial launch velocity. It may seem paradoxical to use the Earth for such a purpose, but if the spacecraft can be placed in orbit round the Sun, then the Earth's orbital velocity is available for boosting, via a suitable gravity-assist manoeuvre.

For a manned mission to Mars, the length of time required between Moon and Earth encounters would be a disadvantage. However for the *return* journey a flyby of Venus can have an advantage – not by reducing the velocity change, but by reducing the waiting time on Mars for the Earth and Mars to be in the correct relative positions. This would reduce the overall duration of the expedition and hence the mass of supplies required. A fly-by of Venus forms part of one of the preferred scenarios for a first Mars expedition, and this technique will feature in some of the favoured Mars mission plans (to be described later). Essentially, on departure from Mars the spacecraft is placed in a fast elliptical orbit that passes inside the orbit of Venus. On return past Venus, gravity assist is used to change the orbit to one encountering the Earth. This actually takes longer than the 259-day Mars–Earth transfer orbit, but has the important advantage of allowing an earlier departure from Mars, only 30 days after arrival. For the first human expedition this might be a major advantage. Even so, the explorers would still be away from Earth for 1 year 9 months, for a 30-day stay. Used in this way, gravity assist shortens the expedition but, of course, does not reduce the total velocity change requirement.

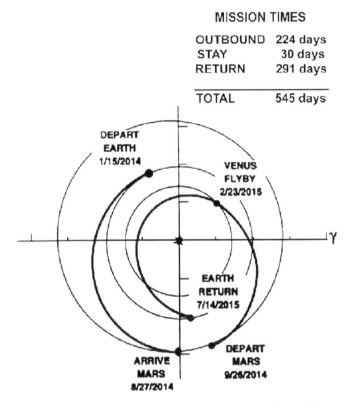

MISSION TIMES

OUTBOUND	224 days
STAY	30 days
RETURN	291 days
TOTAL	545 days

Figure 3.21. The 'short-stay' mission scenario involving a fly-by of Venus to bend the free-flight path back towards Earth. (Courtesy NASA.)

3.17 THE FINAL DELTA-V REQUIREMENT

The value of 19 km/s for the total delta-V, previously calculated, can be modified by the use of aerobraking and aerocapture. To be captured, the velocity has to be decreased to just less than the escape velocity. For Mars, the approach velocity at 500 km altitude is 5.83 km/s, the circular velocity is 3.31 km/s, and the escape velocity is $\sqrt{2}$ times the circular velocity: 4.68 km/s. The maximum aerocapture saving would then be just less than the escape velocity, minus the circular velocity: 1.37 km/s. If landing can be accomplished entirely by use of aerobraking, then the rockets must be fired to reduce the velocity from the circular velocity to 3.2 km/s in order to enter the landing ellipse: a delta-V of 0.11 km/s. The whole of the 3.66 km/s landing velocity change would then be saved by aerobraking. The total saving in delta-V is then $1.37 + 3.66 - 0.11 = 4.92$ km/s.[4] This produces, for the entire mission,

---

[4] These manoeuvres can of course be combined in a single trajectory for an aerobraked capture and landing on Mars. The total delta-V would be the same.

a delta-V requirement of 14.08 km/s – a considerable saving, which can be realised by designing a mission for maximum use of atmospheric braking, and acceptance of some increase in the time spent in orbit around Mars for the aerocapture; plus, of course, the extra risks to the expedition posed by the reliance on accurate guidance of the approaching vehicle.

Aerocapture combined with aerobraking is almost certain to be used for the return to Earth, provided, as before, that the safety requirements can be met; this is how all astronauts, including the Apollo lunar explorers, have so far returned to Earth. The velocities are all much higher than in the case of Mars, because of the Earth's higher gravity. The hyperbolic velocity on return to Earth is 11.16 km/s, and the circular velocity is 7.62 km/s at 500 km altitude. Again, the maximum saving would be just less than the escape velocity of 10.77 km/s minus the circular velocity: 3.15 km/s. If this approach is used for the returning expedition, then the full delta-V requirement for the mission drops to 10.93 km/s. This returns the expedition to low Earth orbit – the location from which we assume they set out to Mars. We have chosen to calculate requirements based on departure and return to low Earth orbit because it is convenient for comparing different ways of accomplishing the expedition. In an actual expedition, however, it is highly probable that aerocapture and landing would be combined in a single manoeuvre, exactly as was done in the Apollo missions.

The journey scheme presented here is the simplest, and is the scheme first devised by von Braun during the 1950s. It is not the only way of going to Mars, and there are other possible combinations of departure times and orbits that may be advantageous, in certain circumstances. These will be considered in a later chapter.

3.18 A HUMAN EXPEDITION

The objective in carrying out these simple calculations has been to begin to illuminate the issues surrounding a human expedition to Mars by setting a firm basis for the technological requirements. The physics of orbits is simple, and the numerical conclusions inescapable. The velocity change requirement may seem to be not all that daunting, given that we have been sending unmanned missions to the outer Solar System, requiring total delta-V delivery of the same order. A bigger issue, however, is the ability to send human beings out into the Solar System and to bring them back safely. This was achieved with spectacular success in the Apollo expeditions to the Moon, and our task is to achieve the same level of success for a much more distant objective. We have now to examine the burden placed on our technology by the peculiar needs of a human expedition. There are two: the sheer mass of equipment and supplies required to keep the explorers alive and to enable them to explore Mars; and the need to bring the explorers back alive and in good health. Neither of these requirements needs to be met by an unmanned probe.

The above calculations demonstrate how the velocity change is calculated, and how much delta-V is required for the expedition. The required velocity change is independent of the mass of the spacecraft, and is the same for a 100-kg probe as for a 1,000-tonne spacecraft. However, the mass of the spacecraft influences the amount of fuel required for a given velocity change. So, how much will the expedition weigh? Estimates will vary with the taste and inclination of the estimator, but very clear physics relates the mass of fuel required for the expedition to the delta-V requirement and the all-up mass. This we can examine securely by using Tsiolkovsky's rocket equation.

3.19 THE ROCKET EQUATION AGAIN

Tsiolkovsky's rocket equation relates the total velocity change to the fraction of the loaded spacecraft mass, which is allocated to fuel, and to the exhaust velocity of the rocket engine. If the payload is doubled, then the amount of fuel has to be approximately doubled. Earlier references in this chapter to Tsiolkovsky's work have indicated that a mass ratio of 7 or 8 would be typical, and this implies a tonnage of fuel that is six or seven times the tonnage of payload; remember that the payload here includes the whole dry mass of the spacecraft, loaded with the expedition, but without fuel. The simplest procedure is to calculate the mass ratio necessary to achieve the required total velocity change. The rocket equation is very useful, in that it can be applied to a single burn of a rocket or to a series of burns. All we have to do is to insert the actual changes in mass due to the fuel exhausted through the nozzle, for each case. Here we can take the total velocity change for the whole expedition – assuming the use of aerobraking, and so on – and calculate the necessary mass ratio. Note that it is the mass ratio we are calculating. This does not require a knowledge of the expedition mass, although it does require an estimate of the exhaust velocity of the rocket engine, because Tsiolkovsky's equation gives $\Delta v = v_e \log_e(M_0/M)$ delta-V in terms of the mass ratio and the exhaust velocity. The exhaust velocity of a rocket engine is a major performance indicator, given that it defines how much payload can be accelerated to a given velocity. (This will be a major subject of discussion in later chapters). For a correctly designed engine, it depends almost entirely on the type of fuel used. For the moment we can cite the Space Shuttle main engine as typical. It has the highest efficiency of any present-day rocket engine, it uses liquid hydrogen and liquid oxygen as fuel, and it develops a vacuum exhaust velocity of 4,550 m/s.

Substituting this and the required delta-V of 10.93 into the rearranged rocket equation,

$$\log_e \frac{M_0}{M} = \frac{\Delta V}{v_e}$$

we derive a mass ratio of 11.05. The typical value for a single stage of a modern launcher is about 3, but by combining rockets of this mass ratio in three-stage launchers we can achieve Earth orbit as noted earlier. For a single-stage launcher – the earnest desire of modern rocket engineers – a mass ratio of about 8 is thought to be technically possible. But how should we consider the achievability of a mass ratio of 11? It is true that for a vehicle assembled in orbit, the static force of gravity is absent, and it would only need to withstand the much smaller forces due to acceleration by the rocket engines themselves. This would reduce the mass of the basic structure of the vehicle, and so help towards a higher mass ratio. On the other hand, an expedition vehicle will be more complicated than a simple launcher, as it has multiple roles which cannot be combined in one vehicle without some cost in terms of additional mass. Many of these issues will be considered in later chapters, and here we should simply note that a mass ratio of 11 is a significant technological challenge.

The mass ratio, as it appears in the rocket equation, is mathematically convenient, but it is perhaps more illuminating to express it in a different way. The initial mass M_0 of the departing Mars spacecraft is made up of three elements: the true payload, comprising the explorers, their life support, their supplies and their equipment; the vehicle itself, its rocket engines, fuel tanks and equipment holds; the all-important structure; and finally, the fuel that will be expended on the journey there and back. At the end of the mission, the explorers, their samples and their vehicle with empty fuel tanks is represented by the final mass M.

$$\frac{M_0}{M} = \frac{M_p + M_d + M_f}{M_p + M_d}$$

The subscripts p, d, and f refer, respectively, to the payload, the 'dry' spacecraft, and the fuel. But what should be the relative ratios of these separate masses? We know already that the initial mass should be eleven times the final mass. We can also guess at the dry mass of the spacecraft in the following way. A mass ratio of 11 implies that the mass of fuel should be 91% of the mass of the spacecraft. This means that we can, to a first approximation, ignore the mass of anything else in determining how much structure we need to hold the whole spacecraft together and to contain the fuel. We could imagine that the mass of the structure will be a fixed fraction of the fuel mass; perhaps 10%. Is this reasonable for a spacecraft? Let us consider the third stage of the Saturn V vehicle that launched the Apollo Lunar Transfer Vehicle – the only true spaceship we have ever built – on its way to the Moon. The mass of the payload – the fully loaded LTV – was 47 tonnes; the dry mass of the third stage itself was 13.3 tonnes; and the all-up mass of the stage was 119.9 tonnes. Of course, in this case the third stage was really a 'booster' that was discarded by the transfer vehicle; but it does provide us with an idea of the structural mass of a basic vehicle. This is just the ratio of the dry mass of the stage to the all-up mass of the stage itself: 13.3/119.9, or 11%. It would not be unreasonable, therefore, to assume that the structural mass should be about 10% of the fuel mass. In fact, as can be verified from the above equation, this would produce a mass ratio for the vehicle, without payload, of 11. With this type of vehicle, therefore, we could send a rocket to Mars and back, but empty, without a payload. This means that considerable work would need to be accomplished on the structural design of the Mars transfer vehicle and its booster, to reduce the structural mass. But this is not impossible – particularly if the vehicle is constructed and, more importantly, fuelled in orbit. Some of the mass in the Saturn

V upper stage was included to hold the fuel against the force of gravity; but gravity is not applicable in Earth orbit, and only the much smaller acceleration of the rocket itself need be considered.

Assuming for the moment that we can achieve a structure 'factor' of 5%, we can rewrite the mass ratio as

$$\frac{M_0}{M} = \frac{1.05M_f + M_p}{0.5M_f + M_p}$$

Evaluation with the payload mass set to zero will confirm that the basic mass ratio is now 21 rather than 11, which provides for payload capability. In fact, setting the mass ratio to 11 and the new structure factor of 5%, we can rearrange the equation to produce $M_f = 20\,M_p$, so that for every tonne of payload, 20 tonnes of fuel will be required, based on our simplifying assumptions. This ratio is useful in assisting us to estimate the magnitude of the technological challenge. We may note, in passing, that the Apollo payload of 47 tonnes would require 940 tonnes of fuel to carry out the Mars expedition.

3.20 THE MASS OF THE PAYLOAD

The actual mass of the payload depends, to some extent, on the intended activities of the explorers on Mars; but it depends mostly on the number of crew, and the length of time that they are away from Earth. It is clear that the expedition has to take everything with it in the spacecraft: food, oxygen, water, and expedition requirements such as housing, transport, and communication and scientific equipment. Clearly, the larger the crew and the longer the expedition, the greater will be the load for the spacecraft. Estimation of this is a major concern for any space agency planning a human expedition to Mars. Here we shall make some preliminary estimate, in order to appreciate the magnitude of the associated technical challenges. A basic datum is the mass needed to support a single astronaut. Depending on the level of austerity, this is 5–10 kg per person per day in life support consumables: oxygen, water and food – the basic necessities. For approximately 1,000 days away from Earth this amounts to 5–10 tonnes of consumables per crew-member. Each Apollo mission carried three explorers, one of whom remained in orbit round the Moon; and the Space Shuttle has a crew of seven. If a crew of seven were selected to go to Mars, then we could be looking at 70 tonnes of consumables just to keep the astronauts alive for that length of time. Remembering that 20 tonnes of fuel are required for every tonne of payload, we already have a spacecraft carrying 1,400 tonnes of fuel. Every tonne of fuel has to be carried up into low Earth orbit, and the present maximum payload of the Space Shuttle is about 24 tonnes. This then involves a very large number of Space Shuttle flights, just to deliver the fuel into low Earth orbit.

At present we need go no further with this estimation process. It is clear that a vast amount of fuel will be required for a human expedition to Mars, and that anything we can do to reduce this will help to make such an expedition possible. So far we have made many simplifying assumptions, and some of these can immediately be discarded. The mass of the payload taken on the journey to Mars is not the same as the mass when returning, as only the crew and the samples gathered from Mars have to be returned, and all the equipment, and the Mars expedition quarters, can be left on the surface of Mars. Proper waste disposal would result in only half the consumables needing to be on-board the return vehicle. Intelligent design of the mission could make use of a much smaller return vehicle. The vehicle used to travel to Mars could be left in orbit, and only a small lander could be used to travel to the surface and back. Some of the equipment, stores, and perhaps the expedition quarters, could be sent to Mars on an unmanned cargo flight that would not have to return at all. Shortening the expedition duration by using the Venus fly-by manoeuvre will reduce the mass of the payload at the expense of a much shorter stay on Mars. Many such refinements to the mission could reduce the total payload to be accelerated in the different manoeuvres, and so reduce the overall fuel requirement. These refinements are the subject of intense debate, and many are included in the various mission scenarios that are being prepared by space agencies, and by those, outside the agencies, who are interested in the human exploration of Mars.

It is the task of technology to meet the requirements outlined here, and later chapters will relate how new space technology will play its part in enabling human exploration of the Solar System, and of Mars in particular.

3.21 SUMMARY

In this chapter we have examined in detail the physics of an expedition to Mars: orbits, propulsion, space flight dynamics, and the motion of the planets that control the duration and scheduling of the expedition. These set the ground-rules for the specification of the technology that will enable the human mission to Mars, and much of this technology will be described in later chapters. This chapter has, in part, been mathematical; but without the ability to calculate the necessary requirements it is impossible to see how the technology will enable the expedition. The calculations set forth here are simple, but provide an accurate idea of the requirements, and can be checked by the reader with the aid of a pocket calculator, as all the methods are included. In Chapter 4 we shall see how an expedition could be mounted using just present-day existing rocket technology; and this will lead, in Chapters 5 and 6, to the simplifications that could be made by the use of emerging new technology for spacecraft propulsion.

4

Human exploration using chemical rockets

Human beings have so far explored space by the exclusive use of chemical rockets; that is, rockets that derive their energy from chemical reactions, in the same way that a car engine operates. From the earliest Chinese gunpowder rocket to the latest Space Shuttle engine, they all produce their thrust by 'burning' something. They are chemically-powered heat engines, just like the steam engine. There are other kinds of rocket at the leading edge of space propulsion technology, and it may well be that their use will bring benefits for a Mars expedition. These will be discussed in detail in later chapters, but for the moment – taking a leaf from von Braun's book – we shall examine how humans could visit Mars using exclusively present-day, proven, technology. An expedition powered exclusively by tried and tested chemical rocket technology will not be easy (as foreshadowed in Chapter 3). Nevertheless, it would be foolish to trust our first human interplanetary expedition exclusively to new and untried technology, without examining the possibility of using the accumulated engineering experience of more than fifty years of rocket design. The only humans to set foot on another celestial body and return safely – the Apollo astronauts – were transported on chemical rockets. But how did they do it, and what lessons should be applied to a Mars expedition?

4.1 APOLLO

The landing of six two-man teams on the surface of the Moon, and their safe return to Earth, stands as the greatest human achievement of the twentieth century, and perhaps of the millennium. It has been so well-documented in books and films that little of the grand scheme needs to be said here. The series of missions began with little more than twenty hours on the lunar surface during the Apollo 11 mission, and progressed to just over three days with Apollo 17. Exploration equipment, on the surface, began at 100 kg and rose to more than half a tonne, while returned lunar samples rose from 20 kg to 100 kg, through the programme. The Apollo 11 astronauts spent eight days away from the home planet, and just over two hours walking on the surface of the Moon. The last men to stand on the Moon – the

Apollo 17 astronauts – spent more than twelve days away, with three seven-hour explorations of the surface. Compared with the requirements – on men and technology – of a Mars expedition, these times and masses are tiny; but for the first attempt by human beings to explore another celestial body it was magnificent. From our point of view, we need to understand how the enterprise was achieved, what technology was used, and what strategies were employed. As the first successful human venture into the Solar System, Apollo should be our first guide as to the means to achieve a human expedition to Mars.

Just as in the concept for the Mars expedition, outlined in Chapter 3, the requirement was to leave Earth orbit, make the transfer to orbit around the Moon, to land the explorers on the Moon, bring them back to lunar orbit, and then make the successful transfer to Earth orbit, prior to landing back on the Earth. The mighty Saturn V rocket, conceived by von Braun, was the means for sending the astronauts safely on their way to the Moon. Compared with today's launchers it was amazingly powerful. The Space Shuttle has a capability to put 24 tonnes into low Earth orbit, while the Ariane V can place 18 tonnes, and the Russian Proton and American Delta IV can place 22 tonnes. The Saturn V, built in the 1960s, could place 118 tonnes in low Earth orbit, and 47 tonnes in lunar transfer orbit. As we shall see later, the whole problem of launching the expedition into Earth orbit is one of the major issues for a Mars expedition; the fact that we no longer have the Saturn V, and that the only current launchers have less than 20% of its capability, is part of the problem. Delaying consideration of the Earth launch to a later chapter, for the time being we shall, in this chapter, concentrate on expeditions leaving from, and returning to, Earth orbit. Early concepts for lunar exploration made use of assembly in Earth orbit, prior to departure for the Moon; but for Apollo this was abandoned in favour of a more or less direct launch and a direct return using aerobraking.

4.2   APOLLO MISSION STRATEGY

Figure 4.1, taken from a NASA press release of the time, shows the series of orbital manoeuvres that comprise the mission. Unlike the Mars mission, the spacecraft never enters an orbit round the Sun, and is instead confined to the Earth–Moon system. Nevertheless, there are many elements of the Apollo programme that are analogous to a Mars exploration programme. From the low Earth 'parking' orbit, the spacecraft enters a high elliptical orbit that intersects the Moon's path, and from this it enters orbit around the Moon. The astronauts transfer to the Lunar Module (LM) for descent to the Moon's surface, and return to lunar orbit using the smaller ascent vehicle, which docks with the Command and Service Module (CSM). For the return home, this module enters an elliptical orbit that passes close to the Earth, and from that it uses aerobraking to enter a flight path that allows a direct landing. This sequence is shown in Figure 4.1. In interpreting this figure it is as well to remember that the Moon is continuously moving in its orbit, and that it is moving faster than the spacecraft is moving at apogee, so that the Moon's motion dominates the encounter.

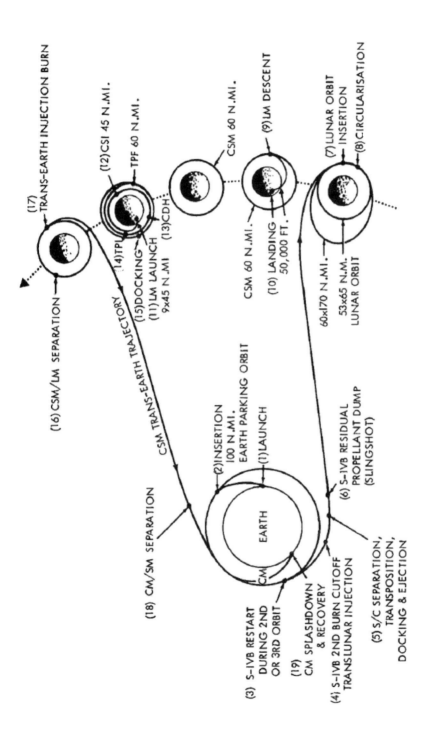

Figure 4.1. The sequence of manoeuvres used during the Apollo lunar exploration missions. The Moon, with an orbital period of 28 days, moves as shown during the 8 days of the mission. The elliptical transfer orbits are here convolved with the Moon's motion to produce the illustrated shape. (Courtesy NASA.)

The third stage of the Saturn V places the expedition in a translunar orbit. It is powered by a single J-2 engine fuelled with liquid oxygen and liquid hydrogen – the highest energy propellant combination (as we shall see later). The J-2 on the third stage is also restartable, which is unusual for engines powered by cryogenic propellants, but was vital for the lunar mission. On separation from the second stage of the Saturn V, it contains 107 tonnes of propellant, of which it uses 30 tonnes to acquire a circular low Earth orbit at an altitude of 190 km. The engine then shuts down while the upper stage makes 1 ½ complete orbits of the Earth in order to place the spacecraft in the correct position, relative to the Moon, to begin the injection into translunar orbit. The engine then restarts, and accelerates the spacecraft from the 7.79 km/s of the low circular orbit to 10.83 km/s for the journey to the Moon. In the process it burns 74.7 tonnes of liquid oxygen and liquid hydrogen fuel.

About an hour into the journey to the Moon a rather complex and, at first sight, extraordinary manoeuvre takes place. Figure 4.2 shows that the CSM is mounted at the tip of the upper stage, and that the LM and ascent vehicle is mounted *below* it. The crew are in the conical tip of the CSM for the launch – and, incidentally, for the last stage of the return to Earth. This is the safest place for them. The LM is larger in diameter than the CSM, and could not sensibly be placed in the nose of the Saturn V. It is also very fragile, being designed to operate only in zero g and in the low lunar gravity. It is therefore safer, during the rigours of launch, for it to be placed inside the protection of the tapering section, between the CSM and the Saturn V's third-stage tanks and engine. The tanks in this stage are now empty and need to be discarded, together with the now useless J-2 engine. These components would simply be an extra mass burden when the spacecraft has to be slowed down for lunar capture. So they must be separated, and the LM must be attached to the CSM, which contains the engine needed for all later manoeuvres.

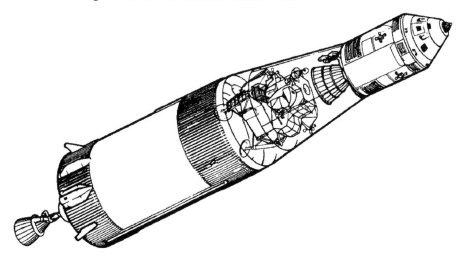

Figure 4.2. The upper stage of the Saturn V, early in the lunar transfer orbit. It carries the lunar transfer vehicle – also called the Command and Service Module – and the Lunar Module. The explorers are in the nose cone of the CSM. (Courtesy NASA.)

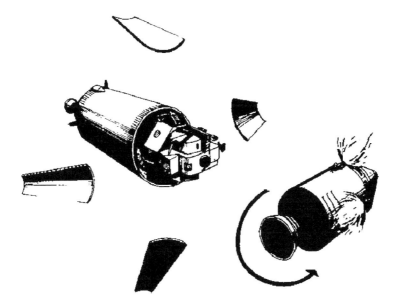

Figure 4.3. Safely in lunar transfer orbit, the Command and Service Module separates from the Saturn V upper stage and the Lunar Module, turns round, and docks with the Lunar Module. (Courtesy NASA.)

The astronauts first separate the CSM from the Saturn V upper stage by releasing clamps that hold the two together, against compressed springs. Once the clamps are released, the springs drive the two apart with a relative velocity of about 30 cm/s; the entire system, being in orbit, is weightless. Using the manoeuvring jets of its reaction control system, the CSM is turned round and docked with the LM, which was exposed when the stages separated. Once the docking is complete, spring-loaded clamps between the LM and the third stage are released, and they separate under the force of the springs, again at about 30 cm/s. Once separated there is a danger of a subsequent collision between the CSM–LM composite and the now discarded third stage of the Saturn V, as they are in the same lunar transfer orbit and very close together. This eventuality is prevented by two further manoeuvres. The CSM–LM spacecraft first burns its main engine for a few seconds, to place itself at some distance from the third stage. Once the path is clear, the remaining propellants in the third stage are vented through the nozzle, unburnt, to provide enough thrust to send the stage into an orbit past the Moon, out of the path of the spacecraft carrying the astronauts. The third stage eventually enters orbit around the Sun.

The result of this complex procedure is just what is required: the spacecraft is on its way safely to the Moon, and the third stage, which injected it into the required lunar transfer orbit, has been discarded and is safely out of the way. It seems a remarkably complicated process, especially for the early days of space exploration; but it was necessary because even with a rocket as powerful as the Saturn V, weight was a major problem. Had the LM been launched attached to the nose of the CSM, a

Figure 4.4. The Command and Service Module, with the Lunar Module attached, separates from the empty upper stage of the Saturn V. (Courtesy NASA.)

large-diameter and very heavy shroud would have been needed to protect it while passing through the Earth's atmosphere, and this would have used up the mass allowance required for other vital components of the mission. This was avoided by combining the necessary inter-stage section between the third stage and the CSM, with a protective shroud for the LM, lower on the rocket, where the diameter was larger. Manoeuvring and docking was less unsafe than it might appear, because it had been a feature of several earlier space missions. At one time, the plan to put a man on the Moon had involved the building of the lunar craft in Earth orbit, and this resulted in an earlier development of docking than might otherwise have been the case.

The next element of the strategy to be considered is the series of manoeuvres at the Moon, which enable landing on the surface and a safe return. On its journey to the Moon, the CSM was in a 'free return' orbit; that is, if nothing more were done, then the spacecraft would pass behind the Moon and return to Earth, where the astronauts could land safely using aerobraking. This was what happened on Apollo 13 after the sending of the well-known message, 'Houston, we have a problem'. It was an important safety feature of the Apollo programme, and the crew was able to safely return to Earth by means of this free-return orbit. If all went well, however, while the spacecraft was passing behind the Moon the rocket engine on the SM could be fired to slow it down and enable the Moon's gravity to capture it. The first orbit to be entered round the Moon had a pericynthion[1] of 113 km above the lunar

---

[1]   This lunar orbital term was used at the time of the Apollo programme, but it seems not to have been used since then.

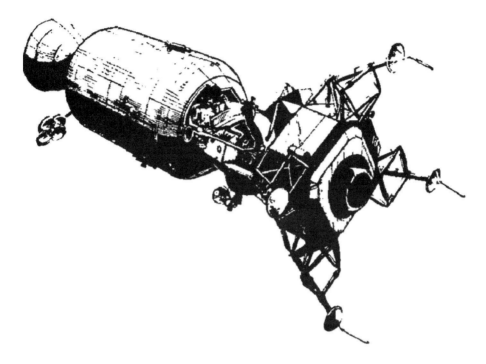

Figure 4.5. The Command and Service Module and the Lunar Module on their way to the Moon. Mid-course corrections, lunar capture, and Earth return manoeuvres, are carried out using the engine of the CSM. (Courtesy NASA.)

surface. If this was satisfactorily achieved, then the orbit could be made circular by another burn. This is the safe parking orbit around the Moon, from which a descent can be made when all is prepared.

There was only one chance for the initial burn with the SPS to bring the spacecraft into lunar orbit. The burn was timed to be at the point on the safe return orbit closest to the Moon's surface, which was naturally on the far side of the Moon from Earth, so the astronauts were out of radio contact. The spacecraft, being in an elliptical orbit around the Earth, was moving more slowly than the Moon in its near circular orbit, and so the Moon was overtaking it. The Moon moves at about 1 km/s in its orbit, while the spacecraft, having lost most of its velocity in the climb up from Earth, is travelling at only about 200 m/s, and the net approach velocity of the spacecraft to the Moon is about 0.8 km/s. This would be sufficient to place it in an hyperbolic (non-capture) orbit round the Moon, and the time of departure from the Earth, and the speed of the spacecraft were all arranged for this to happen in such a way that on leaving the Moon's gravity, the spacecraft entered a safe return orbit to Earth. As in Chapter 3, we can calculate the orbital speed at closest approach to the Moon: about 2.6 km/s; the Moon's gravity has accelerated the spacecraft as it falls in towards the Moon. The velocity necessary for the spacecraft to enter and remain in a circular orbit at NASA's required altitude of 60 nautical miles (111 km) is 1.63 km/s.

Figure 4.6. Safely in a circular orbit round the Moon, the Lunar Module separates from the Command and Service Module prior to descending to the lunar surface. (Courtesy NASA.)

The spacecraft had to be slowed down by firing its main (SPS) engine. The quoted velocity change is 2,924 feet/s (0.891 km/s) for it to enter a 314 × 111-km elliptical orbit. This was altered to a circular orbit at 111 km by a further burn of 48 m/s. The Moon weighs only about one eighteenth of the Earth, and so all the velocities are much smaller than they would be in Earth orbit. Once this burn of the SPS engine was completed, the spacecraft – the CSM, with the LM attached – was safely in a circular orbit about 100 km above the Moon's surface.

The next operation was for the pair of surface explorers to enter the LM. Once it had been been checked out, it was undocked from the CSM, ready for descent to the lunar surface. Using a short downward burn of its main engine, the CSM separated itself from the LM by 4 km to enable the CSM to view the LM during its descent to the surface, without overflying it. With a short burn of its descent engine, which slowed the spacecraft down by 22 m/s, the LM now began the descent to the surface by entering an elliptical orbit with a pericynthion altitude of 15 km. The pericynthion point was, of course, selected to be up-range of the chosen landing site. If nothing was done at pericynthion, then the LM could return freely to the altitude of the CSM – another safety feature. At pericynthion, the descent engine was fired to begin the descent from the approach orbit to the surface. This is a complex manoeuvre. Initially, the engine fires parallel to the surface to slow the spacecraft in its orbit, after which the spacecraft tilts, engine-downward, towards the landing attitude, as the horizontal velocity of the spacecraft reduces and it approaches the surface. At 2

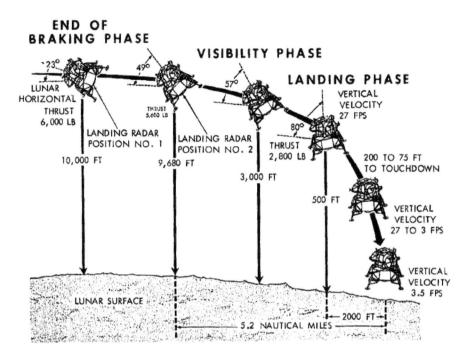

Figure 4.7. The Lunar Module landing sequence. As with all NASA missions, the speeds and distances are given in imperial units and nautical units. (Courtesy NASA.)

km above the surface the LM was fully upright in its landing attitude, with the windows forward, and the horizontal velocity was small. It essentially fell to zero by the time the spacecraft was at an altitude of 45 m; and from there, with manual guidance, it settled on to the lunar surface at about 2.5 m/s. To prevent the exhaust from stirring up dust around the spacecraft, probes on the feet of the LM warn the astronauts to shut down the engine when it was within a few metres of the surface.

Because of the Moon's rather 'lumpy' internal mass distribution and gravitational field, the first landing on the Moon was not exactly a copybook landing. This property of the Moon was noted on Apollo 8, in circular orbit, and was confirmed on Apollo 10, which actually followed the descent ellipse down to 15 km and back up to the CSM. The calculated orbits and trajectories were based on the best knowledge; but the best knowledge was limited to what could be derived from Apollo 8 and Apollo 10. Armstrong realised that his approach trajectory had been disrupted by 'lumpiness' in the Moon's gravity, so that he was aiming for a point 8 km beyond the designated landing site where there was a large crater, strewn with debris. Using manual control and cutting the rate of decent, he was able to fly over this, and a second smaller crater, to find a clear landing site beyond. This is an important lesson. It is tempting to believe that with modern computing techniques and automatic systems, human beings are more or less redundant in space exploration. We often forget that while our computers can calculate and control our creations to a very high degree of precision, our knowledge of the real world, from which come the data input for our calculations, is always limited, and sometimes wrong. This was the case for the first landing on the Moon. An automatic landing would almost certainly have been a disaster. As it was, the human pilot was able to realise the difference between prediction and reality, and take appropriate action. Even so, it was nearly a disaster – but for another reason: the amount of fuel required for the landing had been calculated for the expected trajectory. Armstrong's intervention lengthened the course, and at the moment when he felt it was safe to shut down the engine, he had sufficient fuel for only 30 more seconds.

The LM, resting on the Moon's surface, was really a two-stage spacecraft: the lander which had brought the astronauts safely to the surface, and the ascent stage which was to take them back to the CSM in orbit round the Moon. The complete module is illustrated in Figure 4.8. The philosophy of a two-stage vehicle results very simply from the requirements. The lander has to be powerful enough to bring to the surface the explorers, all their exploration equipment, and the necessary life support supplies for the duration of their stay on the surface. It must also be their base on the surface, with eating and sleeping accommodation, and protection from local hazards. On the Moon, these are limited to space radiation, the Sun's rays and the solar wind; but on Mars, weather and dust storms would also need to be taken into account. In both cases, the lander would need to provide a pressurised environment where the astronauts, out of the space suits, can breath freely, and it therefore has to be robust and fairly large to meet all these requirements. The ascent module, on the other hand, is only required to transfer the astronauts, together with the samples that are to be brought back to Earth back, to lunar orbit and the CSM. It can be rather small and light, as the journey is short and the life support requirements are minimal. It would not be sensible to use the same module for descent and ascent.

Given this two-vehicle philosophy, what are the specific requirements and features of the lander and the ascent vehicle? The lander has to bring to the surface of the Moon not only itself, but also the ascent vehicle, and so it must clearly have a more capable engine and be provided with more fuel. Because of the difficulties of landing on the surface, with its obstructions and uncertainties, the control of the engine must be much more precise and variable. However, docking the ascent vehicle to the CSM in space is much easier, and requires a much less powerful engine. As far as propulsion systems are concerned, the lander requires a powerful and flexible engine, while the ascent vehicle requires a light and simple engine. As we shall see, the Apollo approach to the crew habitat was extremely spartan. The astronauts were only on the Moon for a period ranging from a few hours to two days, and they did not require much in the way of space and comfort. In fact, the only habitable area of the LM was the crew space in the ascent vehicle, which doubled as the crew space for descent and during the stay on the Moon's surface. Certainly, for Apollo 11, there was not even provision for the crew to lie down!

The descent vehicle is a platform upon which sits the ascent vehicle with its crew accommodation. The two are firmly attached by explosive bolts. The structure of the descent unit has an 'egg-box' design of lightweight beams. In the central cavity is located the main engine, while fuel tanks containing a 50:50 mixture of hydrazine

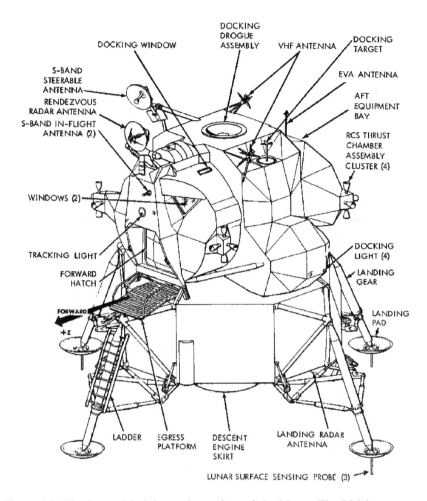

Figure 4.8. The Lunar Module on the surface of the Moon. The LM is a two-stage vehicle: the lander is the part to which the landing gear is attached, and the ascent vehicle and the explorers' cabin is the part above the leg attachment. (Courtesy NASA.)

and UDMH, and the oxidant nitrogen tetroxide, are located in the four cubic outer cavities on the arms of the double cross. Other spaces are used to store life support supplies. The landing struts that ensure a soft landing, support the LM on the Moon, and are attached to the main lander structure. The ladder, used by the astronauts to reach the Moon's surface from the crew cabin, is also attached here, together with the exploration equipment bay, or MESA. The explorers exit through the hatch in the ascent module, and onto the platform above the ladder, before climbing down.

The upper unit – the ascent module – is the home of the crew. In addition to the ascent engine and its fuel tanks, it contains the crew cabin, the tunnel and hatch that allows them to transfer to and from the CSM, and the docking system. The pressurised crew cabin is very small: a cylinder, 2.3 m in diameter and 1.3 m long.

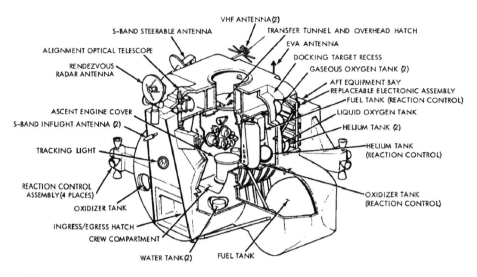

Figure 4.9. The ascent vehicle and crew compartment of the Lunar Module. (Courtesy NASA.)

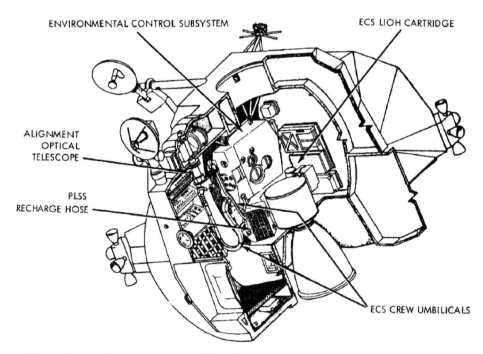

Figure 4.10. The crew accommodation in the Lunar Module. Note how the ascent engine protrudes into the crew space. The black pad, at lower left, is where one of the crew stands during the mission. (Courtesy NASA.)

This cylinder is mounted transversely in the module, and the top of the ascent engine protrudes into it. The crew standing-bays are at one end, where the headroom is larger. There are restraints and arm rests, as well as windows for piloting, but there are no seats or couches. In later Apollo missions, the crew was provided with hammocks on which to sleep, and were allowed to remove their space suits; but on Apollo 11 they kept their suits on, and did not lie down to rest.

The pictures of the cabin interior show that the crew accommodation was very cramped. This restriction was the result of careful calculations based on the amount of mass that the Saturn V could send to the Moon, and the precise aims of the mission. Initially, it was simply, to quote from President Kennedy, 'to land a man on the Moon and bring him back safely', and the idea of serious exploration of the surface was a later development. The LM cabin was adequate for the key aim, although perhaps less than adequate for the later exploration missions, Apollos 12–17. But this is not the place to dwell on the comfort of the Apollo explorers, and perhaps the glory was sufficient. For us it is enough to note how the difficulties of bringing sufficient mass to the Moon had a dramatic impact on the environment of the crew.

Once the exploration was complete and the crew had rested, it was time to return to lunar orbit and to dock with the CSM. With check-out complete, the explosive bolts separating the two halves of the LM were fired, and the ascent engine was ignited. The vehicle rose with its crew and samples – vertically at first, and then pitching over to follow an ascent ellipse towards the CSM orbit. The thrust of the ascent engine was 15.6 kN for a take-off mass of 4.554 tonne. The descent engine could be throttled, and had a maximum thrust of 43.9 kN. The velocity change for the ascent module was 1.84 km/s, followed by rendezvous and docking manoeuvres at 25.4 m/s. The ascent vehicle docked with the CSM, and the astronauts and samples were transferred. The ascent stage was then undocked, and at the correct moment the 91.2-kN-thrust SPS main engine was ignited to send the spacecraft along the return ellipse to Earth.

After up to four mid-course corrections, the final adjustments were made for a fully aerobraked landing on Earth. This saved an enormous amount of fuel which would otherwise have been carried all the way to the Moon and back. The aim point in the upper atmosphere of Earth is at an altitude of 122 km at a minimum angle of 5°.2 to the local horizontal – maximum angle, 2° more. Before atmospheric entry, the SM – which had carried out all the major manoeuvres with its 92-kN-thrust engine and large fuel tanks – was discarded. The CM separated and prepared to enter the atmosphere, base forward. Tremendous temperatures were generated as the 11 km/s that it had gained in falling down from the Moon towards Earth was all removed by the atmosphere and converted into heat. The base of the module was covered with a thick layer of *ablative* insulating material. Ablation is the process in which a solid material is charred and vapourised to produce a layer of relatively cool gas that flows around the vehicle to protect it from the much hotter gas further out. Ablative insulation does not last for many minutes, but is sufficient to allow the module to pass safely through re-entry heating. Once slowed down, at about 60 km altitude, the drogue, and then the main parachutes, opened to drop the module gently into the

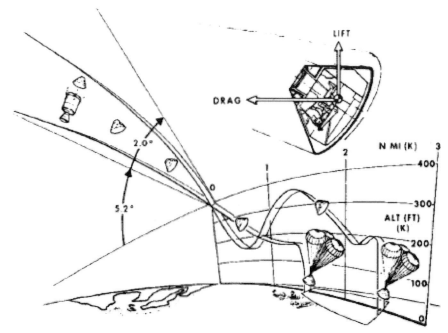

Figure 4.11. The Apollo Earth return sequence. The crew are in the conical crew compartment, which enters the atmosphere with the thermally insulated base of the cone forward. The vehicle is slowed by atmospheric friction. The high temperatures generated burn away the insulation to create a layer of relatively cool gas round the module. (Courtesy NASA.)

sea. The objective had been achieved. Men had walked on the Moon, and had returned safely to Earth.

The Apollo 11 mission was followed by five further successful visits to the Moon; but that was the end of human exploration of the planets in the twentieth century. Over a period of just four years, humans walked on another celestial body – and then it stopped. Public interest had waned, and government interest – which was never very high after the first successful mission – would no longer support the expenditure on a continuing or expanding programme. The budget was cut, and NASA had to abandon its plans. The Apollo programme was the greatest human adventure, and seen from the space-starved beginning of the twenty-first century it seems to have been ahead of its time. Perhaps no single nation could have hoped to sustain the effort and expenditure needed to continue it. Nevertheless, at some time in the future, space explorers will follow the Apollo astronauts and attempt journeys to the planets – and Mars will be first. From our point of view, the Apollo programme is a model from which to determine how we, as human beings, will travel to Mars.

The journey to Mars described in Chapter 3 considered only the crudest mission concept, using rocket power alone to achieve all the necessary manoeuvres. This was to expose the naked facts about getting to Mars. Now we have seen how Apollo managed the lunar mission, we can develop a more efficient approach; and such an

Figure 4.12. The Lunar Module in orbit. (Courtesy NASA.)

approach is needed because of the huge amount of rocket fuel demanded by a Mars mission, compared with the lunar mission. Apollo had a number of key features that improved the efficiency with which rocket power was used, and indeed made the mission possible with early space technology. Among these were the use of three separate rocket vehicles to accomplish the mission – four, if the upper stage of the Saturn V is included, as it should be; the exclusive use of aerobraking for Earth return; and the transfer of the crew to different vehicles at different stages of the mission. In the next section, a simple mission to Mars that makes use of these techniques is developed. From this we can calculate the actual fuel requirements and compare these with those of the basic concept in Chapter 3.

4.3   A BASIC MARS MISSION

Unlike the Apollo mission, which began and ended on the Earth's surface, the Mars mission, to be developed here, begins and ends in low Earth orbit. There are several reasons for this. We do not have a rocket like the Saturn V that launched the astronauts directly to the Moon. We do, however, have the Space Shuttle, which could easily transport the astronauts to and from LEO. We do not, as yet, possess the heavy lift capability to transport the necessary vehicles and fuel for a Mars

Figure 4.13. An artist's concept of an ascent vehicle lifting off from Mars, leaving behind the lander with its empty fuel tanks. Note that in this concept, the expedition quarters on Mars are in a separately landed vehicle (not visible here). (From the NASA Reference Mission, courtesy NASA/JPL.)

expedition to LEO, but this is a challenge that is not specifically connected with a Mars mission. Its want is felt in the building of the International Space Station, in the lack of heavy launchers for the biggest communication satellites, and in several other areas. There is hope that the activities of the space-faring nations will in time result in significant improvements in this area, independent of the needs of a Mars expedition. We should note here that there are Mars exploration proposals that use direct launch from Earth, as with Apollo. Among them is Mars Direct, which will be considered later in this book.

The requirement, then, is to transport the expedition from LEO to Mars, to land it on the surface, to support the explorers during their stay on Mars, and to bring them back safely to LEO. Following the Apollo philosophy, the first decision concerns the number of separate rocket propelled vehicles to be used. The logic of this is to minimise the total quantity of fuel required for the whole journey. *The basic rule is to not carry anything further than necessary, and not to accelerate, at any given time, more mass than is necessary.* For the moment we shall not include aerobraking in the assessment; it will be brought in at the end.

Assessment of the mission requirements begins on the surface of Mars, at the moment that the explorers leave on their journey back to Earth. The reason for this will become apparent as the argument develops. Minimising the mass to be accelerated, and not carrying anything further than necessary, results in the need for a separate ascent vehicle, which carries the explorers and their samples back into Mars orbit. This is exactly the same as in the Apollo missions, in which there was a separate, specialised ascent vehicle. The reasons are perhaps obvious. The vehicle, its fuel and the astronauts have to be transported from Earth orbit to Mars, to be slowed down for capture by Mars, and to be landed on the surface – all before a single engine on the ascent vehicle has been fired. A kilogramme of fuel on the ascent vehicle has already been accelerated three times by rocket engines before it is used in earnest. Efficiency requires that the number of these kilogrammes is minimised, and the mass of the ascent vehicle should therefore be as small as possible. The payload comprises only the explorers themselves and their samples from the martian surface. There is no point in bringing back any equipment, and the life support requirements are just those necessary for the very short journey into orbit and on to docking with the orbiter.

The Apollo ascent vehicle doubled as the crew quarters on the lander and the 'home' of the lunar explorers while on the Moon. Enough has been said of the privations which they suffered to make it clear that this cannot be the case for a Mars expedition. The ascent vehicle should provide crew space for both landing and take-off, but a separate permanent home for the explorers has to be provided that will support them during the long stay on the surface, with some degree of comfort. This will, of course, be left behind when they return to orbit. In terms of the Apollo concept, this would be the equivalent of the lander. Thus we have evolved the requirement for a two-stage vehicle, similar to the Apollo LM, which has an ascent vehicle on top of a permanent surface habitation for the expedition. It could be suggested that these might be separated once on the surface, or even delivered separately; but for the simple mission considered here they should be part of the same assembly, delivered to the surface by the lander engines, and using fuel contained in the lander's tanks. In this way all the used engines, fuel tanks and structure can be left on the surface, to minimise the mass to be brought back into

Figure 4.14. An artist's concept of the expedition habitat landing on Mars. (From the NASA Reference Mission, courtesy NASA/JPL.)

space. The crew habitation will carry the expedition equipment, life support requirements, and all other materiel necessary to support the explorers while on Mars. These can be unloaded onto the surface, to provide a comfortable living space once the expedition is established. It is as well to mention here that the expedition quarters will be pressurised, and will also provide protection against cosmic rays, extremes of heat and cold, and dust storms. The atmospheric pressure on the surface of Mars varies with the seasons, and is, on average, about 6 millibars – the equivalent of the pressure at an altitude of 35 km on Earth, at four times the height of Mount Everest. The explorers will have to wear space suits, and so any activity on the surface will be hampered by clothing. On the first expedition, any type of building work should be avoided. This is why the habitation unit is part of the lander–ascent vehicle unit, standing on its landing legs, and requiring no heavy lifting or external manual activity to prepare it. Having defined requirements on the lander and the ascent vehicle, it will later be possible to calculate the minimum amount of fuel required to land on Mars and to take off again.

The next requirement to be considered is the transport of the explorers to Mars and back to Earth. The journeys are similar in duration, the requirements are the same, and the same vehicle can be used for both. The crew will live in it during the journey to Mars, it will be left in Mars orbit while they descend to the surface, and they will return to it for the journey back to Earth. The Apollo transfer vehicle was the CSM, and a similar concept applies to the Mars expedition. The journey is much longer, and the cumulative amount of time that the crew spend in the module is similar to the amount spent on Mars, so that the size and comfort level in the transfer module should be about the same. The main difference is that they will be in zero-g conditions in the transfer module, which means that the whole volume, rather than just the floors, can be used by the crew.

There has been considerable discussion about the need for some type of artificial gravity during the long journeys between Earth and Mars. From the point of view of comfort, most astronauts would prefer zero g; but the main problem is the physiological effect of long exposure to zero g on the human body. To simply spin the spacecraft to provide artificial gravity is not sufficient, because on a small spacecraft the created gravitational field varies from zero on the axis to a maximum at the circumference. The feet of a person would be on the outer wall at, say, 1 g, while his or her head would be at zero g, near the axis. This would probably be very uncomfortable, and would produce considerable problems in the heart and circulatory system. A proper approach would be to have the spacecraft attached to an outrigger of some kind, so that it could rotate about a circle of much larger diameter. The laws of physics require that the mass on the end of the outrigger would have to be similar to the mass of the spacecraft. Such a problem could be solved, but in this chapter, in which we are considering a simple mission, it will for the moment be put to one side, and zero g assumed for the journey. The general conclusion is that the mass of the Mars lander/habitation vehicle and that of the Earth–Mars transfer vehicle will be about the same.

Since we have defined the journey as beginning and ending at low Earth orbit, all that are required are the three separate crew vehicles: the Earth–Mars and

Mars–Earth transportation vehicle; the lander, combined with expedition head-quarters; and the crew ascent vehicle. For the purposes of this study it is assumed that the crew of explorers use the Space Shuttle to carry them to and from low Earth orbit.

4.4   ROCKET ENGINES

The next factor to examine is the propulsion of the vehicles. Which rocket engines should be used? Which fuel should be used? How many different rocket units will be required? And how should they be used? The fuel to be used is a simple decision. The most secure propellant combination should first of all be *hypergolic*; both components should also be liquid at normal temperatures; and they must be storable for long periods. These requirements are all related to the safety of the crew and the security of their return to Earth. Hypergolic propellant combinations are those that ignite on contact, and it is sufficient simply to allow the propellants to enter the rocket combustion chamber for the rocket to start. These propellants have been used from the early days of the space programme. The fuel is based on *hydrazine* (N_2H_4), a chemical relative of ammonia, and the oxidant is based on nitric acid; indeed, nitric acid itself was used in early rockets. Derivatives of hydrazine are often used: monomethyl hydrazine (MMH – CH_3NNH_2), which has a lower freezing point than hydrazine – 274 K^2 – is used in the Space Shuttle orbital manoeuvring engines, and in the Ariane V's upper-stage engine. Unsymmetrical dimethyl hydrazine (UDMH) is used in many applications, and is sometimes mixed with ordinary hydrazine because it has a wider range of temperature as a liquid (216–336 K). The early rockets used the exotically named *red fuming nitric acid* as an oxidant, but the much safer and less exotic nitrogen tetroxide (N_2O_4) is used in modern rockets; and, incidentally, was used for Apollo. For our model Mars mission, MMH and nitrogen tetroxide are the appropriate choice of storable hypergolic propellants. This choice is not the same as was used for Apollo, for which a 50:50 mixture of hydrazine and UDMH was used, but MMH now seems to be the propellant of choice for the Space Shuttle and for the Ariane V upper stage. As we shall see later in this chapter, hypergolic, storable, propellants do not produce the highest performance when used in a rocket engine, although they are easy to store for long periods, and the engine can be guaranteed to start at any time.

Rocket engines of different sizes are required for the various stages of the mission. The Mars lander will require a high-thrust engine that is steerable and capable of being throttled back so that the pilot can land the spacecraft – the same requirement as for the Apollo lander. The ascent vehicle will use an engine with a lower thrust, as it has much less mass to lift, and does not need to be either throttleable or steerable.

---

[2]   In this book, temperatures are expressed either in the familiar degrees Celsius (centigrade), or in Kelvin where appropriate. The Kelvin scale (K) has the same graduation as the Celsius scale, but the zero point is at absolute zero: –273° C.

For the in-space manoeuvres, Mars capture, and injection into the Earth return ellipse from Mars orbit, the engine of the Earth–Mars transfer vehicle will be used; it is not steerable, and because it is used only in space, the thrust is only determined by the maximum time that can be allowed for a manoeuvre. The thrust can be much less than the *weight* of the spacecraft, if enough time can be allowed for the manoeuvres. For the initial injection into the Earth–Mars transfer ellipse, the transfer vehicle engine is not used. Instead, a separate liquid-fuelled booster is used, and then discarded after the transfer ellipse is entered. This is directly analogous to Apollo, in which the second firing of the Saturn V upper stage injected the CSM into the transfer ellipse, after which the upper stage was detached.

4.5   THE MISSION PROFILE

We now have a complete set of engines, and manoeuvres that will take the expedition from Earth orbit to the surface of Mars and back. The mission manoeuvres are represented in the diagram shown in Figure 4.15.

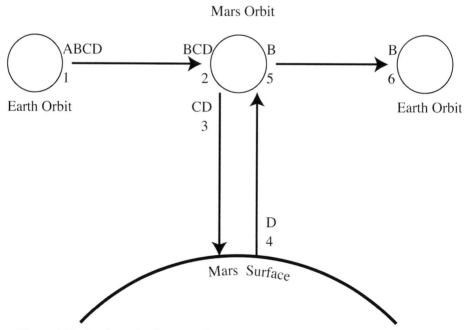

Figure 4.15. A schematic diagram of a basic mission to Mars. The elements of the mission are: A, discarded booster; B, Earth–Mars transfer vehicle; C, Mars lander; D, ascent vehicle. The manoeuvres proceed as follows: **1**, the combined vehicle leaves low Earth orbit, and **2**, is captured into Mars orbit; **3**, the lander–ascent vehicle complex leaves Mars orbit for the surface, and after exploration, **4**, the ascent vehicle returns to Mars orbit; the explorers enter the Earth return vehicle, **5**, leave Mars orbit, and **6**, are captured into Earth orbit.

Having transferred the explorers from Earth to the waiting Mars spacecraft complex using, for example, the Space Shuttle, the mission begins with manoeuvre **1**, the boost from low Earth orbit into Mars transfer orbit. The vehicle, ABCD, consists of all the stages described above, joined together into a multi-stage spacecraft, as shown schematically in Figure 4.15. It is boosted into Mars transfer orbit using the engine and fuel in section A. This section is discarded some time during the transfer to Mars, in the same way as the Saturn V upper stage was discarded during the Apollo missions. On arrival at Mars, the remaining vehicle complex has to slow down to enter low Mars orbit by using the engine and some of the fuel in section B, the Earth–Mars transfer vehicle. This is manoeuvre **2**. The explorers enter the Mars module, CD, which then separates from the Earth–Mars transfer vehicle. The EMTV is left in Mars orbit to await the return of the explorers, and to carry them back to Earth orbit. The Mars module, CD, executes manoeuvre **3**, which lands the explorers on the surface of Mars. When the time comes for the return to Earth, the explorers, who have been living in the lander, C, enter the ascent vehicle, D, which executes manoeuvre **4** to carry them back from the surface of Mars into Mars orbit, and on to the Earth–Mars transfer vehicle. The EMTV, B in the diagram, then executes manoeuvre **5** to enter Mars–Earth transfer orbit. On arrival at Earth, it executes manoeuvre **6** to slow down and enter low Earth orbit to complete the mission. The explorers are then collected by the latest version of the Space Shuttle, and returned to Earth.

The mission is described in this schematic way so that the fuel required for each manoeuvre can be calculated using the correct velocity change and vehicle mass for each case. Some vehicles in the complex make only one major manoeuvre, while the EMTV makes several. It will also be obvious that the fuel for each manoeuvre *forms part of the mass to be accelerated by previous manoeuvres*. This means that there is a cumulative effect whereby any increase in fuel used in a late manoeuvre increases the amount of fuel required for the earlier manoeuvres. It may also be clear that the path through the mission scheme, over which this cumulative effect works, divides in two, beyond entry to Mars orbit. Thereafter, the landing on Mars and the return to Earth follow separate paths. The meaning of this will become clear as we calculate the manoeuvres and the quantities of fuel required.

4.6 CALCULATING THE FUEL REQUIREMENTS

As a first step in assessing the feasibility of a basic Mars mission as postulated here, we can use the rocket equation to calculate the mount of storable propellants that would be required for the mission. A schematic of the vehicle complex in Earth orbit, prior to departure, is shown in figure 4.15. To calculate the fuel requirements it is necessary to make an estimate of the dry mass of the three vehicles involved, and the dry mass of the booster (the mass of the engine and empty fuel tanks), so that the mass ratio can be established. The smallest vehicle is the ascent vehicle, which has to carry a crew of six,[3] the samples from Mars, and enough life support for the short journey. Here a

---

[1]   A crew of six has been assumed for the NASA Reference Mission and is base-lined here.

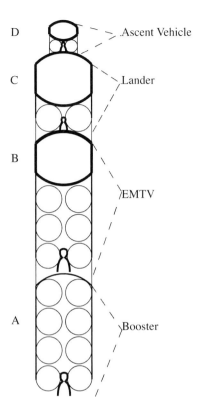

Figure 4.16. The Mars expedition vehicle complex, as assembled in low Earth orbit: A) the booster that sends it on its way to Mars, and is then discarded; B) the EMTV carries the crew to Mars and back; the engines are used both to enter orbit round Mars and to enter the Mars–Earth transfer orbit; C) the lander that delivers the expedition to the surface of Mars; D) the ascent vehicle that brings them back up.

mass of 5 tonnes is assumed, including the engine, the empty fuel tanks and, of course, the crew. A rough estimate of the size is based on the size of the necessary fuel tanks, which suggests a diameter of 4 m and a height of 2 m for the crew cabin. The lander has to be much bigger, both because of the larger quantity of fuel required, and because it will be the base for the explorers on the martian surface. The assumed diameter of the habitation is 8 m, with a height of 4 m to allow two floors, for use on the surface of Mars. The mass of this vehicle, including the engine and empty fuel tanks, is assumed to be 70 tonnes, including all that is necessary for the support of the six explorers on the surface of Mars. Since the time spent in the journey to and from Mars is similar to the time spent on the surface, it is reasonable to assume the same dimensions and dry mass for the Earth–Mars transfer vehicle: 70 tonnes.

The velocity changes necessary for manoeuvres 1–6 were calculated in Chapter 3, and are repeated here for convenience: **1**, 3.55 km/s; **2**, 2.45 km/s; **3**, 3.42 km/s; **4**,

3.42 km/s; **5**, 2.52 km/s; and **6**, 3.55 km/s. We used the rocket equation earlier, to calculate the velocity of a rocket from its mass ratio. What we need to do now is to *calculate* the mass ratio and, more usefully, the amount of fuel required, for a given *velocity* change. To do this, the rocket equation can be inverted to give the *multiplier*, which can be applied to the mass of the payload, to calculate the required mass of fuel.

$$V = v_e \log_e \frac{M_0}{M}$$

The rocket equation gives the vehicle velocity in terms of the mass ratio. What we want to know is: given the required vehicle velocity, and its mass, how much fuel is required? To do this the equation needs to be turned round and rewritten, using the definition of the natural or base e logarithm

If $y = \log_e x$, then $x = e^y$; $e = 2.71828 \ldots$

Applying the same procedure to the rocket equation, and writing the initial mass as the sum of the required fuel, and the final mass, produces:

$$\frac{M + M_f}{M} = e^{\frac{V}{v_e}}$$

$$\frac{M_f}{M} = e^{\frac{V}{v_e}} - 1$$

The ratio M_f/M is the multiplier by which the fuel mass is greater than the empty vehicle mass. It depends on e raised to the power: the ratio of the required velocity change to the exhaust velocity. This is a very important result, because it shows the impact of the rocket engine performance, represented by the exhaust velocity, on the fuel requirement. It is immediately obvious that a high rocket exhaust velocity will result in a lower fuel requirement. To see this more clearly, the function is plotted in the graph in Figure 4.17.

The exponential function, e^x, is a very strong function of its argument, as can be seen in the graph. At a velocity ratio of 1 – when the exhaust velocity of the rocket, and the desired velocity change, are the same – the mass of fuel that must be used is already 1.72 times the mass of the dry vehicle. When the desired velocity change is twice the exhaust velocity, the mass of fuel required is 6.4 times the dry vehicle mass; and at three times the exhaust velocity it is 19 times the dry mass. This graph shows the importance of rocket engines with high exhaust velocity. A small increase in exhaust velocity can introduce a very large improvement in the ratio of fuel to payload in a rocket vehicle. Unfortunately, the hypergolic, storable, propellants that most people favour for manned spaceflight do not produce a very high exhaust velocity. For the present exercise it will be appropriate to take the exhaust velocity of the Ariane V Aestus upper stage engine – an engine designed to work in vacuum, and incorporating the latest design features for a storable propellant engine. The vacuum exhaust velocity is 3.24 km/s, and the mass of the engine is 1.2 tonnes.

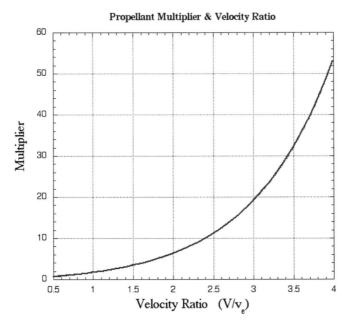

Figure 4.17. The fuel multiplier plotted for different ratios of the vehicle velocity change delta-V to the exhaust velocity.

In calculating fuel loads we can treat a series of manoeuvres, using the same rocket vehicle, as independent manoeuvres, provided the starting and finishing amounts of fuel are properly accounted for. This will become clear as we calculate the fuel requirements of the model mission. Since the fuel required for later manoeuvres forms part of the payload for the earlier manoeuvres, we must begin with the latest operations, working back to the early ones.

The last burn of the engine of the EMTV is the burn that slows it down to enter low Earth orbit. In principle this is the burn that empties the tanks of the remaining fuel. The payload for this manoeuvre is just the vehicle itself, with the crew and samples, defined as 70 tonnes for vehicle B. The velocity change for manoeuvre **6** is 3.55 km/s, and the velocity ratio is 3.55/3.24 = 1.096. Applying the equation, this gives, for the multiplier, 1.99, so that the amount of fuel required for this manoeuvre is 1.99 × 70 = 139.4 tonnes. The preceding manoeuvre (**5**) for this vehicle is the departure from Mars orbit with a velocity change of 2.52 km/s, which produces a ratio of 2.52/3.24. Applying the equation in the same way gives a multiplier of 1.178. Now the mass to be accelerated is the 70 tonnes of the vehicle *plus* the fuel required for manoeuvre 6, which is 139.4 tonne. The mass of fuel required for this manoeuvre, with the much heavier load, is 1.177 × (70 + 139.4) = 246.5 tonnes. This is the fuel required for the Mars departure manoeuvre. It is much larger than the amount required for the Earth arrival manoeuvre, even though the velocity change for the former is larger. This shows the cumulative effect, where some fuel has to be used, just to accelerate other fuel, which will be required for later manoeuvres.

The procedure has so far dealt with the Earth–Mars transfer vehicle, which waits in Mars orbit during the time that the explorers are on the surface. Working backwards, its previous manoeuvre is the capture into Mars orbit – manoeuvre 2. However, between arrival at and departure from Mars orbit, there is the landing on the surface and the return of the explorers. The payload of the EMTV for Mars orbit capture is not only the vehicle itself and the fuel required for the two manoeuvres already discussed, but is also the lander with the ascent vehicle and the fuel required for both of them. This is calculated using the same procedure. The last manoeuvre of this section is the return of the astronauts to the EMTV. The dry vehicle mass of the ascent vehicle is 5 tonnes, and it makes only one manoeuvre (**4**), from Mars surface to orbit, with a velocity change of 3.42 km/s. The multiplier is 1.87, giving a fuel requirement of 9.4 tonnes. This fuel, together with the combined lander and ascent vehicle, is the payload for the lander engine in the journey from Mars orbit to the surface – manoeuvre 3. This has a velocity change requirement of 3.42 km/s, and the fuel multiplier is 1.87, the same as for the take-off. The combined mass of the fully fuelled ascent vehicle and the (dry) lander is 84.4 tonnes, leading to a fuel requirement for the lander engine of 157.8 tonnes. The all-up mass of the Mars excursion composite is then 242.2 tonnes.

For Mars capture – manoeuvre **2** – the total mass of the combined Mars excursion composite, and the EMTV with fuel for the Earth return manoeuvres already detailed, is 698.4 tonne. This is the composite vehicle BCD in the diagram. Most of this mass is the fuel required for subsequent manoeuvres. The velocity change is 2.45 km/s, and the multiplier is 1.13. This produces, for the fuel to be expended in Mars capture, the high value of 789.2 tonnes. All of this fuel, plus the vehicles, have to be inserted into Mars transfer orbit by the booster, A, which is just an assemblage of engines and fuel tanks with no accommodation or cargo hold; the crew will be in the EMTV habitation. An estimate of the dry mass of the booster has not yet been included, but it will not be far from 7% of the total fuel mass, which is a typical 'structure factor' when dealing with liquid-fuelled rockets. Its mass can be approached by approximation. The mass of the fully fuelled composite BCD is 698.4 + 789.2 = 1488 tonnes. The velocity change from Earth orbit to MTO is 3.55 km/s, and the multiplier is 1.99, and the fuel required for this manoeuvre, without the dry mass of A, would therefore be 2,963 tonnes. The structure and rocket engine mass for A is then estimated, using the structure factor, to be 207 tonnes. We add this to the payload mass BCD, 1,488 tonnes, to produce a new estimate of the dry mass, 1,695 tonnes. We then recalculate the fuel requirement using this new value, to produce 3,373 tonnes. This procedure could be repeated to produce a more accurate value, but this approximation is good enough. Strictly, we should apply this method to calculate the mass of tankage on BCD for the Mars capture and subsequent manoeuvres. This would add 50 tonnes to the 1,488-tonne mass of BCD, with a consequent ripple-through effect on the earlier manoeuvres, although here this is ignored.

The all-up mass of the departing Mars expedition vehicle can now be summed up to be 1,695 tonnes of ABCD plus the fuel for A amounting to 3,373 tonnes, producing an all-up mass of 5,068 tonnes. This is the mass which has to be lifted into low Earth orbit to enable the expedition. Compare this figure of more than 5,000 tonnes in Earth orbit with the equivalent Apollo orbital mass of just under 120 tonnes. It is a truly daunting requirement – especially given the current 24-tonne limit of capability to low Earth orbit in a single launch. This is clearly too great a discrepancy to allow for an early Mars expedition. We must seek an optimisation of the mission itself and improvements in the current rocket technology to bring the human exploration of Mars closer in time to the present day.

4.7   OPTIMISING THE BASIC MARS MISSION

With any optimisation task it is necessary to identify the 'pressure points' where small changes in design can make a big impact on resources. It should be clear from the above that the mass of the expedition and its vehicles is only a small fraction of the total mass to be lifted into low Earth orbit; and most of the 5,000 tonnes is fuel. It will also be clear that the majority of this fuel is used to accelerate or decelerate other fuel that will be used later in the expedition. Given the preponderance of fuel in the total resource requirement, optimisation of fuel use will be the dominant route for improvement.

The first consideration is the 'cost' of fuel used for different parts of the mission. To illustrate this, consider a tonne of fuel used in the ascent vehicle to lift the explorers from the surface of Mars. What is the 'cost' of this tonne of fuel to the expedition? First, it has to be placed in Earth–Mars transfer orbit; the velocity change is 3.55 km/s, and the fuel multiplier is 1.99. Second, it has to be slowed down for Mars capture; the multiplier here is 1.13. Third, it has to be landed on Mars, where the multiplier is 1.87. Then it is ready for use. A tonne of fuel on the surface of Mars requires 1.87 tonnes to land it there; it requires 1.87×1.13 tonnes to enable Mars capture; and $1.87 \times 1.13 \times 1.99$ tonnes to place it in Earth–Mars transit. The total 'cost' of a tonne of fuel on Mars is $1.87 + 1.87 \times 1.13 + 1.87 \times 1.13 \times 1.99 =$ 8.19 tonnes. Save a tonne on Mars' surface, and 8.19 tonnes are saved in low Earth orbit. Table 4.1 shows the 'cost' of fuel used for all the manoeuvres required for the Mars expedition, calculated in the same way. The cumulative effect is clear from the Table. The early manoeuvres have a low fuel cost, and the late manoeuvres have a high cost. The highest value is for the departure from Mars surface at 8.19, and the lowest is for Mars capture at just 1.99. Between these are the other manoeuvres: **3**, 4.25; **5**, at 4.25; and **6**, at 5.16.

Table 4.1. Fuel required for manoeuvres

Manoeuvre	Delta-V (km/s)	Multiplier	Vehicle	Fuel cost	Fuel mass (tonnes)
1	3.55	1.99	ABCD	1.00	3,373
2	2.45	1.13	BCD	1.99	789.2
3	3.42	1.87	CD	4.25	157.8
4	3.42	1.87	D	8.19	9.4
5	2.52	1.18	B	4.25	246.5
6	3.55	1.99	B	5.16	139.4

To reduce these fuel costs, the multiplier can be reduce by improving the rocket engines, for example; or the mass of fuel that needs to be transported can be reduced. Already, the most expensive item – the ascent from Mars surface – has been minimised by using a small vehicle that requires a relatively small amount of fuel to be transported. Savings for the lander fuel, the departure from Mars orbit, and arrival at low Earth orbit, would next be examined because of their high 'cost'.

Where a velocity change is a deceleration and is associated with a planetary atmosphere, then aerobraking can be considered. But how much of this slowing down could, for example, be achieved by using Mars' atmosphere? It is tenuous, but it can be used to slow down a spacecraft. Indeed, it has been used many times to slow down satellites, for example, in converting elliptical orbits to circular orbits, and for robotic landers. If aerobraking, in combination with parachutes, could be used to reduce the velocity to a small value, then most of the fuel used in manoeuvre **3** could be saved, and this has a fuel cost of 4.25. The departure from Mars' surface requires rocket power; and here the atmosphere cannot help. The other potential use of aerobraking is for Mars capture, during which a significant saving could be made, but against a lower fuel cost of 1.99. Another very significant saving in fuel could be made if, instead of using rocket power to enable capture by Earth gravity into LEO and then transporting the explorers back to Earth in a separate vehicle, aero-braking were used to both circularise the orbit and to land the astronauts directly back on Earth. This latter approach was, of course, used for the Apollo missions. The astronauts would need to remain in the EMTV for the journey back from Mars, but would transfer to a small re-entry vehicle and separate from the EMTV. The re-entry vehicle would then skim the Earth's atmosphere to slow down and enter a highly elliptical Earth orbit, while the EMTV would continue in the Earth–Mars transfer ellipse. Several passes through the fringes of the atmosphere might then be required to lower the velocity sufficiently to enable safe atmospheric braking and landing. Indeed, if an Apollo-style return can be used, then there need be no elliptical orbits; just a direct capture and atmospheric entry manoeuvre. Taking the optimistic view of these savings, only a small amount of fuel would be required to enter aerobraking manoeuvres. Then the fuel used in manoeuvres **2**, **3** and **6** could be saved. Taking into account the fuel cost factors, this amounts to 1,571, 671 and 719 tonnes respectively, reducing the all-up mass of fuel by 2,960 tonnes.

Not all of this saving could be realised, however, because some fuel would be required to land safely on Mars' surface, and for trajectory changes in order to encounter the planetary atmospheres at the correct altitude and angle to the horizontal. Errors here would be disastrous, and plenty of fuel and thrust capability will be required to ensure the possibility of escape from a wrong re-entry corridor. Nevertheless, these quantities will be small compared with the large fuel expenditure involved in thrust-only major velocity changes. The other consideration is the mass of the aerobraking thermal shield. The spacecraft's kinetic energy is converted into heat by the atmospheric encounter, and it must be protected from the hot gas created by its passage through the atmosphere. This is well-established technology dating from the Gemini and Apollo re-entry vehicles, and involves *ablative* shielding. The material of the leading surface of the shield is made up of layers of composite, fibrous refractories such as silica, phenolics and carbon–carbon, which burn away slowly without cracking. The products of the burning form a layer of relatively cool gas that flows round the capsule, insulating it from the much hotter gas outside. The vehicle is a blunt cone that enters the atmosphere base forward. This generates a shock front that travels ahead of the capsule, protecting it from the hottest atmospheric gases. Designing a shielded capsule for Earth return is relatively simple, because it will be small, containing only the astronauts and the samples. It would be similar to an Apollo return vehicle. For the Mars aerocapture and landing, however, the task is more onerous because the vehicle involved is the Mars landing composite consisting of the habitation and ascent vehicle, the explorers, and all their stores. The shield will have to be large, and therefore heavy; and since it has to be delivered to Mars it will have the same mass cost factor as the fuel it saves. A careful trade-off study will be necessary to determine the exact saving that can be achieved. It will be large, however, because of the huge mass of fuel assigned to Mars capture and landing. Proper studies assign about 15% of the vehicle mass to the aerobrake shield. The mass of the composite BCD is 698 tonnes, without the fuel required for Mars capture. Assuming both capture and landing will be achieved by aerobraking, we can subtract from this the fuel required for landing, to produce 541 tonnes. At 15%, an 81-tonne shield would be required. This saves an actual fuel mass of 947 tonnes in Mars transfer orbit – which is well worth the effort.

The use of planetary atmospheres for manoeuvres that would otherwise require huge quantities of fuel saves about 3,000 tonnes of our 5,000-tonne basic Mars mission; but 2,000 tonnes is still far beyond the capability of present-day launchers. We must look to new mission scenarios and improvements in rocket technology in order to bring a Mars mission closer. The new mission scenarios will come later. Here, we will now examine rocket technology. Referring back to the inverted form of the rocket equation, the key to the reduction of fuel multipliers is to increase the exhaust velocity – and to understand this we need to examine how chemical rockets work. What are their limitations? And how can they be improved?

4.8 ROCKET SCIENCE

This common phrase is today interpreted as anything advanced, and difficult to understand or achieve. Rocket science is, in fact, very old science, originating in the late nineteenth and early twentieth centuries. Tsiolkovsky, Oberth, Tsander, Glushko and Goddard were the pioneers; and we should not forget the ancient Chinese, who first invented the rocket. The reason that rockets changed from being fireworks and rather useless military missiles into awe-inspiring and frankly terrifying modern machines lies in engineering, not in science. Returning for the moment to Chapter 3, Tsiolkovsky and his successors recognised that the gunpowder rocket was simply not powerful enough to enable space travel. The key to more powerful rockets was in the chemistry of combustion and the development and use of high-strength and high-temperature alloys, together with high-power turbines – all twentieth-century engineering.

While we look to rocket engineering to provide us with the technology for Mars exploration, it is the science of rocket engines that will provide us with an insight into their limitations, and opportunities for improved performance. This is really nineteenth-century science updated to the twentieth century.

4.9   ROCKETS AND STEAM ENGINES

At first sight there seems little in common between Stephenson's 'Rocket' and the Space Shuttle; and yet they are, in fact, very similar in concept, and rely on the same science: thermodynamics. Thermodynamics should not scare us; it is only the science of steam engines. C.P. Snow considered that any educated person should know the second law of thermodynamics; which, interestingly, is the basis for the steam engine and the rocket engine. In essence it states that heat can be converted to work, or useable energy, only if it is taken into a machine at a high temperature and released at a lower temperature. There must be a temperature difference in order to extract energy from heat, and it must be positive. This is the basis of modern mechanical society. We make electricity, which can be used to make things move, by using a source of heat – the burning of fuel, or nuclear fission – and by using a machine to take in energy from it at a high temperature and release it at a lower temperature. The work extracted is used to turn a turbine and rotate a generator. More simply, in a motorcar, heat is generated at a high temperature, in the cylinders, by burning petrol; and is then released at a lower temperature through the exhaust. The difference appears as the kinetic energy of the car, moving along the road.

Stephenson's 'Rocket' used this same principle. Steam, produced by heating water in a boiler, was introduced at high temperature into the cylinder, where it pushed the piston down while expanding. At the end of the stroke, a valve opened, connected to a condenser that sucked the steam out of the cylinder by condensing it to form a partial vacuum. The rotating wheels then easily pushed the piston back up the empty cylinder, ready for a new introduction of high-pressure steam. The two temperatures involved were that of the steam 'chest' fed by the boiler, and that of the condenser, cooled by cold water. The energy extracted, to keep the locomotive moving, was replaced by the coal, heating the boiler to make more steam. The actions involved here – heating, expansion, cooling and exhaust – are all present in the rocket engine, as well as in the 'Rocket'.

The main difference is that there is no piston in the rocket engine, and the work is extracted by other means. The combustion chamber of a rocket engine replaces the cylinder. If anything, the piston is represented by the exhaust from the nozzle – as we shall see. The process in the steam engine is cyclic, whereas in the rocket engine it is continuous. Of course, the steam in a steam engine does not return to the boiler, and is instead exhausted to the air. In this respect, the two engines are similar: the cyclic element is in the rotating wheels and the oscillating piston, not the steam. It is easier to appreciate the working of a rocket engine by thinking about a 'lump' or portion of fuel passing through the engine. It is preceded and followed by other 'lumps'; but here we concentrate on one lump. This is then much more like the cyclic operation of the steam engine. The lump of fuel enters the combustion chamber and immediately ignites. A chemical reaction – burning – takes place between the fuel and oxidant, and heat is produced. The products of combustion are raised to a high temperature, similar to the situation in which hot steam has just been introduced to the cylinder of the steam engine. There is a pressure difference between the combustion chamber and the outside world beyond the nozzle, and the hot combustion products begin to move along the chamber towards the nozzle. The cross-section of the chamber is constant, and so there is as yet little expansion of the hot products. However, when they enter the nozzle, the cross-section increases rapidly, and the gases expand and cool as they pass down the nozzle. At the exit of the nozzle, the combustion products have cooled considerably, and their pressure has dropped, both because of the expansion. They then leave the nozzle as the rocket exhaust.

So far, this appears to be very much the same process as in the steam engine. But where is the work? In fact, the energy gained by the cooling and expansion of the hot gases in the nozzle now resides in the kinetic energy of the exhaust stream, leaving the rocket at hypersonic velocity. We know that, in the combustion chamber, the same materials are moving very slowly. In the exhaust, they are moving very quickly; they have been accelerated as they pass through the nozzle. The rocket engine has converted the heat stored in the hot combustion products into the kinetic energy of a stream of gas leaving the rocket. It may therefore be said to have acted as a heat engine in the same way as the steam engine. In the latter case, the heat stored in the hot steam has been converted into the kinetic energy of a moving locomotive. The reader will have spotted that one is more useful than the other. A hypersonic rocket exhaust is no doubt remarkable – but is it useful?

This becomes clear as we examine the forces that accelerate the exhaust stream, as these are the key to the remarkable properties of a rocket. The combustion chamber is normally kept as short as possible: just long enough for the fuel to vapourise and burn. The pressure here is more or less constant. The interesting things happen when the nozzle is reached. Our 'lump' of gas has other lumps of gas in front of it and behind it. As it enters the nozzle, the gas in front expands and is therefore at a slightly lower pressure, while the gas behind is still at the high chamber pressure. There is thus a net force acting on our lump, impelling it towards the exit. The lump itself is always at a higher pressure than gas further down the nozzle, and so it is constantly pushing this gas towards the exit, just as the gas behind it is pushing the lump. It should be clear by now that the plug of gas downstream of any selected lump of gas acts like the piston of the steam engine, receiving the force of the expanding gas upstream of it, and converting that force into motion of the downstream gas. We have, in fact, a kind of continuous 'steam' engine in which the 'steam' is entering the cylinder continuously and the piston is always being pushed downwards.

This explains how the exhaust stream is accelerated down the nozzle. To understand how this causes the rocket to accelerate we need to remember the nature of pressure in a gas: it acts in all directions equally. While the pressure of the upstream gas is pushing on the downstream gas, it is also pushing on the walls of the nozzle, and so there is a force acting on a square centimetre of the nozzle wall equal to that acting on a square centimetre of the downstream gas. In a rocket, it is this force that accelerates the vehicle; the seat of the accelerating force is in the nozzle of the engine. The nozzle has to be anchored securely to the rest of the vehicle to prevent the engine tearing itself out. If this seems strange, compared with the steam engine, consider the different construction of the two. The steam engine cylinder is *fixed* to the crank-case, as is the piston through the connecting rod and the crank. If it were not for the offset of the crank, the piston would be unable to move at all relative to the cylinder As it is, the only motion possible is tangential to the crankshaft, allowing rotation, which can then be transmitted to the wheels. The cylinder itself cannot move at all. In the case of the rocket, the 'piston' and 'cylinder' are not connected to one another at all. Both are free to move – the 'piston' out through the nozzle, and the 'cylinder' in the opposite direction. The operation of a rocket engine results in two moving 'parts': the exhaust stream, and the rocket engine itself. The engine moves in the opposite direction to the exhaust stream, impelled by the pressure forces acting on the internal walls of the nozzle.

It can be shown that the force accelerating the rocket is identical in magnitude, and opposite in direction, to the pressure force accelerating the exhaust stream. This is an illustration of Newton's third law of motion: *action and reaction are equal and opposite*. It is perhaps superfluous to remark that the exhaust does not need to 'push' on any external object in order to propel the rocket; indeed, it travels at hypersonic speed, while pressure forces can only travel back up the stream at the speed of sound. There is no physical connection between the exhaust and the rocket once it has left the nozzle.

4.10 FORMULAE FOR ROCKET SCIENCE

We have so far tried to illustrate the similarity between a steam engine and a rocket engine. Both are what are known as 'heat engines': devices that convert heat into work. They convert the chaotic motion of the molecules in a hot gas into the ordered motion of a stream of gas in the rocket, and the ordered motion of a piston in the steam engine. So much can be illustrated without calculation. But our quest is to explore the limiting performance of rocket engines, and to identify ways toward improved versions; and this can only be achieved by studying quantitative performance. The full discussion, using the mathematical formulations of thermodynamics, is elementary but tedious. Here we shall simply quote some principles and results.

As might be expected for a heat engine, the temperature difference is an important parameter. The upper temperature is that of the combustion chamber, while the lower one is the temperature of the exhaust as it leaves the nozzle. It is a rule of thermodynamics that the work obtained from a given amount of heat depends on the temperature difference: the larger it is, the more work is extracted per unit of heat – which in practice means per given quantity of fuel. There are two ways to maximise this temperature difference. The most obvious is to increase the temperature of combustion, and the other is to obtain the lowest exhaust temperature by sufficiently expanding the hot gases. Both of these have limitations, either from physics, in the case of the expansion, or from the high temperature strength of the materials used for the combustion chamber.

The other factor that affects the performance of a rocket engine is the molecular weight of the exhaust gases. At first surprising, this is, in fact, fundamental, and relates to the physics of the heat engine. The molecular weight – the mass of individual molecules in the gas – determines how the energy stored in the hot gas is converted into the velocity of the exhaust stream. The lighter the molecules, the higher will be the exhaust velocity; and we already know from the rocket equation that the exhaust velocity is the most important measure of an engine's performance.

It is worth recalling here that thrust and exhaust velocity are related to one another, but are quite separate measures of the rocket's performance. Thrust is mainly related to the diameter of the 'throat' (the narrowest part) of the nozzle and to the combustion chamber pressure, while exhaust velocity is mainly related to the temperature in the combustion chamber and the molecular weight of the exhaust products. The thrust of a rocket engine can be doubled by doubling the combustion chamber pressure, or by doubling the cross-sectional area of the nozzle throat, without at all affecting the exhaust velocity. It is also worth remembering that the velocity change, calculated using the rocket equation, does *not* depend on the thrust of the rocket, only on its exhaust velocity. In principle, an engine with a very small thrust is capable of achieving a very high velocity change. Thrust is considered only when the time taken for a given manoeuvre is critical, as a low-thrust engine will require a long time to reach a given velocity change. It is clear that when lifting off from a planetary surface, the thrust must be greater than the weight of the vehicle, otherwise it cannot lift the vehicle.

While the complete equations for rocket engine properties are complicated, rocket engineers often use simpler ideas, which we can adapt here. They define three parameters of a rocket engine: the *specific impulse*, the *thrust coefficient*, and the *characteristic velocity*. The value of each of these parameters can be calculated from the laws of thermodynamics, and, indeed, can also be easily measured for real rocket engines. For a given engine, the quality of the design can be assessed by comparing the theoretical and actual values.

Specific impulse is the exhaust velocity divided by the acceleration of gravity at the Earth's surface. In SI units, g is about 10 m/s/s, so for most purposes the specific impulse is just 1/10 of the exhaust velocity. It is arguable that it would be better to use the exhaust velocity itself, but this definition is so deeply ingrained among rocket engineers that it is always used. It derives from an attempt to represent the fuel efficiency of the rocket. The specific impulse is the momentum imparted to the rocket per kilogramme of fuel expelled through the nozzle. Useful as this definition is, however, it is doubtful if specific impulse as a concept would have become universal if it were not for the fact that its numerical value for a rocket is independent of the units used for mass. Be they kilogrammes, pounds or quintals, the specific impulse is the same. Its units are quoted as *seconds*, because when a velocity is divided by an acceleration, the resultant cancellation of units leaves 'seconds', as can be easily verified. Thus, for example, the exhaust velocity of a particular rocket engine may be quoted as I_{sp} = 300 *seconds*. To calculate the true exhaust velocity, this number should be multiplied by 10.

The *thrust coefficient* is a measure of the efficiency of the nozzle in expanding the hot gases as they travel down it. As a coefficient it has no units; just a numerical value. It is really a measure of the temperature difference and hence efficiency of the heat engine. The bigger the thrust coefficient, the more the gases expand before leaving the rocket, the bigger is the temperature difference, and the greater the exhaust velocity. The thrust coefficient depends partly on the actual shape of the nozzle, and partly on the ambient pressure. It is well known that a rocket engine is more efficient in a vacuum, which shows itself in a larger value of the thrust coefficient. For operation at sea level on Earth, with the optimum nozzle shape, the coefficient has a typical value of about 1.2, while in vacuum an optimum nozzle would have a value of around 2.0. These numbers show the improvement in thrust in vacuum. In designing a rocket engine, the nozzle expansion should be tailored to the pressure of the atmosphere in which the engine will have to work. In fact, the exhaust pressure – the pressure in the exhaust when it leaves the nozzle – should be equal to the ambient pressure. To understand this it is enough to consider what would happen if the exhaust pressure were not equal to the ambient pressure. If the nozzle were too long then the gases would expand to a pressure lower than the outside pressure, and there would be a partial vacuum in the nozzle that would reduce the accelerating force. In the same way, if the nozzle were to be too short, then the gases would not have expanded as much as they should have done, and there would be energy locked up in the temperature of the gases that could have been converted into exhaust velocity. For engines that have to give of their best at sea level, the nozzle should be short and the expansion should be small, just to reach sea-level atmospheric pressure when the gases leave. For vacuum operation, the ideal exhaust pressure should be zero; but this would need an infinitely long nozzle, and so a compromise is reached with an expansion ratio of about 80, to produce an exhaust pressure of about 0.001 times the combustion chamber pressure. This produces a thrust coefficient of about 2.0. The optimisation of the nozzle length for use in a vacuum depends on the mass of the vehicle. For the large vehicle sintended for the Mars expedition, the extra mass of a very long nozzle is negligible, and expansion ratios up to 250 can be used. For most manoeuvres in the Mars mission, the engines operate in a vacuum, and so the long nozzle is appropriate, with a high expansion ratio and a thrust coefficient of about 2.2. Ascent from the surface of Mars, where the pressure is about 6 millibars, would require a slightly shorter nozzle.

The characteristic velocity is a measure of the efficiency of the propellant combination and the combustion chamber, and is independent of the thrust coefficient and hence of the nozzle properties or the ambient pressure. As its name suggests, it is measured in velocity units: m/s. It is not the exhaust velocity, because this also depends on the shape of the nozzle. In fact, the product of the thrust coefficient and the characteristic velocity is equal to the exhaust velocity. It is very useful to have a measure that is independent of the rocket nozzle design and only depends on the fuel combination, because it allows the propellants to be considered, apart from the mechanics of the nozzle, and optimised separately. The characteristic velocity depends on only the properties of the propellant combination: the temperature of combustion, and the molecular weight of the exhaust gases.

The simple definitions and relationships between these parameters can now be given:

$$C_F = \frac{F}{p_c A^*}$$

where C_F is the thrust coefficient, F is the thrust, and p_c and A^* are the combustion chamber pressure and the nozzle throat area respectively. It can be seen that the thrust coefficient is just the ratio between the actual thrust and the notional thrust represented by the product of the combustion chamber pressure and the throat area, remembering that pressure times area is equal to force:

$$c^* = \frac{p_c A^*}{m}$$

where c^* is the characteristic velocity, and m represents the mass of fuel used per second – the same as the rate at which the exhaust stream carries away mass from the rocket engine. Here, p_c and A^* appear again as the notional thrust. The characteristic velocity is then the notional thrust, per unit mass of propellant used, per second. (The units are in m/s because of cancellation in the equation.) Because the thrust of the rocket engine appears here, it is possible by substitution to represent the thrust in terms of these parameters as follows:

$$F = mc^* C_F$$

The thrust can also be derived from Newton's laws of motion, to produce the following:

$$F = v_e m$$

This is simply an algebraic representation of the statement by Newton that 'the rate of change of momentum is equal to the force applied'. In this case, the mass expelled per second, multiplied by the velocity with which it is expelled, is indeed the rate of change of momentum. Note that here, v_e is the same exhaust velocity that appears in the rocket equation, and is a quantity on which much depends in a Mars mission. Again substituting in the equations, we can derive another useful expression:

$$v_e = c^* C_F$$

The specific impulse is, of course, simply given by

$$I_{sp} = \frac{v_e}{g} = \frac{c^* C_F}{mg}$$

These are the relationships between the useful parameters and the performance of the rocket engine (the values of which will appear in later discussions). It is important, however, to see how their values depend on the actual properties of the rocket engine and the propellants that are used. These are derived from the thermodynamics of the rocket engine, and may be expressed approximately as

$$C_F = 2.25 \sqrt{\left[1 - \left(\frac{P_e}{P_c} \right)^{\frac{1}{6}} \right]}$$

$$c^* = 1.54 \sqrt{\frac{8.31T_c}{M}}$$

where $\left(\frac{p_e}{p_c}\right)$ represents the ratio of the exhaust pressure – at the end of the nozzle – to the combustion chamber pressure, and $\frac{T_c}{M}$ is the ratio of the combustion chamber temperature to the molecular weight of the exhaust gases.

These two equations express the performance of the engine in terms of the physics; and perhaps these are the equations of 'rocket science'. The thrust coefficient depends on the ratio of the pressure in the combustion chamber to the exhaust pressure. Since the pressure and temperature of an expanding gas are locked together by the gas laws, this is really an expression of the temperature difference with which we began this discussion; this temperature difference defines the efficiency of the rocket as a heat engine. Decreasing the ratio of temperature or pressure – which in most cases is the same as increasing the combustion chamber pressure – will increase the thrust coefficient. If the ratio becomes infinite – that is, if the exhaust pressure is zero or the chamber pressure is infinite – then the coefficient only reaches 2.25, which is the theoretical maximum value. This will be approximately true in vacuum with a very high chamber pressure. The characteristic velocity depends on the ratio of the combustion chamber temperature to the molecular weight of the exhaust gases. The molecular weight is the number of protons and neutrons in the molecule. For water – the product of the burning of hydrogen and oxygen – it is 2 + 16 = 18.

This completes the formulae for the science of rocket engines. But before proceeding to consider how their performance could be improved, we need to examine some practical examples to understand some of the engineering limitations.

4.11 ROCKET ENGINES

All liquid-fuelled rocket engines have common characteristics. They require a combustion chamber in which the two propellants are burnt; at one end this is attached to the nozzle, and at the other end there is an injection assembly to introduce the propellants into the combustion chamber. We know that the temperatures are very high in the combustion chamber, and over the inner part of the nozzle, where the exhaust gases are still hot. The combustion temperature in an engine fuelled by liquid oxygen and liquid hydrogen is about 3,500 K. Most metals lose a good deal of strength even at temperatures around 500°, and many melt between 1,000° and 2,000°. Alloys using cobalt, chromium and molybdenum keep their strength to higher temperatures, but there is no metal that can function satisfactorily at temperatures above about 3,000°. Since strength is required to contain the pressure in the combustion chamber and to transfer the thrust forces on the nozzle to the rocket itself, a solution to this problem had to be found. If the metals cannot take such high temperatures, then they must be cooled and insulated so that they never reach their softening temperature. Liquid propellants lend themselves to the idea of cooling, because they can be used in a jacket round the hot parts of the engine to carry away the heat and keep the casing cool. This applies *par excellence* to liquid hydrogen and liquid oxygen because they are so cold, but it also works well with room-temperature liquids such as hydrazine and nitrogen tetroxide. The classic approach is to route one of the propellants into an array of tubes welded onto the outside of, or integral with, the combustion chamber and the upper half of the nozzle. The liquid flows through the tubes, absorbs heat from the walls, and then enters the combustion chamber. The flow rate needs to be adequate to keep the engine cool; and in practice this requires the correct cross-section and number of tubes. The technique whereby warm propellant is introduced into the combustion chamber is called 'regenerative cooling', because the heat extracted from the walls is put back into the combustion chamber in the form of warm fuel. It will be appreciated that cold propellants need to be warmed before they can evaporate and burn, and that most of this heat has to come from the hot gas in the combustion chamber, lowering the overall temperature. If the propellant can be warmed with 'waste' heat, there is an improvement in efficiency.

In very simple engines, cooling can be enabled by less complicated methods. The cooling liquid is a small part of the propellant supply that is simply routed through the tubes and then allowed to flow out of holes in the nozzle wall end, to be carried away with the exhaust – a process known as 'dump cooling'. Another approach is to introduce some of the liquid propellant into the combustion chamber so that it sticks to the wall and flows along it, held there by surface tension. As it flows along it is heated, and some of it evaporates to form a layer of cool gas between the hot exhaust and the walls of the chamber and nozzle. This is essentially a combination of cooling and insulation, because the evaporation of the liquid film cools the walls and, at the same time, the layer of evaporated liquid acts as an insulator. This is called 'film cooling'. These techniques are sometimes combined, with film cooling or dump cooling being used for the outer part of the nozzle, and regenerative cooling for the combustion chamber and the inner part of the nozzle.

Once the cooling problem is solved, the next major issue is propellant delivery. Modern large engines shift several hundred kilogrammes of propellant per second through their nozzles, and this has to be supplied to the combustion chamber at the same rate. There are two approaches, depending on the magnitude of the thrust. For low-thrust engines, in which the propellant flow rates can be modest, pressurisation of the fuel tanks is used to force the fuel into the chamber. But for high thrust engines, turbo-pumps are used; indeed, they were used for the first successful liquid-fuelled rocket, the German A4, or V2. The pumps are usually powered by burning some of the propellant to generate hot gas, which is then used to drive turbines, and pump the fuel into the combustion chamber. It may seem obvious, but the fuel has to be delivered at a pressure that is higher than the combustion chamber pressure, otherwise it will be blown back up the pipe. The pump system therefore has to be properly sized to cope with the pressure; and the more pressure required, the more propellant will have to be bled off from the main supply, decreasing the amount available for propulsion.

Figure 4.18. The Aestus storable propellant engine used for the upper stage of the Ariane V. The nozzle and the fuel tanks can be seen here. (Courtesy Arianespace.)

The injector design is also important. The need is to allow a high flow-rate into the combustion chamber, but to encourage evaporation by using very small holes that produce a fine spray. But these requirements are contradictory, and the only solution is to use hundreds of small holes to produce an adequate flow-rate. Mixing is also important, and so the individual holes for the fuel and the oxidant have to be adjacent – an exercise in three-dimensional topography.

An example of the pressure-fed system is the Aestus engine used on the upper stage of Ariane V, which uses the hypergolic storable propellants MMH and nitrogen tetroxide. Its thrust is 29 kN, and the exhaust velocity in vacuum is 3.24 km/ s. This engine is designed as an upper stage, for use in the vacuum of space. It is used to inject satellites into geostationary transfer orbit, and, apart from its smaller size, is analogous to the type of storable propellant engine that could be used for a Mars mission. The fuel is contained in two spherical tanks, and the oxidant in two quasi-spherical tanks visible in the photograph (Figure 4.18). The delivery pressure is provided by helium gas that is stored under pressure in the two spherical tanks. Concern for the elimination of dead weight results in the use of this, the lightest inert gas as the pressure provider. The upper part of the nozzle and the combustion chamber are cooled with the flow of MMH in a surrounding jacket. The warm liquid, together with the nitrogen tetroxide, then enters the combustion chamber. The main part of the nozzle is made of high-temperature alloy, and is cooled simply by radiation; in use it glows red-hot. This engine has a rather low thrust, but it is designed to produce the best mass ratio when attached to the satellite that is being injected into orbit. Because of the low thrust it takes some time – about 1,000 seconds – to reach the required velocity, but the dead weight of the engine is small, so that the ultimate velocity is higher than would be the case for a higher-thrust and heavier engine. The expansion ratio of the nozzle – only 30 – is not very high for vacuum use, but this still achieves a thrust coefficient of 1.87. This engine is designed to sit between the main stage of the Ariane V and the spacecraft. Any length that can be saved on the engine can be used for the spacecraft, and so the engine is designed to be very compact, and the fuel tanks are clustered round the nozzle (as shown in the photograph). Extending the nozzle length to obtain a greater expansion ratio would in this case reduce the available spacecraft space for only a small gain in thrust coefficient. The fuel combination of nitrogen tetroxide and MMH results in a characteristic velocity of 1.72 km/s. The temperature in the combustion chamber is 3,398 K, and the mean molecular weight of the exhaust products (mostly water) is about 20.

It would be wrong not to consider an example of the high-thrust, high-energy pump-fed chemical rockets that are so commonly used in launchers, as there are features of these rockets that can be used in interplanetary thrusters. Large payloads require a high thrust if reasonable manoeuvre times are to be achieved, and a high exhaust velocity is vital for all interplanetary manoeuvres to maximise the payload. The principle is simple: high thrust requires a high combustion chamber pressure combined with a high fuel delivery rate or mass flow rate in the exhaust. To do this, the propellants have to be delivered by pumps to the combustion chamber; and the alternative, to increase the pressure level in the tanks, would necessitate unacceptably thick walls. The pumps must be driven by a power source, and it is usual to use some of the propellants to provide it. A fraction of the propellant flow is diverted to a *gas generator*, which burns them to produce a supply of hot gas. This gas then drives one or more turbines, producing shaft speeds of 10,000 rpm or more. The turbine shafts then drive the propellant delivery pumps. If a single turbine is used, then one of the pumps will be driven via gears to take account of the differing flow rates of fuel and oxidant. These have to be different because of the chemical needs of the combustion process; for instance, the reaction of hydrogen and oxygen requires 8 kg of oxygen to be delivered for every kilogramme of hydrogen.

The propellant used in the gas generator is not available for propulsion, and therefore decreases the mass ratio. This, it will be remembered, is the ratio of the total propellant mass to the empty mass of the vehicle, and the velocity change achieved by the engine depends on the mass ratio. The propellant used in the gas generator decreases the achievable velocity change for a given mass ratio, and various approaches are taken to minimise this effect. In some cases, the exhaust from the turbines flows to the outside world through a rocket nozzle, to create a small additional thrust. In others, the hot exhaust from the turbines enters the combustion chamber, so that the energy contained in it contributes to the main thrust.

The Space Shuttle main engine uses the latter technique, and has separate turbines and gas generators for its liquid oxygen and liquid hydrogen propellant pumps. Furthermore, all the turbine exhaust enters the combustion chamber. These factors contribute to the very high efficiency of this engine in converting the chemical energy in the fuel into thrust, and a very high exhaust velocity of 4.55 km/s. The combustion chamber pressure is 205 bar – vastly greater than that of the Aestus pressure-fed engine – and, indeed, the turbo-pumps generate pressures of more than 400 bar prior to the rather complex flow of the propellants through the system. Some of the liquid hydrogen leaving its turbo-pump is used to cool the nozzle and the combustion chamber, while the rest passes through the gas generators for both propellants, here called *pre-burners*. To keep the generated gas reasonably cool, the mixture is very fuel rich. Only a small fraction of the necessary oxygen is provided, so that most of the hydrogen remains unburned. This 'hydrogen-rich steam' then enters the combustion chamber to combine with the liquid oxygen provided by the oxidant

Figure 4.19. A test firing of the Space Shuttle main engine. Cold, foggy air flows down from the cooled nozzle and propellant pipes. The converging flame is caused by atmospheric pressure. The nozzle is designed for use in a vacuum and here, in air, the exhaust takes the inverted cone shape characteristic of over-expansion. (Courtesy NASA.)

turbo-pump and pre-burner system. The relative quantities of hydrogen and oxygen entering the combustion chamber are controlled by varying the amount of oxygen supplied to the pre-burners, which changes the speed of the turbo-pumps.

The complexity of such a system is clear in the photograph (Figure 4.19), as is the long, high-expansion-coefficient nozzle. It has an expansion ratio of 78, and a correspondingly high thrust coefficient of 1.91. With this engine, everything is done that *can* be done to maximise the exhaust velocity. The use of the cryogenic propellants hydrogen and oxygen produces, rather surprisingly, a combustion temperature lower than that of the storable propellants. But the low molecular weight of the exhaust gases provides for a much more efficient conversion of heat energy into exhaust velocity. The characteristic velocity is 2.39 km/s, compared with the Aestus engine's lower value of 1.72 km/s. Although the molecular weight of water is 18, the presence of additional hydrogen in the exhaust reduces the mean molecular weight to about 12. The additional hydrogen arises from two causes. There is some dissociation of the water molecules caused by the high temperature and pressure in the combustion chamber, but most of the excess hydrogen is deliberately added to lower the mean molecular weight. This is a very important principle. Adding fuel that will not take part in the burning, because of oxygen starvation, reduces the combustion temperature, but it happens that the effect of reducing the temperature is much less than the effect of the lower mean molecular weight. So, the SSME, emitting an exhaust that is nearly 40% unburned hydrogen, has the highest exhaust velocity of any current in-service rocket engine. Three of these engines power the Space Shuttle, and each develops a thrust of 2.3 MN in vacuum. But this is still puny, compared with the thrust of the liquid oxygen–kerosine F-1 engine, used in a cluster of five to power the first stage of the Saturn V. Each one developed a vacuum thrust of 7.9 MN, although the characteristic velocity was only 1.78 km/s – the same as for storable propellants.

These two examples of chemical rocket engines – the Aestus and the SSME – cover the range of current types, and it is to these types of engine, and to future improvements in them, that we must look for a chemically-propelled mission to Mars.

4.12 IMPROVING CHEMICAL ROCKETS

While thrust is an important parameter for the manoeuvres required during a Mars expedition, it is the exhaust velocity which really controls the achievability. The ratio of the exhaust velocity to the required velocity change defines the fuel multipliers used earlier in this chapter to determine the mass of fuel required. The exhaust velocity of 3.24 km/s that we used to calculate the requirements of the basic mission is that of a storable system, based on the Aestus engine. But what would happen if we could increase this, and what are the problems? Moving to liquid oxygen and liquid hydrogen, basing our performance on the SSME, would produce a significant reduction in the fuel multipliers. The multiplier for the departure from Earth orbit, for example, is 1.99 with storable propellants; but it would be reduced to 1.18 for the cryogenic propellants. This would result in a very significant reduction in the mass of fuel required for the whole mission.

4.13 STORAGE OF CRYOGENIC PROPELLANTS

A very significant problem lies in the nature of the propellants: they are not 'storable'. Liquefied gases have what is known as a 'critical temperature', above which no amount of applied pressure will keep the gas liquid. They therefore cannot be contained in sealed tanks, as the pressure would simply continue to rise indefinitely, as all the liquid turned to gas. A sufficiently strong tank could, of course, contain gas at this very high pressure. Such tanks are used by welders and are very heavy for the amount of gas they contain. Liquid oxygen has a critical temperature of $-118°$ C and a boiling point at atmospheric pressure of $-183°$ C, while liquid hydrogen has a critical temperature of $-241°$ C and a boiling point of $-253°$ C. If kept below their critical temperatures these gases can be stored in sealed tanks; and if kept below their boiling points they can be stored in open tanks. The practice for vehicles such as the Space Shuttle or Ariane V is to use lightweight, vented tanks to store these propellants. The boiling-off of some of the propellant cools down the rest by taking away latent heat. This compensates for the heat that leaks in from outside while the rocket is on the launch pad. To replace propellants lost by boiling off they are topped up until a few hours before launch.

It is clear that this process cannot be used, as it stands, for a deep space mission. One possibility would be to load as much propellant as possible, and to hope that enough would be left by the time it was required. This would play havoc with the mass ratio, which would decrease continuously with elapsed time, irrespective of whether the engines were firing. The conditions would not be quite as bad as on the launch pad, because the tanks would be in the vacuum of space, and not surrounded by the warm Florida air. The natural temperature of space, in shadow, is about 3 K ($-270°$ C) – well below the boiling point of either propellant. This is the temperature of the microwaves that permeate space, left over from the Big Bang. If the tanks could be kept in shade, then they would cool down to a suitably low temperature, with hardly any evaporation of the liquids. The problem is, of course, sunlight. There is no hope of executing a Mars mission in shadow, and sunlight would fall on the fuel tanks and raise their temperature.

The optimum shape for the fuel tanks is a sphere, because this has the greatest volume for a give surface area. The sunlight heats the tank on one side only, and the tank radiates its heat away on all sides. The natural temperature of a tank in space, covered with the appropriate solar reflecting white surface, is about 155 K. Unfortunately the efficiency of cooling to space decreases with the fourth power of the temperature. As the temperature of the tank drops, so the cooling ability is reduced – a case of diminishing returns. Using the best insulation, including fifty to a hundred layers of thermal foil, it is still not possible, using passive cooling to space, to keep the tanks below the boiling point of liquid oxygen at 93 K, let alone that of liquid hydrogen. The fuel tanks will therefore need to be refrigerated – and this will require electrical power. The mission is rapidly becoming more complicated. The question, then, is: can the increased performance of cryogenic fuels more than compensate for the extra complexity and mass of the cryogenic system? The improved exhaust velocity will enable a smaller mass of propellant to be used; but the mass of the insulated tanks, the cooling system, and the power supply for the cooling system will increase the dry mass of the rocket and hence reduce the mass ratio for a given amount of propellant.

The insulation for the tanks will add to their mass, but not by very much: it will mainly be MLI or multilayer insulation, made up of many layers of aluminised plastic film, each only a few microns thick. This is the standard material used to protect spacecraft from extremes of heat and cold.

There will be an increase in the mass of the power supply. Refrigeration at low temperatures is inefficient, and considerable electrical power is required to extract a few watts from the propellant tanks at the necessary low temperature. This power will have to be supplied by the solar panels with a significant mass increase.

The savings in fuel achieved by using cryogenic propellants are to be offset against the more complex tanks, the mass of the sunshields and radiators, and the amount of fuel that will boil off during the mission as a result of the irreducible leakage of heat into the tanks. There must also be an allowance for the lower density of the cryogenic propellants. The mean density for liquid oxygen and liquid hydrogen in the correct ratio is 0.32, compared with 1.2 for nitrogen tetroxide and MMH. The tanks will therefeore have to be four times bigger in volume for the same mass of propellants; which translates to about 2.5 times the dry mass of the tanks. All of these factors have to be taken into account in calculating the advantage of cryogenic propellants. They will appear as a reduction in mass ratio to offset the higher exhaust velocity. In general, the increased mass of the tanks for the low-density propellant, together with the insulation mass, can be represented by a structure factor of 10% rather than 7% needed for storable propellants.

4.14 REFRIGERATED STORAGE

The power and mass cost of cooling the propellant can be calculated based on the solar input and the heat radiated to space. The main difference between the cooling requirements for hydrogen and oxygen lies in the critical temperature. As previously mentioned, the cooling process becomes more difficult as the temperature drops, so more refrigeration power is needed. For reasonable assumptions the heat leakage into the liquid oxygen tanks is about 1.3 W/tonne of propellant, while for liquid hydrogen, at a lower temperature, the heat leakage is 3.0 W/tonne. These numbers seem quite small, but when the efficiency of the refrigerators at these low temperatures is taken into account, the power requirements are seen to be rather high: 32 W/tonne for the liquid oxygen, and 524 W/tonne for the liquid hydrogen. These electrical power requirements can be translated into additional mass using data on solar cells; the most efficient modern solar cells produce about 340 W/m^2, with a mass of 8 kg/m^2. This produces a mass cost for electrical power of 24 kg/kW.

For the Space Shuttle cryogenic engine, which we use for a model, the oxidant/fuel ratio is 6; and for each 7 tonnes of propellant, 1 tonne is hydrogen. Using this ratio we can calculate the cooling requirements per tonne of propellant, taking into account the greater power required for hydrogen cooling. This produces a rather modest 102 W/tonne of propellant. The solar panel mass and area for each tonne are 2.43 kg and 0.3 m^2 respectively. This is negligible compared with the 3% increase in structural mass coefficient when changing from storable to cryogenic propellants, and is equivalent to 30 kg per tonne. The mass penalty is thus due to the increase in tank mass, and not the cooling of the cryogenic propellants. The cryogenic propellants can therefore replace the storable propellants with only a change in the structural mass factor of 7–10% of the propellant.

4.15 THE ADVANTAGE OF CRYOGENIC PROPELLANTS

The cryogenic system is, of course, much more complicated and more risky to use; but it results in a major reduction in propellant mass – as shown in Table 4.2, where the new propellant multipliers are used to calculate the amount required for each manoeuvre of the Mars mission, taking into account the increased tankage mass.

Table 4.2. Fuel requirements using cryogenic propellants

Manoeuvre	Delta-V (km/s)	Multiplier	Vehicle	Fuel mass (tonnes)
1	3.55	1.18	ABCD	828.1
2	2.45	0.72	BCD	249.6
3	3.42	1.12	CD	90.3
4	3.42	1.12	D	5.6
5	2.52	0.73	B	113.4
6	3.55	1.18	B	82.6

Comparing with the mass of propellant previously required, we see that the use of cryogenic propellants results in a large decrease. For manoeuvre 1, the reduction is from 3,373 tonnes to 828 tonnes. So, to send the mission on its way to Mars requires only about 800 tonnes of propellant. The all-up mass of the mission becomes 1,597 tonnes compared with 5,068 tonnes – about a third of the original mass. This is equivalent to fifty-three Space Shuttle payloads – still rather daunting – although the number of Space Shuttle payloads required to build the International Space Station is larger, so it is not all that unreasonable.

This is about the limit with conventional rocket engines, as the exhaust velocity generated by liquid oxygen and liquid hydrogen is the highest practically achievable. The physics and chemistry are at their limits, and there is nowhere else to go in the search for greater chemical energy and heat engine efficiency. To assemble a Mars expedition based purely on chemical rockets it would be necessary to initiate an intensive Space Shuttle launch programme together with in-orbit assembly and fuelling of the vehicle. Or, a launcher of much greater mass capability has to be designed and built. Both of these are possible, but a Mars mission would be more problematical because of the investment required. Could alternative means of propulsion produce a further major fuel saving over the use of liquid oxygen and hydrogen? This will be discussed in the following chapter.

5

Electric thrusters: propulsion of the future – now

The vital importance of exhaust velocity for a Mars expedition is clear from the preceding chapter. Changing from the value of 3,240 m/s using storable propellants, to 4,550 m/s using liquid oxygen and hydrogen, resulted in a propellant mass reduction from 5,000 tonnes to 1,600 tonnes, for a 40% increase in exhaust velocity. If we could increase the exhaust velocity by a factor of, say, 2, then we should obtain a much larger reduction in the propellant required, and perhaps bring the expedition within the capabilities currently available. However, the liquid hydrogen–liquid oxygen propellant combination is the best that can be provided for a chemical rocket, and 4,550 m/s is about the highest exhaust velocity that can be achieved.

Are there alternatives to chemical rockets? Yes – and one alternative is currently in use on a number of space missions. Surprising as it may seem, there are already robotic spacecraft traversing the Solar System, powered by beams of particles; and they even emit a blue glow like the starship *Enterprise*! These are spacecraft powered by electric thrusters. Long a dream, and worked on in the laboratory for decades, a simple commercial advantage led to their adoption, first by communications satellites for station-keeping, and more recently by interplanetary probes. For a given spacecraft velocity change they use much less propellant than a chemical rocket. This is a commercial advantage for communication satellites, because more mass is available for equipment, and more business can be handled. The number of parallel communications channels that can be handled by a satellite depends on the number of separate transponders on board. Any mass saved in onboard fuel can be used for more transponders, and therefore increased revenue for the telecommunications company. For robotic planetary probes, a reduction in the amount of propellant again increases the size of the available payload, allowing for more scientific equipment. A Mars expedition requires a huge amount of propellant (as we saw in Chapter 4), and anything that reduces the amount required is a step towards making such a mission practicable. Electric propulsion fulfils this requirement.

5.1   THE PRINCIPLE OF ELECTRIC PROPULSION

In the previous chapter we discussed the vital importance of exhaust velocity, and how the laws of thermodynamics and chemistry put a practical limit of about 4,550 m/s on this parameter. From the equations for rocket science, we saw that the factors that control the exhaust velocity are the temperature in the combustion chamber and the molecular weight of the exhaust products. The mixture – liquid hydrogen and liquid oxygen – produces a high temperature and, more importantly, a low molecular weight for the exhaust gases. This ensures the most efficient conversion of the chemical energy in the combustion into a high-velocity gas stream, through expansion in the nozzle. There are, indeed, reactions that produce a higher temperature in the combustion chamber; and, in fact, the simple propellant combination of liquid oxygen and kerosene does so. But the exhaust products are quite heavy molecules such as carbon dioxide, and the resultant exhaust velocity is smaller than for liquid hydrogen and liquid oxygen. It is difficult to surpass hydrogen in terms of low molecular weight.

There is one propellant combination that outperforms liquid hydrogen and liquid oxygen: liquid fluorine and liquid hydrogen. It has a combustion temperature of 4,260 K compared with 3,250 K for the former combination; the molecular weight of the exhaust is also quite low; and the exhaust velocity is 4,790 m/s. The only problem is that fluorine is a highly corrosive and toxic material which dissolves most metals and even glass. It has therefore not been used in any practical rocket vehicles.

It might legitimately be asked why it is not possible to make the exhaust stream, and hence the rocket, go faster, by supplying more propellant. After all, a motor vehicle is accelerated by depressing the accelerator, which increases the amount of fuel–air mixture entering the cylinder. Here there is a distinction between power and ultimate velocity. In a motor vehicle, the engine is working against all sorts of loss mechanisms, including friction in the transmission and the road wheels, and air resistance. Increasing the power output of the engine will increase the speed, because more energy is available to combat the loss of kinetic energy from the vehicle caused by friction and drag. For a rocket in the vacuum of space, however, none of these mechanisms apply, and the speed is simply determined by the mass of the rocket, the exhaust velocity, and the amount of propellant burned. Increasing the flow of propellant will increase the thrust but will not increase the ultimate velocity, and increasing the thrust simply shortens the time taken to reach maximum velocity. A rocket engine burning more fuel per second is increasing the energy release in the combustion chamber, but there is also more propellant to eject from the nozzle, and the exhaust velocity therefore remains exactly the same.

To increase the exhaust velocity it is necessary to increase the energy released in the combustion chamber *without* increasing the amount of propellant flowing through it. In such a case there is more energy per molecule of exhaust gas, and the exhaust stream will flow faster. This cannot be achieved by chemical means, because the energy released is simply proportional to the mass of propellant flowing through the chamber, and there is always the same amount of energy per kilogramme of propellant. It can, however, be achieved by supplying *electrical* energy to the

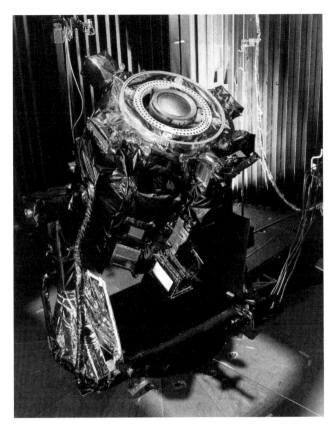

Figure 5.1. Deep Space 1, the first interplanetary spacecraft to be powered by an ion thruster, pictured here during ground testing. The ion thruster is uppermost in this picture. (Courtesy NASA/JPL.)

propellant, and the energy supply and the propellant flow are then decoupled. This is the principle of electric propulsion. The energy is supplied electrically, and the flow of propellant is separately controlled. By turning down the flow of propellant, or increasing the electrical energy supplied, the energy per molecule is increased, and the exhaust velocity is therefore increased. This can produce thrusters with exhaust velocities much higher than those of chemical rockets, and perhaps make a Mars expedition easier. There are many different forms of electric thruster, but the simplest serves as a practical demonstration of the principle.

5.2   THE ELECTRO-THERMAL THRUSTER, OR RESISTOJET

This simple electric thruster finds a wide application in *station-keeping* on communication satellites. These satellites are positioned in geostationary orbits above the equator, where their orbital motion exactly matches the rotation of the

Earth. They therefore appear to be permanently stationary over the equator; but in actuality, small disturbing forces act on them and they gradually drift away from their optimum location. To restore them to the correct position, small thrusters are used to correct the drift. The propellant for this station-keeping can amount to quite a large quantity – perhaps 500 kg – on a satellite that operates for ten years or so. The use of electro-thermal thrusters reduces the amount required, and allows for a larger and more capable satellite.

The construction of an electro-thermal thruster is very simple. There is a combustion chamber and a nozzle, and propellant is delivered to the thrusters in the same way as for a conventional engine, except that there is only one propellant, not two. This single propellant is the working fluid of the engine. Inside the combustion chamber there is a heater coil made of resistive wire – hence the term 'electro-thermal' thruster. An electric current is passed through the heater coil, which glows hot and heats the propellant in the chamber. As a hot gas, it expands down the nozzle and generates thrust, in the same way as in a conventional rocket, but with the difference that the flow rate and energy release are separately controllable. As the current through the coil is increased, for the same propellant flow rate, the amount of energy per molecule of the exhaust gas increases, as does the exhaust velocity. This simple device produces a quite remarkable increase in exhaust velocity, over conventional rockets. It can be as much as 10,000 m/s – more than twice the exhaust velocity of the Space Shuttle main engine.

To examine the impact of such a high exhaust velocity on the Mars expedition we can compare the propellant requirements for the liquid hydrogen–liquid oxygen engine with the requirements for an electro-thermal engine producing an exhaust velocity of 10,000 m/s. The results are shown in Table 5.1.

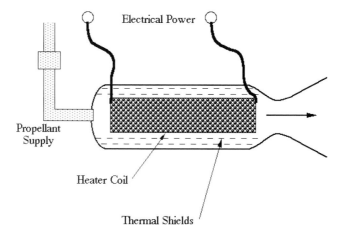

Figure 5.2. A sketch of an electro-thermal thruster. Propellant enters the chamber and is heated by the resistive element; the hot gas then expands through the nozzle, as in a chemical rocket engine.

Table 5.1. Comparative fuel requirements of different engines

Manoeuvre	Delta-V (km/s)	Multiplier	Vehicle	Fuel mass (H_2-O_2)	Fuel mass (electric)
1	3.55	0.43	ABCD	828	137
2	2.45	0.28	BCD	250	67
3	3.42	0.41	CD	90	32
4	3.42	0.41	D	5.6	2.1
5	2.52	0.29	B	113	29
6	3.55	0.43	B	83	30

The total mass of propellant would be 296 tonnes compared with 1,600 tonnes in the case of liquid hydrogen and liquid oxygen, using a conventional engine. About ten Space Shuttle flights would be required to place this amount of propellant in orbit. A Mars mission is theerefore feasible with this level of exhaust velocity.

There are problems, of course, otherwise such a mission would already be in preparation. Some of these problems are surmountable with current technology, but others are much more intractable. But before examining these technical challenges and exploring the full range of electric thrusters currently available, the question of electrical power needs to be addressed.

5.3   POWER AND THRUST

The source of energy in the electric thrusters is electrical, and the measure of the total energy input per second is electrical power in watts. The equivalent in the chemical rocket is the flow rate of the propellant multiplied by the specific thermal energy released in the chemical reaction. This did not previously appear in our considerations, because we have so far confined discussion to the exhaust velocity alone, and this is not the same as the *power* of the engine. For an electric thruster, in which energy is delivered by electricity, power is an issue. It is perhaps obvious that in the case of an electric thruster there has to be an electrical power supply in addition to the combustion chamber, nozzle, pumps and propellant tanks associated with a chemical engine. The size of this supply will depend on the power required by the electric thrusters, and it is of course dead weight in the rocket vehicle, as it does not become lighter as power is used. As dead weight, it decreases the mass ratio of the rocket vehicle, which in turn will reduce the velocity ultimately achievable with the electric thrusters.

The power required by the thrusters will depend on both the exhaust velocity and the thrust. To understand this, consider the energy flow in the exhaust stream. The energy per unit mass of propellant in the exhaust stream is just

$$E = \frac{1}{2} v_e^2$$

while the power, or energy flow per second, is the above, multiplied by the mass flow rate *m*, producing

$$P = \frac{1}{2} m v_e^2$$

where *P* is in watts, *m* is in kg/s, and v_e is in m/s. This power will define the required generating capacity of the power supply. Note how an increase in mass flow rate, *m*, or in velocity, v_e, will require more electrical power to be supplied.

The thrust of the engine is given, as in Chapter 4, by

$$F = m v_e$$

where *m* is in kg/s.

The power is therefore related to the thrust by

$$P = \frac{1}{2} F v_e$$

Until now, thrust has not been a very important consideration in evaluating the Mars expedition. We have assumed that thrust will be adequate, and have concentrated on the final velocity of the spacecraft, which does not depend on thrust. Recalling the rocket equation, we see that the only factors which determine the velocity of the spacecraft are the exhaust velocity and the mass ratio. For a launch vehicle such as the Space Shuttle or Ariane V, thrust is important. It is necessary for the thrust to exceed the weight of the fully loaded rocket sitting on the launch pad, otherwise it could not lift off. Again, a high thrust will help the rocket accelerate quickly against the Earth's gravity, and reduce what is called 'gravity loss'. This is the fuel burned solely to lift the as yet unburned fuel in the rocket's fuel tanks. In space, the thrust is, in general, less important, because most manoeuvres are made by firing the rocket engine at right angles to gravity. There is no 'lifting' involved, as the vehicle remains at the same distance from the Earth throughout the manoeuvre. The effect of thrust here is simply to determine the time taken for the vehicle to reach the desired velocity; and a high thrust will shorten this time.

For electric propulsion, however, the thrust becomes an important issue, for two reasons. However the electricity is produced, the mass of the electrical power supply depends on the power of the rocket engine, which in turn depends on the thrust. Electric thrusters also tend to have a rather low thrust. They can still achieve a high vehicle velocity with great fuel economy, but it may take a long time to do so. The model we have for a chemical thruster – which burns for a short time with high thrust, and then cuts out after the full velocity has been achieved – will not apply to an electric thruster. The vehicle does not move very far in gaining velocity from a high-thrust chemical rocket, and so it does not change its distance from the Earth very much during the burn. As mentioned above, no 'lifting' occurs, and all the fuel is used to accelerate the rocket. For low-thrust electric propulsion, the vehicle travels a long way whilst accelerating, and it cannot help but spiral slowly away from the Earth, rather than, for example, leave the Earth on an hyperbolic trajectory after a

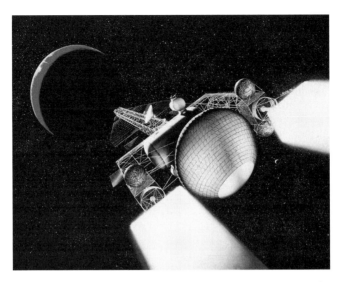

Figure 5.3. An artist's concept of a 10-MW nuclear electric-powered Earth–Mars transfer vehicle. Note the relatively small fuel tanks and the two electric thruster banks on each side of the reactor. In the background is a crescent Mars. (Courtesy NASA.)

short burn of a high-thrust chemical engine. This spiralling involves the lifting of the unburned fuel against the pull of gravity, and some of the propellant is wasted in doing so. Gravity loss has become an important parameter.

For these two reasons we need to know the thrust when considering electric propulsion. In general, the latter problem – the gravity loss – can be dealt with by increasing the amount of fuel provided by about a factor of 2. It can immediately be seen that the mass of a high-power electrical supply, required to produce a higher thrust, will also offset some of the saving in fuel that the higher thrust engenders. There is only an advantage when the saving is greater than the effect of the heavier power supply, and of the extra fuel required to offset the gravity loss..

The above equation for power is not at all specific to electric propulsion, and it is instructive to consider the power output of the chemical rocket engines described in Chapter 4. The Aestus engine, using storable propellants, has a thrust of 29 kN and an exhaust velocity of 3.24 km/s; the power is therefore 47 MW. The SSME, using liquid oxygen and liquid hydrogen, has a nominal thrust of 2.3 MN and an exhaust velocity of 4.55 km/s; its power is therefore 52,000 MW. These are very great power levels indeed – but hardly surprising, considering what these engines have to do. However, seen as electrical power, their demands on a power supply would vastly exceed what can reasonably be carried. From this we can see that if electric propulsion is to be used for a human Mars mission, its performance will be determined by the amount of electrical power that can be supplied. While the exhaust velocity will depend on which types of thruster and propellant are used, the thrust will be determined by the available power.

In space, the options for electrical power are limited. The most common are, of course, solar cells, directly generating current from sunlight; batteries can be used for short-term requirements. Where sunlight is too faint or absent, then power generated using heat from radioactive material is normally used. More exotic possibilities are nuclear fission and, where sunlight is available, mechanical generators powered by sunlight focused on a 'boiler' containing a working fluid. Solar cells are not very effective in generating power (as mentioned in Chapter 4), as one requires 2.94 m^2/kW at Earth, weighing 24 kg. A 1-MW solar panel would need to be about 3,000 m^2 in area (about 55 m on a side), and would weigh 24 tonnes. This is not impossible, but it is certainly not current technology.

We can calculate the thrust of a solar-electric vehicle with a 1-MW solar array and an exhaust velocity of 10 km/s. Rearranging the equation,

$$F = \frac{2P}{v_e}$$

the thrust is just 200 N. It may be noticed that as the exhaust velocity increases, the thrust for a given power decreases. For a given velocity change requirement, the amount of fuel required decreases as the exhaust velocity increases. At the expense of a lower thrust and a longer acceleration period, we can save fuel by increasing the exhaust velocity beyond 10 km/s.

5.4   ELECTRIC PROPULSION TRAJECTORIES

The advantage of electric propulsion is offset by the long acceleration and deceleration times. To appreciate how long these times really are we can calculate the time and distance travelled for the last manoeuvre of our model mission: the deceleration to enable capture into Earth orbit, on returning from Mars. The returning vehicle weighs 70 tonnes, and the fuel required with chemical propulsion – assuming storable propellants and the Aestus engine – is 139.4 tonnes (see Chapter 4).

The rather complicated equations for time and distance are as follows.

$$s = \frac{v_e^2 M_0}{F} \left[1 - \frac{M}{M_0} \left(\log_e \frac{M_0}{M} + 1 \right) \right]$$

$$t = \frac{M_0 v_e}{F} \left(1 - \frac{M}{M_0} \right)$$

Here M_0 is the original mass of the fully fuelled vehicle, M is just the bare mass of the vehicle, and v_e is the exhaust velocity: 3,240 m/s for this engine and fuel. Substituting, we arrive at 5,206 seconds, or about 1 hour 26 minutes. The distance travelled is 7,587 km – about the same magnitude as the final orbital radius. In slowing down, therefore, the spacecraft travels about a third of the way round the orbit. Using liquid oxygen and liquid hydrogen, and assuming the Space Shuttle main engine to be the thruster, these times and distances become much shorter because of the higher thrust.

These numbers can now be compared with those for the electric thruster. The mass of the empty vehicle is 70 tonnes, and the electric thruster – of 200 N thrust and 10 km/s exhaust velocity – requires just 30.1 tonnes of fuel.

Substituting in the equations we determine the thrusting time: 17.4 *days*. This is not an unreasonable time *per se*, but the distance the spacecraft would travel is 2.5 million km. This can only be 'fitted in' if the spacecraft is slowing down while following a tight spiral path down to low Earth orbit. Nevertheless, this is a very long distance indeed compared with the circumference of the desired low Earth orbit of 43,000 km. Without calculating the exact trajectory we can see that the spacecraft must make about fifty-eight circuits of the Earth while slowing down.

The above calculation, however, is not strictly correct, because we should allow for the extra dead weight of the power supply. If we assume solar panels, then it is 24 tonnes. In this case the mass ratio is altered and more fuel is required, basically to decelerate the extra mass of the solar panels. The amount of required fuel can be calculated using the electro-thermal thruster multiplier from the above table, for a final mass of 70 plus 24 tonnes: 40.4 tonnes. Using these values in the equations, we find the time is now 23.2 days and the distance is 3.4 million km, while the number of circuits is now seventy-nine. A final comparison with the propellant required for this manoeuvre, using liquid hydrogen and liquid oxygen, shows that electric propulsion, using solar panels to provide 1 MW of electric power, saves about half the propellant. The former requires more than 80 tonnes, while the latter – with the new mass ratio, taking into account the solar panels – requires just 40 tonnes. The penalty is in the long period spent in slowing down.

The saving in propellant, calculated here, can be applied to most of the manoeuvres during a Mars mission; and it can be very significant. Chemical thrusters, of course, must be used for departure from the surface of Mars and for the landing; there is no substitute for thrust here, where gravity dominates the physics. Electric propulsion is confined to operations in space, where gravity can be partially avoided, by thrusting at right angles to the gravitational pull. Here the long slow manoeuvres with electric thrusters are very different from the short burns of chemical rockets. More time and distance must be allowed if the benefits of reduced fuel requirements are to be realised. On approaching the Earth from the interplanetary voyage, for example, to slow down the spacecraft the electric thrusters are turned on much earlier than when using a chemical rocket. Because of the low thrust, the capture must first be made into a very eccentric elliptical orbit from the approach hyperbola. The thrust must be sufficiently high to enable this capture in the first cycle of the orbit – before the spacecraft passes perigee – and this defines a minimum requirement on thrust. If this is beyond the capabilities of the electric thruster, then a short burst of chemical thrust may be required to ensure capture. The electric thruster then continues to slow the vehicle through a large number of elliptical orbits of ever decreasing apogee until the circular low Earth orbit is achieved. This takes a long time and distance, as roughly calculated above. On the other hand, the saving of fuel is very significant.

This type of trajectory is the method by which all electric propulsion manoeuvres are executed. For departure from low Earth orbit the reverse trajectory is followed, with orbits of gradually increasing eccentricity leading eventually to escape from

Earth along a hyperbolic path. Again, it may be necessary to use a burst of chemical thrust to enter the final trajectory; but this still enables most of the process to be carried out by electric propulsion, thus saving the considerable amount of propellant. Despite the long acceleration times, the saving in fuel is so important that electric thrusters are a serious option for a Mars expedition, and scenarios employing them are included in NASA's studies.

5.5   DIFFERENT TYPES OF ELECTRIC THRUSTER

All of the above discussion has been based on the simple idea of an electric thruster, epitomised by the electro-thermal thruster, which provides a moderate thrust and an exhaust velocity of about 10,000 m/s – about twice that of the best chemical thruster. It is now appropriate to look at the different kinds of thruster, some of which may be more suitable than this simple device, which heats the propellant by an electric current passed through a resistive heater element; not much different in principle from an electric kettle.

A paradox with the electro-thermal thruster is that the temperature that the propellant reaches is lower than that reached in chemical rockets, and yet the exhaust velocity, which, as will be recalled, depends partially on the temperature, is higher. It will be recalled that the exhaust velocity depends on the 'combustion' chamber temperature, and the molecular weight of the exhaust molecules. In fact most of the advantage of the electro-thermal thruster comes from the low molecular weight of the exhaust. Because it is not necessary to use a chemical reaction to generate the heat, the exhaust is free from the products of combustion, which are usually of higher molecular weight than the pure propellants. The propellants *combine* to release energy in a chemical rocket. The most obvious case is when liquid hydrogen is used as the propellant: the molecular weight of water, the product of burning hydrogen and oxygen, is 18 while the molecular weight of pure hydrogen is only 2. This is nine times lower, and because the exhaust velocity depends on the square root of the molecular weight, it gives an advantage of a factor of three in exhaust velocity. This means that the electric thruster, compared with the chemical thruster, would generate a three times higher exhaust velocity, all other things remaining the same. In fact the temperature cannot be as high as in a chemical rocket, and so the advantage is closer to a factor of two.

The fundamental limitation of the electro-thermal thruster is that the propellant must be heated by *contact* with the hot heating element. The propellant is therefore *cooler* than the element, otherwise the heat could not pass from the element to the gas. The element is therefore the hottest component in the chamber of the thruster. This means that, crudely speaking, the gas cannot be hotter than the melting point of the element. (We shall find that this same issue arises in nuclear thrusters, discussed in Chapter 6.) Most metals tend to soften and melt at around 1,500° C; but some – such as chromium, molybdenum, niobium and tantalum – can take higher temperatures, and can be used to make refractory resistive alloys that allow for temperatures in excess of 2,000° C in electro-thermal thrusters. Tungsten can also endure very high temperatures, but is very difficult to work. A typical value for most useful resistive alloys would be about 2,200° C.

This is remarkably low compared with the temperatures in excess of 3,000° C encountered in chemical rocket engines. The reason for this is that the hot gas in the chemical engine never has an opportunity to heat the walls of the combustion chamber, or nozzle, above its service temperature. The walls of the chamber and nozzle are cooled by the flow of cold propellant through channels behind the surface that is in contact with the hot gas. There is therefore a layer of relatively cool gas in direct contact with the walls, which remain below their softening point. This is vital to the integrity of the chamber, and any interruption in the cooling liquid flow results in a catastrophic failure of the engine. Cooling can be arranged in the chemical rocket engine because of a very simple fact. The heat is generated *in the gas itself* by the chemical reaction, so the hottest component in the chamber, in which the chemical reactions are taking place, is the gas itself. The walls can be cooler. This cannot be arranged for an electro-thermal thruster, because the hottest item must be the element, made of metal; and it would be ridiculous to cool it. The walls and nozzle can, of course, be cooled in the usual way, but the service temperature of the heater element remains the limiting factor in the performance of electro-thermal thrusters. Other types of thruster attempt to overcome this limitation. It may be noted that while the electro-thermal thruster could in principle be used in atmosphere, it is not; and none of the other types to be discussed here can be used other than in a vacuum; they are exclusively space thrusters.

5.6   THE ARC-JET

If the heat could be released in the gas itself, then the problem of the service temperature could be eliminated, and higher propellant temperatures could be achieved. This is the process carried out with the arc-jet, in which the gas is heated directly by passing an electrical discharge or arc through it. The general scheme is shown in the Figure 5.4.

The object is to pass current directly through the gas; but instead of a heater element there are two electrodes. The anode (positive) – made of tungsten, with a melting point of 3,244° C – is a cylindrical rod, tapered at the end to a conical point. The (negative) cathode is made of the same material shaped to receive the conical end of the anode, leaving a gap in which the electrical discharge takes place. In this version the anode is also shaped to form the nozzle that will expand the hot gas created in the discharge. The boron nitride insulator, which holds the other components in place, and electrically separated from one another, is also formed into a chamber from which the propellant is introduced into the discharge region. A few hundred volts are applied across the gap between the anode and cathode, and an arc discharge is formed. This heats the gas, which then expands down the nozzle in the usual way to generate a high-velocity exhaust stream. As in the chemical rocket, the thrust acts on the walls of the nozzle. The main advantage of the arc-jet is that up to

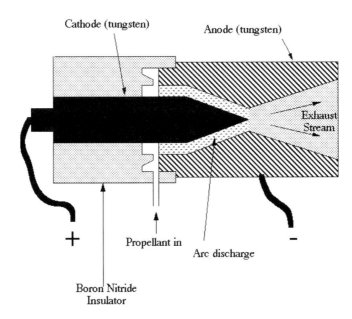

Figure 5.4. A schematic diagram of an arc-jet. The propellant is heated by an arc struck between the electrodes, and then expands through the nozzle in the normal way, to generate a high exhaust velocity.

a hundred times more current can be passed through the arc than can be passed through the heater element of the electro-thermal thruster; and this produces more power in the thruster. Whether or not the thrust will be higher depends on the exhaust velocity. Higher temperatures are possible because the propellant is directly heated by the arc, and cooler regions near the walls can be maintained to avoid softening. Typical temperatures of the gas itself range from 4,000 K to 5,000 K – much higher than for chemical rockets – and exhaust velocities up to 20,000 m/s are achieved. This is a much more powerful thruster than the electro-thermal thruster, and arc-jets have been operated at 200 kN for short periods, although those in practical use have much lower thrust.

There are, however, some problems associated with the use of the arc to heat the gas. The electrodes themselves are made of refractory metal, and do not come into contact with gas at the arc temperature, although they can still be attacked by the ions in the arc. In order to conduct electricity, the gas has to be ionised. These ions are accelerated by the electric field, and some strike the cathode. This process gradually sputters away the surface of the metal, and this limits the operational life of the thruster. It is also less efficient in converting electric power into thrust. The gas is very hot, and radiates heat to the cathode; and this energy is lost. The other effect occurs in the discharge itself. The current, if high enough, can cause the 'pinch' effect where the magnetic field, generated by the current flow, squeezes the ion path into a number of dense streamers, with open areas between. Most of the energy then goes into sputtering the cathode rather than heating the gas. The result of these effects is an efficiency of 30–40% rather than the 60% or so of the electro-thermal thruster. For the same performance, therefore, a larger power supply must be carried.

It might be reasonably asked: why bother with an arc-jet, given these disadvantages? One reason is the higher exhaust velocity, and the consequent saving in propellant. Using the figures above, the propellant required for the vehicle weighing 94 tonnes with solar panels is only 18.3 tonnes. This more than compensates for the increased area of solar panels required. The main reason, however, touches on a different aspect of electric thrusters which has so far not been considered: the size and number of thrusters required to handle the 1 MW of power we have chosen to use. Clearly there must be limits, and it will not be possible to push all that power through a single electric thruster. Electro-thermal thrusters have quite a small power capability because of difficulties in heat transfer from the heater to the gas, and this cannot easily be accomplished for high-power heaters. So, many electro-thermal thrusters would be required, in parallel, to provide the thrust. Arc-jets can handle a hundred times more power, and so reduce the complexity of operating perhaps five hundred electro-thermal thrusters together. Lifetime is an issue, of course, and the arc-jet must not wear out before the slow manoeuvres required for electric propulsion have been completed.

Although it makes use of ions in the gas to carry the current and heat it up, the arc-jet is really an electro-thermal thruster. It gains thrust by expanding a hot gas through a nozzle of conventional shape, just like the resistojet. There is, however, an entire group of electric thrusters that use the direct accelerating force of electric and magnetic fields to generate an exhaust stream: *ion thrusters* and *plasma thrusters*.

5.7   ION THRUSTERS

The ion thruster is a very simple idea. Ions are formed, and accelerated away from the spacecraft by an electric field held between two transparent grids or meshes. They can have a very high velocity indeed – 20–60 km/s – and as we have seen before, such high velocities save a huge amount of propellant for a given manoeuvre. The electric power is being used directly to accelerate the ions, rather than to heat them up, and expand them down a nozzle to create a fast-moving exhaust. Instead of being developed against the inner walls of the nozzle by gas pressure in an expanding exhaust stream, the thrust is developed against the grids that provide the accelerating electric field, by the reactive force of the accelerated ions.

It is a simple concept, and relatively simple in execution. A propellant – usually xenon – is introduced into a chamber, where it is ionised by an electric discharge similar to that in an arc-jet but with a much lower current. An applied magnetic field helps to maximise the number of ions produced. The ions then pass through the first of the two grids that will accelerate them. Once in the region between the grids, the ions encounter the electric field and are accelerated towards the outer of the two grids. They then pass through this grid and out into space at 20–60 km/s. The thrust is produced by the reaction on the grids of the ions that are accelerated. The ions are positively charged, and so in time the spacecraft would become negatively charged,

Figure 5.5. The ion thruster on Deep Space 1, during engine tests in a vacuum chamber. (Courtesy NASA/JPL.)

and this would attract the ions and slow them down. To prevent this, electron guns (similar to those in television tubes) are mounted near the outer grids, and these release sufficient (negatively charged) electrons to keep the spacecraft neutral. Electrons weigh very little compared with ions, and so this procedure has little effect on the thrust. Once again, the main advantage of these thrusters is the high exhaust velocity. They are, in fact, amongst the highest exhaust velocity devices, producing jet speeds of 20,000–60,000 m/s. This provides great fuel economy, but with the accompanying disadvantages of low thrust and protracted acceleration times.

5.8   THE NSTAR ION ENGINE

A good example of a practical ion engine is the NSTAR engine, used with great success both for station-keeping and for interplanetary probes. This engine was fitted to the Deep Space 1 spacecraft, the mission of which was to execute a fly-by of an asteroid and, if possible, several comets. Only 82 kg of xenon propellant provided a total spacecraft velocity change of 4.5 km/s. The diameter of the accelerator grid is 30 cm, and the maximum power level is 1.3 kW, provided by solar panels.

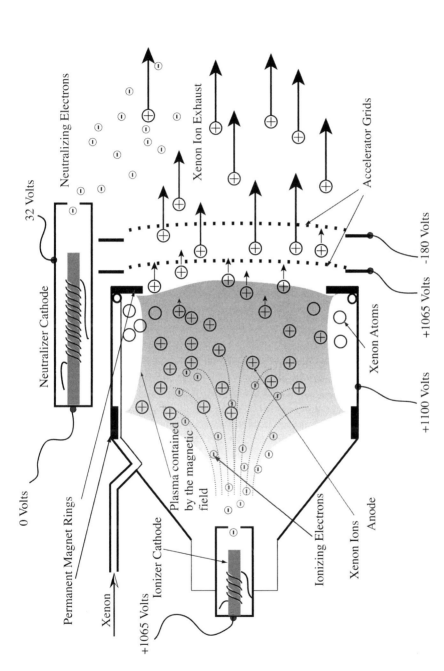

Figure 5.6. A schematic diagram of the NSTAR ion thruster as used on Deep Space 1. The xenon gas enters the chamber from the left, and is ionised in the shaded region, confined by a magnetic field. The ions drift into the gap between the two grids on the right, and are accelerated by the electric field between the grids. They then pass out to form the exhaust stream. Electrons are released from the neutraliser to keep the spacecraft neutral.

The NSTAR ion engine is shown schematically in Figure 5.6. Xenon gas at low pressure is delivered into the discharge chamber, to be ionised. The ioniser cathode releases electrons into the discharge chamber, where they follow the magnetic field lines set up by two rings of permanent magnets. This confines the discharge plasma so as to maximise the number of collisions between electrons and xenon atoms. The actual discharge is between the ioniser cathode and the casing of the discharge chamber, and there is a potential difference of 35 V between the cathode and the casing. Xenon atoms are ionised by the electrons and drift towards the screen grid, which covers the discharge chamber exit. The screen grid is more negative than the casing, and attracts the positive ions. On passing through the screen grid, the ions encounter the strong electric field set up between the screen and accelerator grids, and are accelerated to speeds of up to 30 km/s in the gap between the grids. The grids are only a few millimetres apart, but across the gap they have a potential difference 1,245 V. The ions then pass through the accelerator grid and leave the engine. The thrust is caused by the reaction of the accelerated ions on the grids. The neutraliser cathode releases electrons into the exhaust stream to match the positive charge loss of the departing xenon ions, and this keeps the spacecraft electrically neutral. The conversion of electrical power into thrust is very efficient: more than 90%. If gaining the maximum velocity change from the minimum power and propellant is the main requirement, then ion engines are by far the most efficient.

The NSTAR engine has a thrust that can be varied from 19 mN to 93 mN, with equivalent power levels from 423 W to 2,288 W. This is a very small thrust – much lower than that of the arcjet or electro-thermal thruster – and consequently a long time is needed to achieve a given spacecraft velocity change. If sufficient power were to be available, then a more powerful engine could be used. This is, however, not simply a matter of increasing the ion current in the engine. There is a limit to the operating current, called the space charge limit, which occurs when the number of xenon ions passing between the two grids exceeds a certain value. Since the ions carry the current, this amounts to a current and hence thrust limit. Beyond this limit the charge of the xenon ions in the gap offsets the field between the electrodes, and no more ions can be accelerated. Essentially, the ions closer to the accelerating grid screen its field from those entering the gap, and so the only way to increase the thrust beyond this limit is to make the engine bigger; that is, to have grids of a larger diameter. The current, and hence the thrust, then increases with the area of the grids; but this cannot continue indefinitely, because the grids become difficult to support. They experience a mutual attractive force – the force that accelerates the ions. As the grids become bigger, so this force increases, and at the same time the relative stiffness of the grids decreases. This imposes a natural limit to the size of the grids.

An alternative is to arrange several smaller ion engines in parallel. This is analogous to using several chemical rocket engines – as in the case of the Space Shuttle, which has three. Ganging engines together in this way will reduce the efficiency compared with one large engine, because of the multiplication of neutraliser and discharge cathodes, which do not contribute to the thrust, but still require electrical power. Progress towards higher-thrust ion propulsion will no doubt require both an increase in engine size and the use of multiple engines.

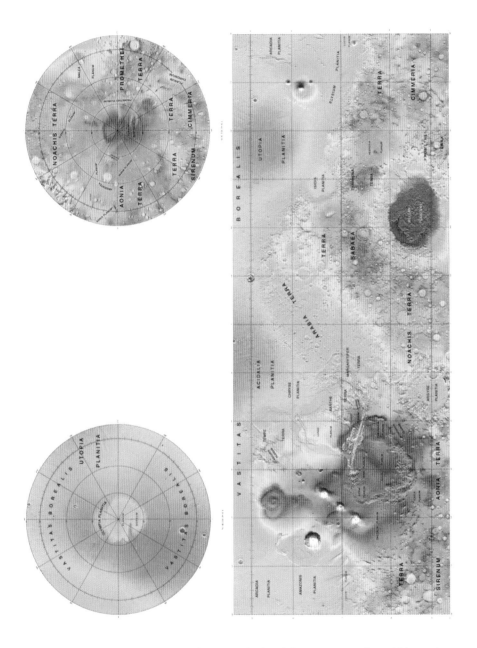

Plate 1. The global topography of Mars. The low-lying areas are coloured blue and green, and the highlands are yellow to red. Valles Marineris, Olympus Mons and the Tharsis volcanoes are very obvious, as is the impact basin Hellas. (Courtesy US Geological Survey.)

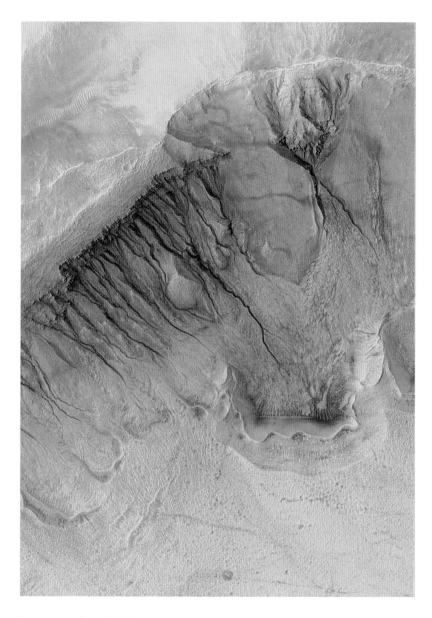

Plate 2. Water-formed gullies on the surface of Mars, in true colour. (Courtesy NASA/JPL.)

Plate 3. A Viking lander image of the martian surface, in true colour, and reprocessed in keeping with the latest information. Note also the colour of the sky. (Courtesy NASA.)

Plate 4. A Viking lander image of the martian surface, in true colour. Ice has formed on the surface over winter. This is water ice condensed out of the atmosphere. (Courtesy NASA.)

Plate 5. The Space Shuttle main engine under test. (Courtesy NASA.)

Plate 6. The RS-68 engine under test. (Courtesy NASA.)

Plate 7. An ion engine firing in a test chamber. The blue colour arises from the recombination of the xenon ions; the glow from the electron gun can be seen at lower left. (Courtesy NASA.)

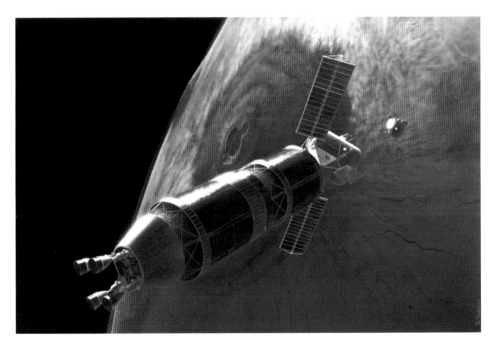

Plate 8. A nuclear-powered Earth return vehicle awaits the arrival of the explorers in the ascent vehicle. It was placed in Mars orbit four years previously, to ensure that a safe return was prepared for the expedition. On the surface below is Olympus Mons. (Courtesy NASA.)

Plate 9. The nuclear rocket, having been assembled in orbit, prepares for departure to Mars. The rear section contains the liquid hydrogen propellant and the three fission engines. (Courtesy NASA.)

Plate 10. The pressurised truck that can be used to take the explorers far from the base. The propellant factory attached to the ascent vehicle has the capacity to produce extra oxygen and methane, which can be used to power the truck. (Courtesy NASA.)

Plate 11. Unloading the Mars rover from the cargo lander. (Courtesy NASA.)

Plate 12. The expedition headquarters, with the attached inflatable extension to increase accommodation for the members. In the distance is a cargo lander. (Courtesy NASA.)

Plate 13. The ascent vehicle is prepared for departure, by removal of the landing systems. It has already been fuelled with methane and oxygen prepared on Mars. (Courtesy NASA.)

Plate 14. The explorers leave the martian surface, to begin their long journey to Earth. (Courtesy NASA.)

Plate 15. Explorers near the Mars base. One of the shallow-pitched but immensely high martian volcanoes is in the background. The base is located amongst the layered islands formed in an ancient water-course. Greenhouses can be seen in the background; the white structure in the centre background is a well that melts the underground permafrost to provide water for the base. (Courtesy NASA.)

Plate 16. Rock-climbing on Mars. Human explorers dare to go to places where robotic explorers would fail, so that no potential location for knowledge of the past history of Mars is omitted. Gravity is only 38% of Earth's gravity - but any fall would still be dangerous. (Courtesy NASA.)

Plate 17. An explorer examines a rock which might contain fossil life. (Courtesy NASA.)

Plate 18. Explorers see the sunrise across one of the water-created features on Mars. The mist is composed of ice crystals condensed out of the atmosphere overnight. (Courtesy NASA.)

Figure 5.7. A close-up view of the ion thruster on Deep Space 1. The curved acceleration grid can be seen in the centre, and the electron gun to the lower middle. The surrounding structure is a plasma shield to prevent space plasma from entering the device. (Courtesy NASA/JPL.)

5.9   PLASMA THRUSTERS

The space charge limit of an ion engine is the direct result of the use of ions as the exhaust stream; it consists of positively charged particles. Between the grids, their positive charge offsets the field, and limits the current that can flow. If the direct acceleration of ions could be achieved, but with an approximately *neutral* exhaust stream, then there should be no space charge limit. This can be done if electric and magnetic fields are used together to accelerate the exhaust. If a current flows through an ionised gas, which has a transverse magnetic field applied to it, then the direction of the force on the electrons, negatively charged, and flowing in one direction, is the same as the direction of the force on the ions, positively charged and flowing in the opposite direction. This accelerates both the electrons and the ions out of the engine. The thrust would be mainly developed by the much heavier ions, as in the case of the ion thruster, but because the plasma is neutral there is no space charge limit.

This *plasma thruster* had been the dream of electric propulsion scientists for decades, but it had not been successfully realised in the West. However, throughout the years of isolation, Russia developed a successful plasma thruster using a somewhat different principle. Used on more than a hundred Russian spacecraft, it was well qualified and reliable. With the renaissance in propulsion technology consequent upon the end of the Cold War, this device is now being made and used worldwide. It is variously known as the Hall effect or stationary plasma thruster. The principle is somewhat complicated, but its construction and operation are very simple.

5.10 ACCELERATION BY CROSSED ELECTRIC AND MAGNETIC FIELDS

The *dynamic* plasma thruster – yet to be realised in practical form – is simpler to understand, and is a useful starting point for grasping the principle of the more practical *stationary* plasma thruster.

The principle is illustrated in Figure 5.8. A current passing through the ionised gas follows the direction of the electric field. Because it is crossing the magnetic field, set up at right angles to the electric field and to the axis of the thruster, the ions and electrons, which carry the current, experience a force *along* the axis. In moving outwards along the thruster axis, they collide with neutral gas atoms and also drive these outwards. So, a stream of ions, electrons and un-ionised gas molecules is driven out of the thruster to form the exhaust. The exhaust is more or less neutral, the thrust is developed against the electric and magnetic fields in the thruster, and these in turn act on the electrodes and coils that set up the field. In this sense the plasma thruster is the same as the ion thruster: the electrons and ions are directly accelerated by electric and magnetic fields. Typical velocities for the exhaust are about 20 km/s, and there is no space charge limit.

Another way to understand this process is to think of the thruster as a linear electric motor. In an electric motor, the current is carried by wires that are set at right angles to the magnetic field. The wires experience a transverse force at right angles to

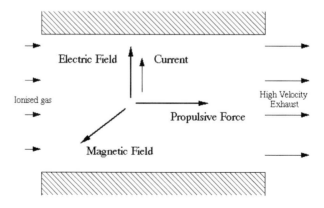

Figure 5.8. Crossed electric and magnetic fields accelerating ionised gas. The electric current driven by the electric field flows between the poles, and the magnetic field drives the electrons *and* the ions out from between the poles. This used to be called magnetohydrodynamics.

both the magnetic field and to the wire itself. It is this force that causes the wire to move and the motor to act. In the plasma thruster, the current-carrying wire is replaced by the ions and electrons in the gas. The transverse force is the same, and the action is the expulsion of the gas from the thruster. In reality, of course, it is all much more complicated, and there is a whole scientific discipline of *magnetoplasmadynamics*; but the principle can be grasped from this simple explanation. The key advantage is that more thrust can be produced from a smaller engine. However, as mentioned above it has proved very difficult to produce a practical dynamic plasma thruster – which is why the stationary or Hall effect plasma thruster must be considered.

5.11 THE HALL EFFECT OR STATIONARY PLASMA THRUSTER

One of the complications of the plasma thruster, as opposed to the electric motor, is that in a gas or plasma the electrons and ions are free to move in any direction, while in the electric motor the electrons can only move along the wire. Given freedom of movement, the ions and electrons will follow a spiral path around the magnetic field lines, as their natural tendency is to orbit the magnetic field lines. The diameter of the orbit or spiral is determined by the strength of the magnetic field and the charge-to-mass ratio of the ions and electrons. Electrons, being very light, move in tight, rapid circles and spirals, while the ions, being heavy, move in large slow circles and spirals. This different behaviour of ions and electrons in a magnetic field has important consequences in a real plasma, because the ions and electrons also collide with neutral gas molecules. Each time a collision occurs, the spiral path is disrupted and the transverse motion, described above, stops momentarily until a new spiral path is established. It is only when the ions or electrons are spiralling round the magnetic field lines that the axial acceleration that leads to thrust can occur. If collisions dominate the process, then no axial thrust is generated; but if spiral motion dominates, then the axial motion occurs, and thrust is generated. This clearly depends on the gas density and the tightness of the spiral path, and the latter depends on the charge-to-mass ratio and the strength of the magnetic field. The factor that determines whether or not collisions dominate is called the Hall parameter, defined as the ratio of the frequency of gyration of the particles in the magnetic field to the collision frequency. If the Hall parameter is large, then axial drift occurs and the thruster works; if it is small, then there is essentially no thrust, and the electrons and ions simply drift across the thruster, not out of it.

It will be clear that under a wide range of conditions of gas density and magnetic field the gyro frequency of the electrons will be much higher than the collision frequency, while that of the ions will be much lower. Even for hydrogen, the charge-to-mass ratio of electrons is 2,000 times higher than for ions; for xenon there is an additional factor of 131, for its atomic weight; and electrons will spiral a few hundred thousand times more rapidly than will xenon ions. There will therefore be a charge separation in the plasma, with the electrons responding to the magnetic field and moving axially; while the ions, with their motion dominated by collisions, simply drift across the thruster. The current generated by the electrons is called the Hall current; and the effect – ions and electrons behaving differently – is called the Hall effect (after the American scientist Edwin Hall, who in 1879 observed it in a flat metal strip placed in a magnetic field). This effect is the key to producing a successful practical plasma thruster.

The general principle of the Hall effect thruster is illustrated in figure 5.9. Note that while this figure is superficially similar to figure 5.8, the applied fields are differently arranged. In the Hall thruster the electric field is *axial*, along the thrust direction, and not transverse as in the dynamic plasma thruster. The magnetic field remains transverse, and the electrons travel in a direction at right angles to both fields, across the cavity. This electron flow – the Hall current – then interacts again with the magnetic field to generate an axial motion, in the same way as the normal current in the dynamic plasma thruster. The net result is an axial flow of electrons, which interact with the ions, and drag them along. The ions collide with neutral gas molecules, and the fast-moving mixture of electrons, ions and neutral gas molecules forms the exhaust. The thrust is developed against the magnetic field, and through that to the coils producing it.

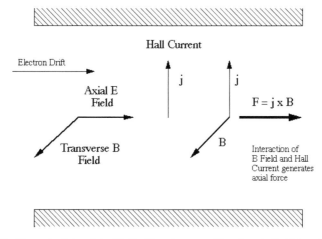

Figure 5.9. The principle of the Hall effect thruster. The electric field is axial rather than transverse, but the magnetic field is still crossed with it. The transverse electron current is the Hall current, on which the magnetic field acts to produce the axial motion that leads to thrust. The ions are dragged along with the electrons.

The figure shows theoretically how the fields and currents are arranged, but in reality, all the forces act together and the system is much more complicated than can be analysed here. In fact, the Hall thruster might have remained no more than a theoretical concept had the Russian scientists not made a breakthrough by creating a thruster with a cylindrical or coaxial geometry. This happened to be eminently practical, and all the Hall thrusters used today follow this idea very closely. For the coaxial Hall thruster the electric field is still axial, but the magnetic field is wrapped round to take up a cylindrical geometry, as shown in Figure 5.10.

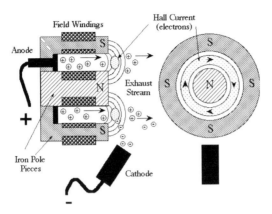

Figure 5.10. The Hall effect thruster as developed in Russia, with cylindrical geometry and a circular Hall current. The cathode both maintains the discharge and ensures neutralisation.

A central cylindrical north pole piece is surrounded by an annular south pole, and there is a gap or cavity between the two. This creates a radial magnetic field across the gap, in place of the linear field in Figure 5.9. The whole of the magnet is lined with an insulating material, and an axial electric field is set up along the gap from the anode to the external cathode. This is just the same as in Figure 5.9, except for the cylindrical shape of the cavity. The difference is that the Hall current – which in Figure 5.9 flows *across* the gap from one electrode to the other – now flows in a circle, without touching the electrodes. In effect, the electrons orbit in the radial magnetic field, and the Hall current then becomes continuous. The electromagnetic forces act in the same way, and there is an axial flow of electrons, ions and neutral gas molecules. This generates thrust, in the same way as before, by the reaction of the accelerated ions on the magnetic field.

To turn the device on, xenon is admitted through many fine holes in the insulator at the bottom of the cavity. At the same time, electrons are produced by the cathode with is located externally, as shown. An electrical discharge then develops down the cavity between the anode and cathode, after which the Hall current establishes itself and stabilises just outside the cavity. The ions created in the discharge flow towards the outside, pulled by the electrons in the Hall current, collide with neutral molecules, and create the more or less neutral exhaust stream. This is the practical form of the Hall thruster, and many examples exist.

The magnetoplasmadynamics are, of course, complicated, and in the case of the cylindrical Hall thruster it is perhaps easier to see how it works by considering it as a form of ion thruster. The electrons in the Hall current ring, just outside the cavity, could be thought of as the accelerator grid in an ion thruster. They attract the positive ions and accelerate them out through the annular cavity. The difference is that in the Hall thruster the stream is neutral, and there is no space charge limit.

Devices of this type can handle up to 20 kW and generate up to 1 N of thrust, and be no more than 220 mm in diameter. Efficiencies are quite high – in the order of

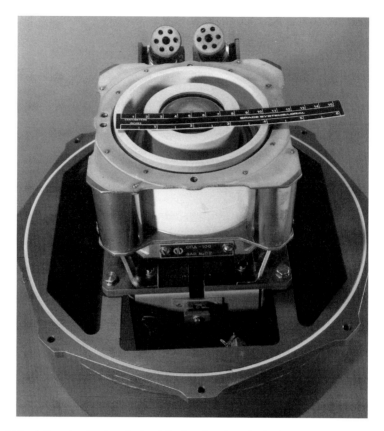

Figure 5.11. A Russian SP-100 device. The insulated cavity is shown with a ruler across the 100-mm diameter. The coils that energise the annular S pole are visible, and the central coil for the N pole is inside the central insulator. Twin neutraliser cathodes are visible at the top.

80% – and they are presently the most versatile electric thrusters available. For a Mars mission, however, much larger units would still be required. Scaling such Hall thrusters up to larger sizes is not as difficult as with the ion thrusters, because there are no large area grids to be supported.

5.12 RADIO-FREQUENCY PLASMA THRUSTERS

Although not yet fully developed, there is another type of plasma thruster that uses radio frequency electromagnetic fields to ionise and accelerate the plasma. The process here begins with the gas – a mix of hydrogen and helium – being exposed to the electromagnetic radiation from an RF antenna that ionises most of the atoms. It then passes into a cavity with strong magnetic fields, where cyclotron resonance is excited by a high-power RF generator. The electrons and ions oscillate within the plasma and raise its overall temperature to a very high value. Temperature here is somewhat different from temperature in an un-ionised gas, because the ions and electrons behave somewhat differently from neutral molecules at high temperatures. Nevertheless, the assemblage of ions and electrons behaves rather like a gas, at high temperature and pressure. It then enters a *magnetic nozzle*, which behaves like a conventional nozzle in that it allows the hot, high-pressure plasma to expand and so to generate a high-velocity exhaust stream and thrust in the conventional way. The difference is that the nozzle is made up of a carefully shaped, strong, static magnetic field, and has no mechanical presence at all. The electrons and ions in the plasma are forced to travel along the diverging magnetic field lines, so that the plasma expands as it emerges from the engine. The thrust is generated by the reaction of the ions on the magnetic nozzle. This type of nozzle has also been proposed for *fusion*-powered rocket engines when controlled fusion becomes practical. The RF plasma thruster partakes of some of the characteristics of the electro-thermal thruster, in the sense that the thrust and exhaust velocity arise from cooling and expansion of a hot substance. The difference is that for the plasma engine, the material forming the exhaust stream is a plasma and not a neutral gas, so that it can be expanded with a magnetic nozzle. Such engines, if they can be fully developed, promise to allow very high power thrusters, because the heating and expansion takes place in the plasma itself, and nowhere does the plasma touch the wall of the engine; and, of course, there is no physical nozzle. This means that high power can be dissipated without risking the integrity of the engine at the very high temperatures achieved.

The modern developed version of this engine is called **VASIMIR**, and a possible scheme for a fully functional thruster is shown in Figure 5.12.

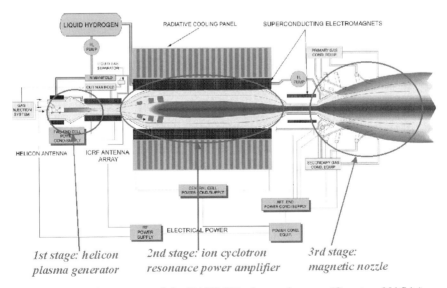

Figure 5.12. The concept of the **VASIMIR** plasma thruster. (Courtesy NASA.)

5.13 ELECTRIC PROPULSION IN USE TODAY

Most space vehicle propulsion is still chemical, despite the rapid advance of electric thrusters. There will always be a need for chemical propulsion in space, mainly for high-thrust applications; and launchers will, of course, always need to be powered by chemical rockets. There are two areas of space activity in which electric propulsion comes into its own: planetary missions, and station-keeping for communications satellites. In both cases it is the reduction in the amount of propellant required to execute a particular set of space manoeuvres or velocity changes that gives electric thrusters their advantage. We have seen that a Mars expedition makes huge demands on propellant, and electric propulsion could well help here. But electric propulsion has only been a practical proposition worldwide for less than ten years. There are a few commercially available electro-thermal thruster systems for communications satellites, and these are in use. There is an increasing availability of ion thruster systems for communications satellites. These systems have low thrust, but this is well adapted to station-keeping applications, and because of their higher exhaust velocity they generate a major saving in propellant over thermo-electric thrusters. The ion thrusters are smaller versions of the NSTAR engine used on Deep Space 1.

The Artemis spacecraft – a joint venture of the Italian Space Agency and ESA – is an example of an ion thruster used for station-keeping and orbital manoeuvres. Intended as an experimental station-keeping device, the ion thruster was used to save the mission after a launcher malfunction placed it in the wrong orbit. Electric propulsion changed a near-disaster into success. The orbital change was very slow, of course, and took many hundreds of circuits of the Earth to achieve; but the high fuel efficiency of the ion thruster enabled manoeuvres that would have been impossible using chemical propulsion. The spacecraft was launched in July 2001, and it was immediately apparent that the satellite had entered the wrong orbit. The upper stage of the Ariane V launcher had cut out prematurely, and the apogee of the orbit was only 17,000 km instead of the 36,000 km required for geostationary operation. Some of the chemical propellant intended for orbit circularisation and station-keeping was used to place the satellite in a 30,000-km circular orbit – the best that could be achieved. The ion propulsion system was then used to raise the apogee to 36,000 km. It was a very slow process, with an increase in the apogee of only 15 km per day; but it worked, because sufficient chemical propellant had thereby been saved to circularise the orbit.

There were two types of ion thruster on board, and one of them is shown in the photograph in Figure 5.13. They operated at a maximum power of 600 W, and had an exhaust velocity of 30 km/s. The thrust was only 2 mN, but acting over six months or so, the orbit change was achieved using 40 kg of xenon propellant.

Another European example is the PPS 1350 Hall thruster, to be launched on SMART 1 – an ESA lunar probe designed to test planetary mission technologies. This operates at 1.5 kW, and has a thrust of 92 mN and an exhaust velocity of 18 km/s. It will be used to raise the spacecraft from low Earth orbit to a lunar flyby, followed by deep space exploration to encounter an asteroid.

Figure 5.13. One of the two types of ion thruster used on the ESA Artemis mission to raise the apogee from 30,000 km to 36,000 km. (Courtesy ESA.)

5.14 USING ELECTRIC PROPULSION FOR A MARS EXPEDITION

While the major saving in propellant that arises from the use of electric propulsion is very promising for a Mars expedition, the low thrust is clearly a major problem. Currently available electric thrusters can be used very acceptably for propelling robotic spacecraft to the planets, and all the major space agencies are urgently developing electric propulsion for this purpose. For these missions the slow acceleration is acceptable; and, moreover, the spacecraft are not very heavy compared with those required for a human expedition. But it would be unacceptable to subject a human crew to an extra, say, six months in a long slow acceleration away from Earth, particularly as most of this period would be spent in the Earth's radiation belts – an environment very dangerous for human beings. Electric propulsion could be used successfully for cargo missions to Mars, during which the extended acceleration times would be acceptable. This seems to be the best way of using electric propulsion; by combining it with the occasional use of chemical propulsion for critical, high-thrust operations. (More details of this approach are included in Chapter 7.)

The main challenge for those advocating the use of electric propulsion for a Mars expedition is one of thrust, and therefore power. Much larger thrusters are required to accelerate the hundreds of tonnes needed for a Mars expedition; but these could be developed, and the Hall thruster shows considerable promise as a high-thrust device. But high-thrust requires high power, and this places demands on the power supply which may not be possible with solar panels, because on moving away from the Sun, the available power decreases with the square of the distance. As observed earlier, some 24 tonnes of solar panel at Earth orbit would be required for 1 MW of electrical power. Apart from the unwieldy nature of 3,000 m^2, the mass of 24 tonnes adds considerably to the dead weight of the spacecraft, and, moreover, has to be raised from Earth's surface into low Earth orbit – a complete Space Shuttle load.

However, there is another option for the provision of electrical power: nuclear fission.

5.15 NUCLEAR ELECTRIC POWER GENERATION

Nuclear energy is already in use for many of the deep space missions that venture beyond Mars orbit, in the form of radioactive thermoelectric generators (RTGs). Fundamentally, they utilise the heat generated by radioactive decay to produce electricity, via the thermoelectric effect. It is a rather inefficient process, but beyond Mars orbit it is currently the only method of generating sufficient electricity to meet the spacecraft's needs. Such power generators have been in use since 1961; early examples used lead–tellurium thermoelectric converters, and produced only a handful of watts of electrical power. The early deep-space missions such as Pioneer and Viking had RTGs that produced 25 W and 35 W respectively; the generators on the Voyagers produced 128 W, using silicon–germanium converters; and the most modern examples produce 285 W. Pu^{238} is used as the heat source, in the form of plutonium oxide, which is a stable, strong and refractory material that can take the high temperatures involved, and will not shatter or release dust on impact. The operating temperature is 1,235° C. To produce electrical power, the thermoelectric converter requires a temperature drop across it, and the waste heat has to be radiated away to space. This is enabled by using a radiator operating at a few hundred degrees centigrade. Because of the inevitable inefficiency of the conversion process, the thermal power produced is much greater than the electrical power; it is rather like a heat engine, in which the efficiency depends on the temperature drop. The latest versions have an efficiency of about 6%: 285 W of electrical power is produced by more than 4 kW of heat. Such 'high-power' generators weigh more than 50 kg, and the specific power is 5 W/kg. Compare this with 44 W/kg for solar panels. RTGs are only used because sunlight is so faint beyond the orbit of Mars.

The requirement for a Mars expedition is a power supply that has a much higher power per unit mass than an RTG or a solar panel; and the only solution is to use nuclear fission. A few kilogrammes of uranium 235 (U^{235}) can generate more or less unlimited power. There are, of course challenges, associated with the use of nuclear fission, that have so far limited its practical use in space to a few experiments during the 1960s. The main technical challenges are radiation levels, safety, and the efficient conversion of the heat generated by fission into electricity. And there is also a more subtle challenge: the moral and environmental argument relating to the use of nuclear power in space.

Nuclear reactors for mobile use are a well-established technology, and they are used in nuclear submarines and specialised surface ships such as ice breakers. Such vessels have to be independent of the shore for fuelling, and nuclear fission is ideal for providing continuous power over an extended period. In space, the mass of the reactor also becomes important. The mass of the core itself, and the fissile material, is only a few kilogrammes, which is very small. The main mass will be the system for extraction of heat from the core, and the conversion of that heat into electricity; and, of course, shielding will also be the necessary to protect the crew from radiation.

5.16 NUCLEAR BASICS

Any kind of nuclear power supply provides energy by converting one type of atomic nucleus into another. In doing so, some of the energy that holds the nucleus together, against the mutually repulsive force of the protons, is released, and can be used as heat. Radioactive decay is a natural process whereby heavy elements, that are unstable, decay into somewhat lighter elements, with the release of some energy in the form of fast-moving particles and photons: the well-known alpha-, beta- and gamma-rays. Typically, the nucleus decays into an element one or two places down the periodic table. This process is characterised by a *half-life* – the time during which half of the atoms in any piece of the element will have decayed. The shorter the half-life, the more the power generated. Plutonium has a half-life of 87 years, against radioactive decay; and enough heat is released to make it 'pleasantly warm to the touch', according to one foolhardy pioneer. It is this heat that is used in the RTG.

Atoms of U^{235}, and of some other elements, can also decay by fission. This is different from radioactive decay in that the nucleus splits more or less in half, rather than releasing one or two protons to shift down the periodic table by a few places. Fission releases much more energy from the split nucleus, because the change in atomic weight, from the original nucleus, to the fission fragments, is much larger. Fission is different from radioactive decay in that a catalyst is needed to cause fission; the half-life of an isolated uranium nucleus *against fission* is more than the age of the Universe. However, in the presence of neutrons the probability of fission becomes very high indeed, and a great deal of energy is released. Thus fission does not take place unless the environment of the uranium nuclei is favourable to the process. The nuclear reactor is a system that creates these special conditions. Outside the reactor, the nuclear material is inert and harmless. The fuel in a nuclear reactor becomes radioactive only after it has been used, as it is the fission fragments that are radioactive, not the uranium itself.

The basic fission process is the absorption by a nucleus of a neutron. Fission arises naturally at a very low rate because of neutrons produced by cosmic rays. In the natural environment the rate is very low. However, it is possible, under special circumstances, to produce a chain reaction in which fission becomes much more frequent. This is because, in splitting, the uranium nucleus releases neutrons. The absorption of a neutron by another uranium nucleus makes it unstable against fission, so that it splits, and in turn releases more neutrons. The number of nuclei that split, at any time, depends on the density of neutrons in the material. Clearly, the bigger the piece of material, the more chance a neutron has of encountering another nucleus and inducing a split. The well-known 'critical mass' is just the amount of material required to make sure that this happens. Neutrons are sub-atomic particles that have the same mass as a proton, but no electric charge. This absence of charge gives them a huge range in matter, and they have no way of interacting with another atom unless they happen to collide with the nucleus itself; and such interactions are rare. This is the reason for the huge shields around atomic power stations. Great thicknesses of material are required to make the area safe.

The large reactors used for power stations contain the neutrons, and secure the reactions necessary for stable fission by interspersing the fuel with a moderator that slows down the neutrons and provides them with more chance to interact with uranium nuclei. The moderator can be made up from carbon blocks or, indeed, be just water. This works well with natural uranium, which contains less than 1% of isotope U^{235}. If enriched uranium, containing more U^{235}, is used, then less moderator is required and a much smaller reactor can be made, because these nuclei have a higher probability of capturing a fast neutron. The escape probability for neutrons from such a small volume is much larger, so a *reflector* is used to bounce the escaping neutrons back into the fissile material. Carbon can be used, but beryllium is more suited to use in space, being of very low density and very efficient at scattering the neutrons. It should be noted that the reactions that make a nuclear reactor work are different from those used in a bomb. A bomb exclusively contains material that will respond to fast neutrons, otherwise it would melt rather than explode; and melting, rather than an explosion, is indeed what happens if a reactor runs out of control.

So, to produce electricity from fission in space, a small reactor containing enriched uranium with a few percent of U^{235} can be used, surrounded by a beryllium reflector and the necessary shielding. But how is the power output from the reactor controlled? The rate of fission depends on the density of neutrons in the reactor. If there are fewer than one per fission, then the reaction will die out; and if there are greater than one per fission, then more and more atoms will split and the power output will increase indefinitely. The reactor will not explode, because the energy release is still relatively slow, but it will melt, as in the case of the Chernobyl accident. To prevent this, materials are used that absorb neutrons efficiently, one of which is cadmium. By arranging holes in the fissile material, through which rods of cadmium or a similar material can be inserted, the neutron flux within the reactor can be absorbed and the reaction stopped. If the rods are removed then the reaction starts up again, and by partially removing the rods the reaction can be stabilised. The reflector, of course, can also be used in this way. If the reflector is removed from the outside of the fissile core, then neutrons escape and the reaction stops. Replacement of the reflector restarts the reaction, and again a partial covering will produce a stable power output.

A very important difference between an RTG and a nuclear reactor is that the reactor can be switched on and off, whereas the RTG starts working the moment it is assembled. There is no need to launch with an *active* nuclear reactor. It can be quite inert, and produce no heat or radiation, up to the moment that it is switched on in orbit. As mentioned above, the fissile material is not radioactive, and so an accident during launch would produce little danger even were the reactor to disintegrate. RTGs have to be constructed in a very robust way to ensure they do not disintegrate in an accident, because the plutonium is both poisonous and radioactive.

There are two other issues to be addressed with respect to nuclear power plants in space. The first of these is extraction of heat from the core. The energy released by fission is initially in the form of fast-moving nuclear fragments, and radiation. These are absorbed in the surrounding material, and their kinetic energy is converted to heat. If this heat were not removed, then the core would melt; besides, the heat is required to generate electricity. Water, air and carbon dioxide can be used, circulated in the spaces between the fuel rods, to extract the heat. For use in space, liquid metals produce a more compact system, because of their high conductivity. Sodium – with a melting point of 98° C – is used on the ground, but lithium is preferred for use in space because of its lower density and high boiling point, 181° C. The cooling system is not unlike the cooling of the cylinder block in petrol engines: the liquid metal flows through cavities in the core and carries the heat to the outside, where it can be used. It may be obvious that the temperature of the liquid metal can be used in a feedback system to control the positions of the control rods. Operating temperatures for the core can be quite high: more than 1,000° C. Similar temperatures arise in the core of an RTG, but the difference is in the power levels. The most powerful modern RTG produces 4 kW of heat, but even in 1965 the SNAP 10-A reactor was producing 40 kW. A more recent Russian space reactor, TOPAZ, used from 1987 onwards, produced 150 kW of heat; and planned reactors for the near future are sized for up to 0.5 MW. Thus the cooling of the reactor core is much more demanding than the extraction of heat from an RTG. Liquid metal is useful for this application because it can be pumped round the reactor core using electromagnetic fields alone, with no moving parts. It is therefore comparatively simple to keep the core cool and bring the heat to the outside.

The conversion of heat into electricity requires either a thermoelectric system or some kind of mechanical generator, converting heat to motion via a working fluid. Thermoelectric systems – even the most modern – have only 6% efficiency. Spacecraft engineers hate moving parts, and until now thermoelectric conversion has been used, with the accompanying low efficiency for both RTGs and for space nuclear power systems. Future designs will have sealed heat engines – such as the Stirling engine or the Brayton turbine – to convert heat to work and hence electricity. With these, efficiencies of 30–40% can be achieved.

Even with the low efficiency of thermoelectric conversion, the high thermal power of reactors can produce much more electric power than RTGs or solar panels. The Topaz reactor produced 10 kW from 12 kg of U^{235}, in a reactor with an all-up weight of 320 kg. At 31 W/kg this is close to the specific power of the most modern solar panels; but the system is much smaller, and can generate power anywhere in the Solar System. As in many areas, the Russian space programme has advanced in different directions from the rest of the world. Reactor power supplies were used in a significant number of Russian spacecraft.

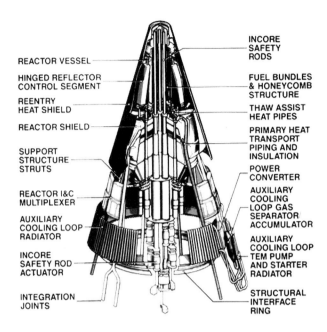

REACTOR VESSEL

HINGED REFLECTOR
CONTROL SEGMENT

REENTRY
HEAT SHIELD

REACTOR SHIELD

SUPPORT
STRUCTURE
STRUTS

REACTOR I&C
MULTIPLEXER

AUXILIARY
COOLING LOOP
RADIATOR

INCORE
SAFETY ROD
ACTUATOR

INTEGRATION
JOINTS

INCORE
SAFETY
RODS

FUEL BUNDLES
& HONEYCOMB
STRUCTURE

THAW ASSIST
HEAT PIPES

PRIMARY HEAT
TRANSPORT
PIPING AND
INSULATION

POWER
CONVERTER

AUXILIARY
COOLING
LOOP GAS
SEPARATOR
ACCUMULATOR

AUXILIARY
COOLING LOOP
TEM PUMP
AND STARTER
RADIATOR

STRUCTURAL
INTERFACE
RING

Figure 5.14. An early US design for an in-space nuclear reactor. The small fissile core is in the nose, while the reflectors, and the radiators, which provide the cold sink for the power converter, form the skirt. It is provided with a heat shield to prevent it from burning up should it re-enter the atmosphere. (Courtesy NASA.)

We will conclude this chapter with a brief description of the SP100 space power reactor, being developed by NASA As yet it has not flown, but many studies have been carried out with a view to making use of the concept for manned and unmanned exploration of the Solar System. The basic design is intended to provide 100 kW of power, equivalent to that produced by 300 m^2 of solar arrays. The system is modular, however, and can be configured to produce 0.5 MW of electrical power. The heat source is a fission core containing enriched uranium in the form of uranium nitride. Enriched uranium has a higher percentage of the rare isotope U^{235} than natural uranium, and the remainder is the common form U^{238}. In uranium that is sufficiently enriched, the chain reaction is further propagated by fast neutrons, and less moderator is needed; which makes for a compact reactor. The fuel is produced in the form of small disks of uranium nitride contained in sealed metal tubes. As in all fast reactors, the quantity of fuel carried is too small to become critical and sustain a fission reaction on its own. Fission is induced by an external reflector, made of beryllium oxide, that bounces the neutrons back into the fuel tubes. When the reflector is put in place, the neutron flux in the core rises, the reaction begins, and heat is generated. When it is removed, the neutrons escape, the reaction dies away, and the reactor cools down. An additional safety feature is the presence of neutron-absorbing control rods, which are in place during launch and whenever the reactor is switched off. These need to be withdrawn before the reactor can become critical,

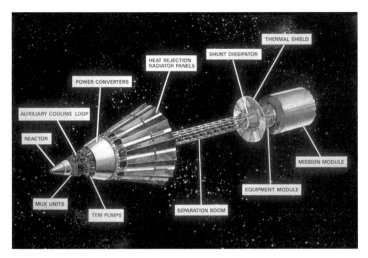

Figure 5.15. An early design for a spacecraft with nuclear-electrical generation. The reactor is like that shown in Figure 5.14, with the radiators forming the extended skirt. It is separated from the spacecraft proper by the boom, to reduce radiation damage to the electronics and equipment. (Courtesy NASA.)

even with the reflector in place, and there is therefore a double lock-out to prevent accidental initiation of the fission reaction.

The heat must be removed from the core, both to generate electricity and to prevent the core from overheating. This is achieved by using liquid lithium circulating in metal pipes that thread through the core of the reactor. The lithium is pumped by electromagnetic pumps, and operates at 1,000° C. When the reactor is turned off it takes several days to cool down, because the high flux of neutrons generates short-lived radioactive isotopes in the material of the reactor. There needs to be an auxiliary cooling system to ensure that this heat can be safely removed from the core. The outer heat exchanger on the lithium circuit contacts the hot junctions of the electrothermal generator, which uses silicon and germanium–gallium–phosphide. The cold junctions of the generator are maintained at about 500° C by a second lithium loop that conducts the heat away to a radiator made up of heat pipes containing potassium as the working fluid. The heat is eventually radiated into space.

The neutrons, and some gamma-rays, which escape from the reactor are, of course, dangerous – as much for the electronics as for any humans in the vicinity – and for this reason the reactor has to be shielded. It is not necessary to shield all of it, however, because in most directions the radiation would pass harmlessly into space. It is only the directions toward where the electronics or the crew will be that need to be shielded. Considerable surface area is also required for the radiators which carry away the waste heat. The optimal mass solution is to place the reactor on a boom that separates it from the main spacecraft. There is then plenty of room for the radiators, and only one side of the reactor needs to be shielded. From the other sides, radiation escapes harmlessly into space.

Such nuclear fission electric power supplies will play a big part in high-thrust electric propulsion systems. Above about 30 kW, fission generators are lighter than solar panels – even with the necessary shielding and safety systems. Beyond the orbit of Mars they are the only high-power solution, as, of course, RTGs can only provide low power. Irrespective of its use for electric propulsion, fission power will be required on the surface of Mars to meet the requirements of a human expedition. It will therefore be impossible for a Mars expedition to proceed using only solar panels to provide electricity. There are, of course, those who quite honestly hold moral and environmental objections to nuclear power in any form; and the same people might also object to a Mars expedition. It is important, however, to realise that these objections – however valid and self-consistent – are not founded on scientific arguments. Nuclear power is safe. The number of human beings killed by nuclear power is infinitesimally smaller than the number who have died as a result of fossil fuel exploitation and wars fought over reserves of fossil fuel. It is, of course, right to insist on the highest standards of safety and public protection; but it is not right to prevent the use of nuclear power in space, based on spurious objections that have no grounding in fact.

6

Fission thrusters: the logical answer?

In the previous chapters it was shown that an expedition to Mars, using chemical propulsion, requires an extremely large amount of fuel – some 5,000 tonnes, using our basic expedition model with storable propellants – and that this can be reduced to 1,600 tonnes by using cryogenic propellants. It was also shown that electric propulsion, with its very high exhaust velocities, can in principle produce major savings in propellant, perhaps to just a few hundred tonnes; and that electric propulsion produces a very low thrust, involving very long acceleration times that may not be acceptable to a human crew.

Given the disadvantage of the low thrust of electric propulsion, is there any other way to achieve a small propellant requirement, and still have a high-thrust system? We know that direct chemical propulsion cannot do this; but we also know that by separating the energy supply from the propellant, as in the electric thruster, high exhaust velocities can be produced. However, the conversion of heat into electricity is inherently inefficient, and the conversion of sunlight into electricity is even more inefficient. The electro-thermal converters currently proposed for nuclear generators have only 6% efficiency, and if mechanical converters can be made sufficiently reliable, this could be improved to perhaps 30%. Thus it is inefficient to use nuclear power to make electricity and then use the electricity to accelerate a very high-velocity exhaust stream, as much of the original energy is wasted as heat radiated into space. We should, however, not forget the importance of exhaust velocity, as high exhaust velocity is vital in order to have a reasonably low propellant mass requirement. It may be worth the inefficiency of conversion of heat into electricity if the 6% can be used to generate a high-velocity exhaust stream.

It is a peculiarity of space propulsion that efficiency and power are sometimes less important than exhaust velocity. The very high exhaust velocity of electric thrusters produces a potentially enormous saving in propellant; but as has been shown, the low thrust may produce such protracted acceleration times as to be a crippling disadvantage. The very high thrust chemical engines require huge amounts of propellant, but their power output is such that acceleration times are negligible. Perhaps something in between would be more useful for a Mars expedition: a moderately high thrust engine, with a moderately high exhaust velocity. The exhaust velocity must be significantly more than that of a liquid hydrogen–liquid oxygen engine, and the thrust must certainly be in the high kiloNewton to megaNewton range.

This requirement was understood as long ago as the early 1950s, before the first satellites were launched. At that time many people were interested in space travel, as it was then known, and the main objective was to explore the planets. The physics of a Mars expedition were the same as now, and the need for fuel economy and short acceleration times was appreciated. At the same time, nuclear power was seen as the solution to most energy supply problems, and we were all going to have reactors in our towns and cities, to provide unlimited free electricity. So, it was not unreasonable that the idea of a nuclear rocket should be advanced, and a programme to put it into practical execution, devised and followed. Development therefore proceeded in the United States and in the USSR. In America it stopped during late 1960s, overshadowed by the success of the Apollo programme, using just chemical propulsion; but in the USSR, design work continued until 1989. Both nations included test firings of their engines, although none were actually flown; and then it all stopped. Nobody would be voyaging to Mars in the near future; and the funding margins – within which such 'blue skies' developments are pursued – were decreasing everywhere. But now – when an expedition to Mars is again on the agenda, and

Figure 6.1. An artist's concept of a nuclear thermal rocket capturing into Mars orbit. The nuclear rocket engine is being used to slow down the spacecraft from its hyperbolic approach orbit. The disk near the engine is the shield for the crew. Other directions can be left unshielded. (Courtesy NASA/JPL.)

chemical propulsion is beginning to be understood as limiting all forms of space exploration – nuclear propulsion is re-emerging as, perhaps, essential technology.

There are two approaches to nuclear-powered propulsion. One of these was described in the last chapter: a high-power electric thruster – perhaps a cluster of Hall effect plasma devices – running at more than 1 MW, and powered by a nuclear generator. This type of system will have the high exhaust velocity required for low propellant consumption, but perhaps insufficient thrust. The other approach is to use fission power to drive the exhaust stream directly, which, in terms of efficiency, saves losses in conversion of heat into electricity, and in the electric thrusters themselves. This latter idea drove the nuclear rocket programme during the 1950s and '60s. This kind of approach is known as *nuclear thermal propulsion*, or NTP, to distinguish it from *nuclear electric propulsion*, or NEP. Both appear in current ideas for a Mars expedition, just as they did fifty years ago.

6.1   THE PRINCIPLE OF NUCLEAR THERMAL PROPULSION

This principle is very simple, and is identical to that of the electro-thermal thruster or resistojet: to heat the propellant directly by the fission reaction. A propellant enters the core of the reactor, is heated by the hot fuel elements, and is then expanded down a conventional nozzle to develop thrust. The nuclear thermal rocket has similar properties to the electro-thermal rocket: the propellant cannot be heated beyond the service temperature of the fuel elements, and will probably be below 3,000° C. Most of the advantage in terms of exhaust velocity will arise from the use of an exhaust of low molecular weight. We may also expect similar exhaust velocity: about 10,000 m/s, in the case of hydrogen as a propellant. The big difference between the NTR and the resistojet lies in the power dissipation. While the electro-thermal rocket will use one or two kW of electrical energy, the nuclear thermal rocket has at its disposal perhaps several hundred MW of power – in the same performance region as large liquid hydrogen–liquid oxygen engines. The NTR develops a thrust similar to that of a cryogenic engine, but with approximately double the exhaust velocity, if hydrogen is used as the propellant. The general principle is illustrated in Figure 6.2.

Hydrogen propellant is used because it provide the lowest molecular weight for the exhaust. The flow of hydrogen cools the reactor core, and is itself heated to a high temperature by contact with the fuel rods. The hot gas then expands down the conventional nozzle to generate a fast exhaust stream. Russian and US ground trials produced exhaust velocities of about 9,000 m/s – twice that achievable with the best chemical rockets.

The hydrogen is, of course, carried as liquid hydrogen in the propellant tanks, and is used to cool the nozzle, as in a conventional rocket, before being injected into the reactor core. In many respects a nuclear thermal engine is therefore similar to a chemical engine: turbo-pumps, to supply liquid hydrogen, at high pressure, to the reactor chamber, are required, and the nozzle and injectors will be very similar to those of, for example, the Space Shuttle main engine. The main difference is that the heat is generated by nuclear fission, not by chemical reactions. Hydrogen flow can be

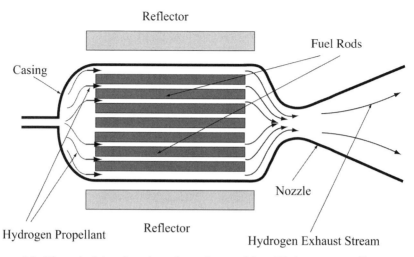

Figure 6.2. The principle of nuclear thermal propulsion. Hydrogen propellant enters the reactor chamber and flows between the hot fuel elements. It then expands in the normal type of nozzle, to create a fast exhaust stream. Also shown is the reflector that controls the escape of neutrons from the fissile core.

controlled, as in a conventional engine, by starting or stopping the turbo-pumps. The fission reaction can be turned on and off by deploying or removing the reflector, and by operating the control rods.

The shutting down of a nuclear thermal engine is different from the situation with a chemical engine. In a conventional rocket engine, shutting off one of the propellants will stop the engine, by preventing the chemical reaction that generates heat. It may also be necessary to flow some of the other propellant through the engine for a few seconds while it cools down. Fission, on the other hand, is not a chemical reaction, as it requires a high flux of neutrons to maintain it; and when the reflector is withdrawn, the fission will stop. Unfortunately, the engine does not then immediately cease to operate; neither does it cool down. Many of the fission products – the fission fragments from the split nuclei – are created as unstable (radioactive) nuclei, while at the same time the high flux of neutrons can cause nuclear transmutation, converting stable nuclei in the structure of the reactor into unstable radioactive variants. These elements invariably have short half-lives, of hours to a few days. While the engine is operating with full power, the decay of these short-lived radioactive nuclei contributes to the heating of the propellant. Once the reactor is shut down, and the fission is stopped by scarcity of neutrons, these nuclei continue to decay until they are all converted back into relatively stable elements. The engine needs to be cooled during this phase, otherwise enough heat would be generated to damage or even melt the fuel rods. The turbo-pumps therefore need to continue to supply hydrogen to the engine for a few days after the shut-down. The quantity required is very much less than that used during operation, but the emerging hot hydrogen will still generate thrust. After a burn of the nuclear thermal engine there is a period during which the engine continues to develop a low level of thrust; and this must be taken into account when planning the manoeuvre and calculating the amount of propellant required for each manoeuvre.

While the engine is operating it produces a large flux of neutrons and gamma-rays, and is unsafe to be near. As in the case of an electrical power fission reactor, shielding needs to be provided on the sides nearest the spacecraft, to protect the crew and the electronics. Unlike the electrical power reactor, which operates continuously, the nuclear thermal engine is only 'on' during the burn. Once a few days have elapsed after the burn, the engine is relatively safe, and extensive shielding is no longer needed. There is a way to take advantage of this fact, and at the same time provide significant safety factors in the shielding of the crew. Liquid hydrogen is a very good neutron stopper. As mentioned before, neutrons, having no electrical charge, do not interact with atoms unless they hit the nucleus. Even then they may just bounce off, with their energy almost undiminished. This is why they have such a long range in matter, and are difficult to stop. It happens that the best way to slow them down is to use a material with a low atomic weight; because, in a billiard-ball type collision between particles, most energy is extracted from the impacting particle if the mass of the target particle is the same. Typical solid shielding materials for neutrons are paraffin wax or polythene, which contain a large amount of hydrogen plus carbon atoms, and lithium hydride, which contains the low-mass element lithium (atomic mass 7) and hydrogen. Liquid hydrogen is therefore the archetypal neutron shield, being only hydrogen with a single proton as its nucleus. One way to protect the spacecraft during the operation of the engine is to place the propellant tanks between the engine and the crew compartment, and leave enough hydrogen in them to provide the necessary shielding after the engine is shut down. This may be the most mass-economic way to provide the shielding. It means, moreover, that the tanks cannot be allowed to run dry during operation, and that sufficient acceleration must be maintained to keep the liquid hydrogen at the base of the tank. In weightless conditions there is no reason for it to do so, and it might all migrate to the middle of the tank or to one side, leaving the crew compartment unshielded.

Before the engine is first operated it will be completely safe, and non-radioactive. This is a very important point, in that the conveyance of the nuclear thermal engine into orbit can be carried out with it switched off, and not a radioactive hazard. After it has been operated there will be some residual radioactivity, but not enough to be a danger to the crew during, say, transit to Mars. Once the engine has become redundant, and nearly all propellant for its use exhausted, it needs to be disposed of safely. It is therefore detached from the rest of the vehicle and, using the remainder of its propellant, is placed in a safe orbit round the Sun. A similar procedure is already used, under international law, for unwanted chemical boosters and spacecraft in Earth orbit. Sufficient propellant must be retained either to place the vehicle in a high or solar orbit, or to direct it into the atmosphere to be burned up. The latter option is not available for vehicles containing any nuclear or radioactive material.

6.2   NUCLEAR ROCKET ENGINE DESIGN

Most of the research and development of nuclear engines was carried out during the 1960s, and we therefore have to examine the work of that period to appreciate how a modern nuclear engine will work. The practicalities of a nuclear engine are simple to understand. The primary determinant of performance will be the maximum temperature to which the propellant can be heated. The exhaust velocity – which determines the amount of propellant required, or possibly the journey time, if there is enough spare delta-V to allow a short transit trajectory – is simply determined by the molecular weight of the propellant and the temperature. Hydrogen is the lowest molecular weight that can be used, and so the temperature becomes the main issue.

The maximum temperature in a chemical reaction – for example, between hydrogen and oxygen – in a chemical rocket, is more or less fixed for any given combination. It partly depends on the pressure, but the primary determining factor is simply the nature of the participating molecules. Typical combustion chamber temperatures are about 3,500 K. Above this temperature, at combustion chamber pressures, the reaction starts to reverse, and the water molecules break down into hydrogen and oxygen, absorbing heat. There is, therefore, a natural limit; which in the case of nuclear fission is so high that temperature hardly has any meaning. The two fission fragments each carry away the best part of 100 MeV. The mean kinetic energy of a molecule at 3,500 K is only 0.3 eV; and the 100 MeV corresponds to about 100 million K. These fragments, of course, slow down, and heat up the material surrounding them; but in practical terms, the temperature in a fission reactor can become arbitrarily high. At some point the structure itself will melt or break down, and the conditions for fission may cease; and so the temperature depends on how much cooling is applied. In a sense, this is like the electro-thermal thruster. If the gas flow is turned down, then the temperature will rise. For the electro-thermal thruster the heating element will eventually melt; for the fission reactor, the structure holding the fuel, and the fuel itself, will eventually melt if the cooling is turned off. This type of event occurred at Chernobyl, where the uranium metal melted and collected at the bottom of the reactor, but still retained critical mass so that fission continued.

Clearly, the highest safe temperature will be the temperature beyond which the structure, holding the fissile material, loses its integrity. Nuclear thermal rocket engines do not incorporate uranium metal, which melts at 1,400 K. For use in rockets, uranium must be in a form that can withstand a very high temperature without melting or softening. Uranium compounds can be quite refractory. Uranium oxide melts at 3,075 K, and uranium nitride can withstand temperatures up to 3,160 K. It does not matter that the uranium is in chemical combination, as fission takes place at the nuclear level, and the combination of atoms is of no consequence. To build the fissile core of a nuclear thermal rocket, therefore, the uranium must be present in a compound that can withstand the high temperature, and be held in a structure with similar thermal resistance. Table 6.1 lists some materials, with their melting points, used in reactor cores.

Table 6.1. Melting points for common reactor core materials

Type of material	Material	Temperature (K)
Fuel metal	Uranium (U)	1,400
Fuel compound	Uranium nitride (UN)	3,160
	Uranium dioxide (UO$_2$)	3,075
	Uranium carbide (UC$_2$)	2,670
Refractory metal	Tungsten (W)	3,650
	Rhenium (Re)	3,440
	Tantalum (Ta)	3,270
	Molybdenum (Mo)	2,870
Refractory non-metal	Carbon (C)	3,990 (sublimation)
	Hafnium carbide (HfC)	4,160
	Tantalum carbide (TaC)	4,150
	Niobium carbide(NbC)	3,770
	Zirconium carbide (ZrC)	3,450

Even if the structure and fuel elements retain their integrity and do not melt, it is also advantageous if they do not shed material into the exhaust stream. The hydrogen passing through the core does not itself become radioactive, and is not affected by neutrons. The rocket exhaust would be safe and non-radioactive if the core material could be prevented from entering the exhaust stream. All of this places a requirement on the fuel elements to be refractory, and not to shed material in the hot high-pressure hydrogen stream that flows past it.

Efficient heat transfer to the hydrogen requires that the fissile material should have a large surface area, and so the core must consist of a structure with many channels, through which the hydrogen flows. Different approaches have been adopted in the various types of nuclear engine developed over the years.

6.3   PROPELLANT DELIVERY

In liquid-fuelled rocket engines, the propellant has to be delivered to the combustion chamber with a high flow-rate and at a pressure sufficiently high to overcome the pressure in the combustion chamber. This is normally achieved by using turbo-pumps, powered by a gas generator that burns part of the propellant in a separate chamber. For high-thrust engines this is the only way to deliver the propellant, in sufficient quantity, to the combustion chamber. For a nuclear engine there is only one propellant, and so it is not possible to operate a separate gas generator. The requirement for turbo-pumps, however, still remains. The solution adopted is to use fission in the main chamber to provide the hot gas to power the turbine, as only one turbo-pump is required in a nuclear engine. Two approaches have been used. In one of these – the topping cycle – the hydrogen is first passed through the cooling channels in the nozzle and the walls of the reactor chamber, with the flow arranged so that it boils and becomes gaseous before leaving. This hot gas is then used to drive the

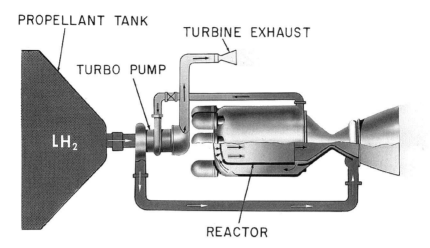

Figure 6.3. The propellant delivery system for a nuclear thermal engine. Hot gas from the reactor chamber drives the turbo-pump, and is exhausted after driving the turbine through an auxiliary nozzle to generate some extra thrust. The bulbous extensions on the left of the reactor chamber are the motors controlling the reflectors and control rods. (Courtesy NASA.)

turbine, before being exhausted to space or into the main reactor chamber. This method can be used for engines of moderate thrust, and is, indeed, used by some chemical rocket engines. Where high thrust is required, gas is bled off from the main reactor chamber at the full chamber pressure, and this gas is used to drive the pump. This is known as the hot bleed cycle. Low-power electric pumps can be used to flow the hydrogen while the reactor is starting up and until the pressure available from the cooling channels or the main reactor chamber is sufficient to operate the main pumps.

6.4   THE REACTOR CHAMBER

The reactor chamber replaces the combustion chamber of a chemical rocket. While the combustion chamber is kept as small as possible, the reactor chamber has to be big enough to contain the fissile core, and to be long enough to allow full heat transfer to the hydrogen flow. At its lower end it is connected to the nozzle, and at its upper end to the turbo-pump and hydrogen delivery system. It is cooled by channels in the wall, through which the hydrogen flows. It also has to contain the pressure at which the engine operates, and this may be quite high: more than 50 bar. This requires thick walls and end pieces. Even though it must contain the fissile core, it is surprisingly small (compared with ground-based reactors): 1 m in diameter, and about 1.5 m long. This is because of the use of enriched uranium containing extra U^{235}. Enriched uranium has a much higher neutron cross-section than natural uranium, and so the neutrons do not need to travel so far before interacting. The result is a very small fissile core.

6.5 THE REFLECTOR AND CONTROL RODS

By itself the core is non-critical – an important safety feature – and becomes critical only when the reflector is in place to bounce neutrons back into the fissile material. In Figure 6.2 the reflector is shown schematically as a cylinder surrounding the combustion chamber; but in engineering terms this is difficult to arrange. The same is true of control-rods, which need to be inside the reactor; and a gas-tight sliding joint is difficult to make, especially at high temperature and pressure. The two requirements are therefore combined in a control drum system. Half-cylinders of neutron absorber, such as boron carbide, and neutron reflector, such as beryllium oxide, are bonded together to make a composite cylinder. These control drums are mounted all round the core, inside the pressure vessel, and can be turned by electric motors. For full power operation the reflecting sides of all the cylinders are presented to the core; and for full shut-down the absorbing sides are presented. By adjusting the degree of rotation of the cylinders, the power level of the core can be controlled. Being so close to the core and inside the chamber, the control drums need to be cooled, and cold gaseous hydrogen is routed up through the shell containing the control drums before entering the core. This warms the hydrogen and keeps the drums cool enough to be safe. Some additional static reflecting and shielding material may be included in the reactor chamber. This, as well as the walls of the chamber itself, is heated by the radiation from the core, and needs to be cooled by hydrogen flowing through channels

6.6 THE CORE MATERIALS

There are many different configurations and compositions of the fuel elements in the reactor. Some designs use fast neutrons for fission and have no moderator incorporated, and require a high proportion of expensive U^{235}. Other designs that use cheaper, less-enriched, fuel make use of slow neutrons and require a moderator. Since everything has to withstand the operating temperature, the moderator is usually mixed into the fuel elements rather than being separate.

Figure 6.5 shows a typical early design using carbon as a moderator and uranium dioxide as the fissile material. The fissile material itself consists of enriched uranium oxide particles a few microns in diameter, suspended in a solid graphite matrix. The composite fuel and moderator is extruded into the hexagonal units, with nineteen holes down the length to carry the hydrogen. It can be seen that this brings the hydrogen into intimate contact with the hot graphite–uranium oxide fuel elements. The elements are made up into clusters of six, as shown, and then packed together to make the reactor core. In the largest core used – generating 4,000 MW of power – 4,086 elements were used to form a core about 1.8 m in diameter.

While carbon in the form of graphite is very refractory, with a sublimation temperature just under 4000 K, it is eroded very rapidly by the hot hydrogen flowing past; the hydrogen combines directly with the carbon to form hydrocarbons. If nothing is done, then the fuel element is quickly worn away and ejected with the

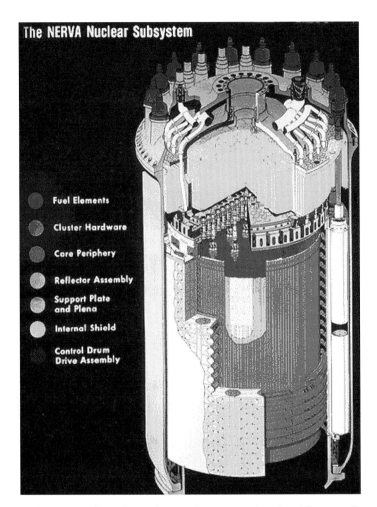

Figure 6.4. A cutaway view of a nuclear engine reactor chamber. The propellant enters through the central port at the top, and passes through fine tubes in the internal shield. It is then distributed to flow through the core. The complex of pipes at the top are for the cooling flow through the outer reflector and control drums. A control drum is shown at right. (Courtesy NASA.)

exhaust. This is a well-known phenomenon, the solution of which is to coat the whole surface of the fuel elements with a material – niobium, or zirconium carbide – resistant to erosion. These materials are refractory, and resist any erosion by the hydrogen.

6.7   EARLY DEVELOPMENTS IN NUCLEAR ROCKET ENGINES

Fuel elements such as this were used in the early Kiwi series, which developed up to 937 MW of power and a gas temperature of 2,330 K. The major problem with these

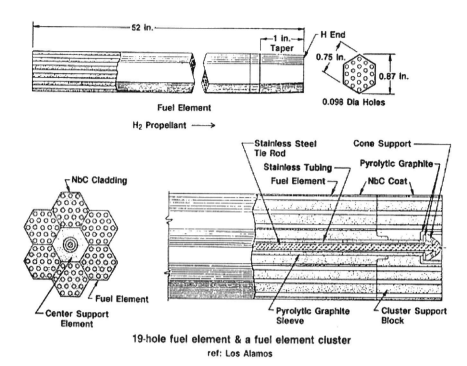

Figure 6.5. Fuel rods as used for the early Kiwi reactor, which was operated on the ground but never flown. Notice the many channels through which the hydrogen flows. (Courtesy NASA.)

reactors was damage to fuel elements caused by vibration set up by turbulence in the propellant flow through the holes. During tests, in fact, whole fuel elements became detached and flew out through the exhaust – which was not very reassuring! Despite the high strength of these elements, the huge amount of power being dissipated in the reactor can cause very powerful turbulence.

The Kiwi was not intended to fly (hence its name), but the later reactors operated with nozzles as ground-tested flight-type engines. In 1968 the Phoebus series operated at 4,000 MW, a temperature of 2,300 K, an exhaust velocity of 8,400 m/s, and a nominal thrust was 93 kN. This engine used enriched uranium carbide as the fuel – dispersed, as before, in a carbon matrix – and the protective coating was niobium carbide–molybdenum. The Phoebus reactor series was, in fact, designed to meet the needs of an interplanetary propulsion system – a human mission to Mars. During the late 1960s, a nuclear engine concept for the Space Shuttle was based on an engine called Peewee, which achieved a temperature of 2,556 K using zirconium carbide to coat the fuel elements.

Under the umbrella of a NASA programme called Nuclear Engine for Rocket Vehicle Applications (NERVA), begun in 1960, these engines were further developed towards a specification for real interplanetary missions. Among the (now more modest) requirements of this programme were a chamber temperature of greater than 2,360 K; a chamber pressure of 30 bar; a thrust greater than 17 kN; a total thrusting time of 600 minutes with sixty restarts; and human rated. In its final form, immediately before the programme was cancelled, most of these requirements had been met. In particular, a vertical test-firing of a more or less flight-representative engine was successful, with a temperature of 2,472 K and a power level of 1,140 MW.

There were also other engine designs developed, including examples using highly enriched uranium, which can become critical without the use of a moderator. The fuel consisted of a mixture of carbides – uranium carbide, niobium carbide and zirconium carbide – which do not act as moderators, as does carbon, but form a refractory fuel element that can be used in the same way.

In 1972 the programme was stopped, and efforts were devoted to the development of the Space Shuttle, with exclusively chemical propulsion.

Figure 6.6. A mock-up of a complete NERVA engine with propellant distribution system and complete nozzle. The turbo-pump exhausts can be seen near the top. (Courtesy NASA.)

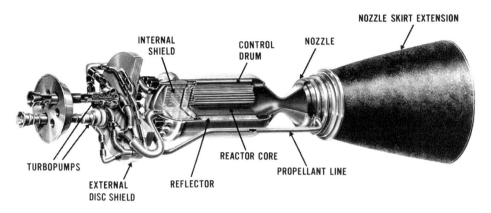

NOZZLE SKIRT EXTENSION

INTERNAL
SHIELD

CONTROL
DRUM

NOZZLE

TURBOPUMPS

REACTOR CORE

PROPELLANT LINE

EXTERNAL
DISC SHIELD

REFLECTOR

Figure 6.7. A cutaway drawing of the NERVA engine shown in Figure 6.6. (Courtesy NASA.)

6.8   A REVIVED NUCLEAR ENGINE PROGRAMME

Although the programme had been stopped, the knowledge was preserved, and was brought out and first reviewed for the 'Star Wars' programme; and now the same information is forming a background for new developments in nuclear engines for non-military purposes – particularly the exploration of the planets. At the same time, Russian technology and expertise has also become available. But much has changed since the tests stopped in 1972. There is little possibility of renewed testing in the open air, as public opinion would prevent it; and at the same time, materials science has advanced remarkably, and our ability to use computer modelling to determine thermal stress patterns and hydrogen flow patterns, among many other essentials, has advanced by an enormous amount. Much of the work carried out during the past seven or eight years has therefore been theoretical; but not in the sense of impractical, because advanced simulations can cover much of the progress that during the 1960s required extensive practical tests. Tests will eventually be required, however, but these will be carried out in enclosed facilities with exhaust scrubbing to remove any radioactive material from the hydrogen exhaust.

The new engine designs are based on the NERVA results, and some details are presented in Table 6.2. The main thrust of the development has been to increase the chamber pressure and the temperature. As far as pressure is concerned, this is a matter of using stronger materials for the pressure vessel. The final NERVA model used aluminium, but the use of high-strength steel, such as that used in the Space Shuttle booster casings, will allow for a much higher pressure. For flight, steel could be replaced with titanium for reduced mass. The other development is to increase the temperature of the fuel elements. The obvious move to all-carbide fuel, in order to prevent erosion at high temperature, leads to operating temperatures slightly above 3,000 K. This should present a significant advantage in exhaust velocity, with values as large as 10,000 m/s being within reach. Control of these engines uses the same

Table 6.2. Details of engine designs based on NERVA results.

Parameters	NERVA 1	Graphite	Composite	Carbide
		NERVA-derivative reactors (NDRs)		
Engine flow cycle	Hot bleed/topping	Topping (expander)		
Fuel form	Graphite	Graphite	Composite	Carbide
Chamber temp. (K)	2,350–2,500	2,500 2,350–2,500	2,700	3,100
Chamber press. (psia)	450	500 1,000	500 1,000	1,000
Nozzle exp. ratio	100:1	200:1 500:1	200:1 500:1	500:1
ISP (sec)	825–850 845–870	875 850–885	915 925	1,020
Engine mass* (kg)	11,250	7,721 8,000	8,483 8,816	9,313
Engine T:W** (w/int shield)	3.0	4.4 4.3	4.0 3.9	3.7

* w/o external disk shield
** Thrust:weight ratios (T:W) are 5-6 at 250 klbf level

Engine type	NRX XE	NERVA 1	New designs based on NERVA		
Fuel Rods	UO_2 beads embedded in graphite	UO_2 beads ZrC coat embedded in graphite	$UC_2 + ZrC + C$ composite	$UC_2 + ZrC$ all carbide	$UC_2 + ZrC + NbC$ all carbide
Moderator	Graphite	Graphite + ZrH	Graphite + ZrH	Graphite + ZrH	Graphite + ZrH
Reactor vessel	Aluminum	High-strength steel	High-strength steel	High-strength steel	High-strength steel
Pressure (bar)	30	67	67	67	67
Nozzle exp. ratio	100:1	500:1	500:1	500:1	500:1
ISP (sec)	710	890	925	1,020	1,080
Chamber temp. (K)	2,270	2,500	2,700	3,100	3,300
Thrust (kN)	250	334	334	334	334
Reactor power (MW)	1,120	1,520	1,613	1,787	1,877
Engine avail. (year)	1969	1972	2006?	–	–
Reactor mass (kg)	3,159	5,476	5,853	6,579	–
Nozzle, pumps etc. mass	3,225	2,559	2,559	2,624	–
Internal shield mass	1,316	1,524	1,517	1,517	–
External shield mass	None	4,537	4,674	4,967	–

Figure 6.8. Tests of a NERVA-type engine at Jackass Flats, Nevada. (Courtesy NASA.)

procedure as in the NERVA programme: cylinders with reflecting material in one half and absorbing material in the other half. They are rotated by electric motors to start and stop the engine.

While, at the moment, there is a good deal of discussion about NTP – and there is apparently a wish to develop such engines, and perhaps even a programme funded by the US government – there is as yet not a practical version ready for flight. But there is no doubt that such a version could be produced within a few years. It only requires a decision to be made. For our purposes, to examine how an expedition to Mars can be mounted, it is sufficient to have a rough idea of how well such an engine would perform, and how it would influence the achievability of a Mars expedition. The data given in Table 6.2 are sufficient to understand how such devices will work, and to enable us to grasp the nature of the special promises and challenges associated with a nuclear thermal engine. It is, of course, sad, from a scientist's point of view, to note that the NERVA 1 engine was ready in 1972, and that now, at the beginning of the twenty-first century, there is nowhere in the world, ready for use, a nuclear engine suitable for a Mars mission.

6.9 THE SAFETY OF NUCLEAR PROPULSION

One reason for the absence of a suitable off-the-shelf nuclear thermal rocket engine is the marked antipathy to all nuclear technology that pervades our society. It would be difficult to go anywhere in the world today and to not encounter people who object fundamentally and vehemently to such technology – in spite of the fact that in many countries a significant fraction of electrical power for domestic use is generated by nuclear power stations, and that there is a wide use of radiation for industrial inspection, medical diagnosis, and the treatment of dangerous cancers. While nuclear-powered submarines are not regarded as a universal blessing, most people regard domestic smoke-detectors as a good thing, saving many lives each year; and yet these contain americium 234 – a radioactive element that ionises the air to detect smoke particles.

The reasons for antipathy are not difficult to find. Many of my generation grew up more or less under the threat of nuclear war. Then, of course, the Chernobyl disaster, preceded by a number of less serious accidents, brought home the reality of the dangers of nuclear technology. We collected rainwater from the roof of the Physics Department in Leicester, in the UK, and easily detected radioactive iodine – fission fragments from Chernobyl. The fear of radiation is rational and deep-seated. The human senses have absolutely no way of detecting its presence; without equipment there is no way of determining whether or not one has been exposed to it; and disease and death, if they result from such exposure, may come as late as twenty or thirty years after the event. In many cases, particularly when low levels of radiation are concerned, the incidence of disease is random in nature. The threat to a specific individual is not quantifiable, and can only be referred to in terms of so many deaths per million, within twenty years, in the population at large. Another factor is the clear separation between those who work with and understand radiation – the scientists and engineers – and those who do not understand it and who do not work with it – the rest of the population. It is difficult to bridge that gap, and to convince those without specific knowledge, of the safety of what is being proposed. Enormous effort has been made by the US government to inform the public and to attempt to convince them that RTGs used in space are not a danger to those here on Earth; but there are still protests whenever a spacecraft containing one is launched, or even passes close to Earth on its journey elsewhere. In Europe, there is presently no hope of being able to use such devices without major public protest. It is clear, then, that there will be an uphill struggle to gain public approval for a nuclear rocket. The conditions of manufacture and of use will have to be strictly controlled.

The first essential is that the engine should not be fired in the atmosphere; it can only be used in space; and it cannot be an active stage on a launch vehicle. Secondly, during launch, when the inactive engine forms part of a payload or orbital manoeuvring stage, the engine must be safe in the event of a launcher accident. Thirdly, it must not be a danger to human beings, specifically the crew, while it is in operation. Finally, it must be disposed of safely when it is of no futher use. All of these aspects must be monitored and controlled to the highest standards of management and quality assurance, and the entire process must be open and subject to public scrutiny at any stage.

A legal and organisational model for the control and monitoring process exists in the system used to permit the use of RTGs in space vehicles. This is one of the reasons why their use outside the United States is very difficult: the safeguards and consensus that exists in the US have yet to be established elsewhere. The

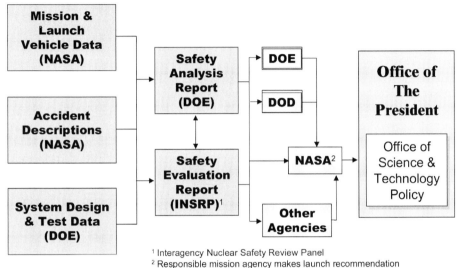

Figure 6.9. The safety diagram for approval of RTGs used by NASA. (Courtesy NASA.)

organisational model, shown in Figure 6.9, ultimately involves the office of the President of the United States. It is, of course, the detail of the engineering and quality assurance that makes the devices safe, rather than the legal procedures.

The principle that makes RTGs safe for use by NASA is the principle of total containment: the radioactive material must not escape from the device in any conceivable accident. The radioactive material – plutonium oxide – is formed into hard, inert ceramic discs. This material is very resistant to abrasion and formation of dust, has a high melting point, and will not burn. The discs are contained in iridium capsules; iridium is strong but malleable, and so the capsules will not burst or rip under impact. The capsules are in turn contained in graphite blocks, and the assemblies are then stacked together inside an aeroshell to protect the graphite during high-speed re-entry into the atmosphere. This 'belt and braces' approach has prevented release of radioactive material in the numerous tests and the few accidents that have involved RTGs.

Can such a process be applied to space nuclear reactors? There are two very important differences between the RTG and the fission reactor. The RTG is strongly and continuously radioactive – this is what provides the electrical power that it generates – and it remains radioactive for more than eighty years after manufacture. On the other hand, the fission reactor is not radioactive before first use, and in the quiescent state between uses it is mildly radioactive. It uses enriched uranium: a mixture of the naturally occurring isotopes, U^{238}, which has a half-life of 4.5 billion years, and U^{235}, which has a half-life of 0.7 billion years. Remembering that the activity depends inversely on the half-life, we see that, compared with plutonium with a half-life of 84 years, the activity of the uranium fuel in a nuclear rocket is negligible. The fissile material itself can be transported quite safely, and any accident resulting in the release of enriched uranium would not result in any significant radioactive contamination of the human environment. The second difference is that the reactor will only become active, to produce thrust and radiation, under very special conditions. It requires the presence of moderator and reflectors to work, and without these it is just an inert lump of material. These facts suggest that the conditions for safe use will be very different from those for RTGs. Provided there is no possibility of the reactor becoming fission-active during launch or any conceivable accident, there is no need to contain the fissile material to prevent release. This would be true in any case; remember that the fuel rods are designed not to melt, or release dust particles under attack by fast-moving hydrogen at pressures of 50 bar and temperatures of 3,000 K, it becomes clear that they would survive most accidents intact.

The key requirement is to prevent any possibility of fission occurring while the engine is still in the atmosphere, or in an accident. This implies that the engine should be launched in a disabled or incomplete state. There are various possibilities, which fall short of building the whole engine in a high orbit. One possibility is to launch the fuel rods separately from the engine, and arrange for them to be inserted in orbit; another is to launch with the control drums and perhaps any passive reflector separated from the fissile material, to be replaced in orbit. In the absence of the reflectors, neutron build-up cannot take place, and fission will not start. It may be safe enough to launch with the engine complete, with just the control drums locked in the 'off' position, with boron carbide facing the fissile core. No fission could occur unless the drums were rotated. Additional safety could be achieved by launching with a neutron-absorbing plug in the engine, which can be removed in orbit. All of these precautions would individually prevent fission, and it is a matter of safety analysis and quality control to decide how, and in what combination, to use them. But the foregoing should be enough to show that safe launch of an inactive nuclear rocket engine is entirely achievable.

The next requirement is safe operation; and here the safety of the crew is paramount. Their safety will in any case depend on proper operation of the engine, in that it is needed to place them in the correct orbit; but here we are concerned with radiation safety. While operating, it will emit large quantities of neutrons and gamma-rays, both of which are penetrative and require significant shielding to stop them. The engines described above are all fitted with internal shields. Shielding is most mass-effective close to the source of radiation, because it can be physically smaller. No shielding is provided in any outward or downward direction – regarding the nozzle as pointing 'down' – and we need only protect the spacecraft, which is forward of the engine. Additional shielding will probably be required for the base of the spacecraft; and we have already alluded to the use of the partially full hydrogen tanks to protect the crew from neutrons. Additional shielding will also be required for protection against gamma-rays – and here the requirement is opposite to that required for protection against neutrons. Dense heavy-element shields will be used to stop them, and tungsten or molybdenum will have to be incorporated in the spacecraft shield. Another possibility, used in some of the earliest designs by von Braun, but still valid, is to place the engine on a boom, well away from the crew compartment to take advantage of the inverse square law. Again, it is matter for detailed design and trade-off study to devise the most mass-efficient method of protecting the crew.

It is, of course, important to ensure that the engine, in use, continues to operate in a safe mode. The fission reaction and the hydrogen flow have to be balanced so that all the heat produced is carried away by the hydrogen. This is a matter of feedback and control loops. The position of the control drums controls the heat produced by fission, and the speed of the turbo-pumps controls the hydrogen flow. These two need to be connected by a fail-safe control loop so that a sudden drop in the hydrogen flow-rate does not cause the fissile core to overheat. In practice, the exit temperature of the hydrogen, as well as the pressure in the chamber, needs to be monitored closely. Any increase in the former or decrease in the latter should result in a rotation of the control drums to reduce the fission rate. During the 1960s, the many hundreds of hours of operation of engines allowed this system to be brought to a high state of development. An example of the wrong procedure is shown in Figure 6.10 – a deliberate opening of the control loop to establish what would happen if the fission was allowed to enter runaway. This did *not* lead to a nuclear explosion, of course, but it did lead to the bursting of the reactor chamber and the release of a

Figure 6.10. The result of allowing fissile runaway on a NERVA programme engine. The explosion was caused by the bursting of the reactor chamber and the release of hot hydrogen into the atmosphere. It was *not* a nuclear explosion. (Courtesy NASA.)

great deal of very hot hydrogen into the oxygen-rich atmosphere, with the inevitable explosive result. In space, of course, there would be no hydrogen explosion.

The final requirement is safe disposal of the engine. The primary route will be to place it in a long-term solar orbit so that it will never again approach the Earth, Mars, or any inhabited region. This is a fairly simple matter to arrange, by keeping a little propellant back for this purpose. Once a few days have elapsed from the last firing, the intact engine will not be a danger. The spent fuel rods will retain their integrity and their high-temperature and abrasion resistance; but it will now contain radioactive fission products in place of some of the safe uranium atoms. It is essential that a used reactor does not re-enter the atmosphere until many years have passed since its last use. The general consensus is that if firings are confined to orbits above 700 km altitude, then the probability of an accidental re-entry is sufficiently small to be safe. For a Mars expedition, the firings will be to enter Mars transfer orbit, or to capture into a high Mars or Earth orbit. So, this condition naturally arises.

The above considerations are objective, and are based on scientific analysis of the issues surrounding the use of nuclear rockets in space. On the basis of this analysis, nuclear thermal propulsion, under the conditions proposed above, is safe. Nevertheless there will still be a substantial body of opinion that will disagree with this conclusion. Such views are honestly held, and probably cannot be changed by scientific argument. Ultimately, the space agencies will have to 'bite the bullet' and proceed with nuclear propulsion – if it is the most effective way to mount a Mars expedition, and can be shown to be safe before the most critical and informed review boards. An accumulation of instances of safe use will eventually ameliorate the adverse public perception.

6.10 NUCLEAR THERMAL PROPULSION FOR A MARS EXPEDITION

The revived interest in nuclear thermal propulsion is partly based on its possible use for a Mars expedition. The attraction is obvious, and has, in the past, been extolled in enthusiastic terms: nuclear rockets more powerful than chemical rockets, short journey times to Mars... and so on. Some of these statements are simply not true, and others are over-confident. For a direct comparison with chemical engines we can work in terms of power. The equivalent power output can be calculated from the thrust and the exhaust velocity using the equation given in Chapter 5, repeated here for convenience.

$$P = \frac{1}{2} F v_e$$

The Space Shuttle main engine, burning liquid hydrogen and liquid oxygen, has a vacuum thrust of 2.3 MN and an exhaust velocity of 4,550 m/s. This leads to a power output of 5,323 MW – somewhat larger than the most powerful reactor tested on the ground: the Phoebus series, operating at 4,000 MW. It is, in fact, the power output that is calculated from the thrust rather than the total thermal power. Chemical engines are specified in terms of the power *output* in thrust and exhaust velocity; nuclear engines mostly in terms of the thermal power released by the reactor. There is the question of the efficiency with which the thermal energy, produced by burning hydrogen and oxygen or by nuclear fission, is converted into the kinetic energy of the exhaust stream. If we treat any rocket engine as a heat engine, then the maximum efficiency is defined by thermodynamics as

$$\eta = 1 - \frac{T_{Cold}}{T_{Hot}}$$

The temperatures are just the combustion, or fission chamber temperature, and the temperature of the exhaust gases. For an engine operating in vacuum, this is simply determined by the nature of the nozzle: the expansion ratio. A high expansion ratio will therefore produce a high efficiency of conversion of heat to exhaust gas flow, and a small expansion ratio will produce a low efficiency.

$$\frac{T_{Cold}}{T_{Hot}} = \left(\frac{P_c}{P_e}\right)^{\frac{1-\gamma}{\gamma}}$$

Here γ is the ratio of specific heats of the exhaust gases, and has a value of about 1.2. The expansion ratio for the SSME is 78:1, producing a pressure ratio of about 0.001. This produces an efficiency – using the above equations – of 68%. If the expansion ratio were to be enhanced to about 500:1, which may not be practical, then the pressure ratio would be 0.0001, and the efficiency would be 78% – independent of whether the engine is nuclear or chemical. Given the actual efficiency of the nozzle, it is clear that in general the largest chemical engines have a greater thermal power output than nuclear engines, by perhaps a factor of 2 or 3.

This digression helps to identify the single property of the nuclear engine that makes it profitable for a Mars expedition. This is the exhaust velocity. As mentioned above, it is a factor of 2 higher than can possibly be achieved by a chemical engine. It arises purely because of the use of hydrogen as the propellant. Its low molecular weight produces a better exhaust velocity than the steam that forms the main constituent of the exhaust of a liquid hydrogen–liquid oxygen engine.

Since the modern nuclear thermal engine has yet to be developed – although, as we have seen, there has been much development work carried out in the past – we may select an engine from Table 6.2: the NDR engine, using uranium carbide fuel elements coated with zirconium carbide to prevent hydrogen erosion. It has a power output of 1,787 MW, a thrust of 334 kN, and an exhaust velocity of 9,250 m/s. The reactor weighs 5,853 kg, the nozzle and fuel pumps weigh 2,559 kg, and the shields, internal and external, weigh 1,517 kg and 4,674 kg respectively. The total mass is then just over 14.6 tonnes. This is deliberately chosen to be conservative (similar to the NERVA engines developed up to 1972), and using slow neutron fission with enriched uranium and a graphite moderator. If a modern nuclear engine is developed, then its performance should be at least as good as this.

For a direct comparison with the chemical rocket mission, we should again evaluate the propellant requirements for the Mars mission. These are shown in Table

Table 6.3. Comparison of nuclear chemical propellant requirements
for a Mars mission

Manoeuvre	Delta-V (km/s)	Multiplier	Vehicle	Fuel mass (LH_2/O_2)	Fuel mass (nuclear)
1	3.55	0.43	ABCD	828	155
2	2.45	0.30	BCD	250	74.4
3	3.42	0.45	CD	90	34.8
4	3.42	0.45	D	5.6	2.3
5	2.52	0.31	B	113	32.2
6	3.55	0.48	B	83	33.6

6.3. This gives a total fuel mass of 332 tonnes, which is not much larger than the electric propulsion total of 296 tonnes. This is totally consistent, because we assumed an exhaust velocity of 10,000 m/s for the electric thruster and 925 m/s for the nuclear thruster. This again emphasises the importance of exhaust velocity in determining fuel requirements. Note that in all these calculations we have not included any mass for the engine. In the case of the storable-propellant Mars expedition we added 207 tonnes to the mass of the booster vehicle, to take account of the tankage mass for the 3,000 tonnes of fuel. Here, and for electric propulsion, this has not been done, because the fuel mass is much smaller. The mass of the nuclear engine of 14 tonnes would certainly have to be taken into account in some manoeuvres, but in the above this has been neglected for simplicity and to enable direct comparisons.

The important advantage of nuclear thermal propulsion over electric propulsion is the thrust – but will the thrust of the nuclear engine be sufficiently large to provide a significant advantage? It is clear that the propellant requirements are very similar, and far superior to any chemical rocket engine. We can use the quoted thrust to calculate thrusting times for two manoeuvres: the capture into Earth orbit on return from Mars, and the boost into Mars transfer orbit, where the mass is much larger. The equation for the duration of the thrust burn is:

$$t = v_e \frac{M_0}{F} \left(1 - \frac{M}{M_0} \right)$$

where the symbols have the same meaning as before.

Substituting: the mass M_0 is 70 tonnes for the returning spacecraft, plus the fuel required for the manoeuvre, 33.6 tonnes; the exhaust velocity is 9,250 m/s; and the thrust is 334 kN. The thrusting time is calculated to be 1,310 seconds, compared with the circular orbital period of 5,247 seconds. In fact, the manoeuvre can be accomplished comfortably within one orbit. This is a crude approximation to the actual trajectory, but the order-of-magnitude estimate is reasonably sound. The Aestus engine – a chemical engine with a thrust of 29 kN – used in a previous calculation, produced just over an orbit period for the same manoeuvre, while the higher thrust of the nuclear engine produced a shorter manoeuvre. The propellant saving is, however, very significant: 33.6 tonnes, compared with 139.4 tonnes for the Aestus.

A more thrust-critical manoeuvre is the departure from Earth orbit, during which the mass of the vehicle-complex is very much greater. We shall first calculate the manoeuvre with a single nuclear engine, assuming an all-nuclear mission: essentially the conditions assumed previously. The all-up mass, M_0, is then 441 tonnes, and the payload for this manoeuvre is 322 tonnes – the total, minus the propellant for the manoeuvre. The time is then 3,296 seconds. This is still less than the circular orbit period, and the manoeuvre can therefore be accomplished with a simple manoeuvre, using a single burn of the engine. Compare this with the same manoeuvre using a single Aestus engine. Referring back to Chapter 4, we find, for the all-up mass, 4,860 tonnes, and for the propellant burnt, 3360 tonnes, producing a payload mass of 1,487 tonnes. The time is then 376,845 seconds, or about 105 hours. This is many orbital periods, and such a low thrust is not sensible for the departure booster, using storable propellants. There are much higher-thrust storable propellant engines – especially some of the Russian developments such as the RD-253, which produces a thrust of 1.67 MN. Six of these are used on the first stage of the mighty Proton launcher, currently being used to ferry major sections of the International Space Station into orbit. The use of just one of these engines would reduce the burn time for insertion into Earth–Mars transfer orbit to 6,544 seconds – a manageable manoeuvre.

These examples show how the different characteristics of engines affect different aspects of the mission. The exhaust velocity determines the amount of propellant required; the thrust determines the time taken over a manoeuvre; and the expansion ratio determines the efficiency of a thermal engine, whether nuclear or chemical. If a high exhaust velocity is available, then this reduces both the propellant mass and also the amount of thrust required to execute the manoeuvre in a reasonable time. In some cases – as with the chemical engine – it is easy to see how sufficient thrust can be generated: these engines are inherently high-thrust. For electric thrusters, although the fuel savings are very great it is difficult to see how sufficient thrust can be generated to execute manoeuvres in a sensible time, and it is instead necessary to accept long manoeuvre times, and slowly spiral out until the required final velocity is achieved. The nuclear engine sits nicely in the middle ground, as it has a moderately high thrust and an exhaust velocity more than twice that of the best chemical rockets. The propellant savings are certainly comparable with those of electric thrusters at the low end of the exhaust velocity spectrum; but unlike the electric thrusters, the thrust of the nuclear engine is high enough to execute these low-propellant-mass manoeuvres in a reasonable time. It is no surprise that those who wish to see a human expedition to Mars look greedily to the nuclear engine, as it offers the best prospect of such a mission within the next twenty years.

6.11 HOW A NUCLEAR THERMAL ENGINE CAN BE USED FOR A MARS EXPEDITION

The nuclear engine cannot be used for all manoeuvres in a Mars expedition. This is dictated by safety considerations, both for the crew and for the local environment.

Protection for the crew from the neutron and gamma radiation produced while the engine is actually operating, is incorporated in the engine itself, in the form of internal and external shields. The exhaust stream of hot hydrogen is itself not radioactive, but it may contain radioactive particles eroded from the fuel elements by the hot hydrogen. Fuel elements are designed to avoid this as much as possible, to increase the operating life of the engine, although some erosion is possible and will make the exhaust dangerous. These factors will decide how and where the engine can be operated. The other factors relate to its launch into orbit, its storage, and its safe disposal. Assuming that it can be placed safely in orbit, then it can be used for the departure: the injection of the expedition vehicle into Earth–Mars transfer orbit. From the point of view of crew safety, the engine could remain attached to the vehicle during Earth–Mars transfer, and be used again for the Mars capture manoeuvre. Similarly, it could remain in Mars orbit and be used for Mars–Earth transfer, and for the final capture into Earth orbit on return from Mars. The engine would then be used once more in a final thrust to place it in a disposal orbit, such that it will not return to Earth. It should not be used for Mars landing or departure, as contamination of Mars with radioactive exhaust would not be acceptable. Neither is contamination of the Earth's atmosphere acceptable, and Earth capture should therefore be into a high orbit, beyond 700 km altitude, followed by disposal of the engine. Subsequent manoeuvres would require chemical propulsion.

This is the most economical use of the nuclear engine; and its dead mass of 14 tonnes would suggest that more engines than necessary should not be flown. It is also the best solution from the point of view of launch safety, and it would be necessary to dispose of only one engine. There may be arguments for using the engine only once, or for having several engines; but these will probably relate to the limitations of the engine itself, such as the total energy available from the core loading of fissile material. The amount of U^{235} will decrease as the engine operates, and some 20–30 kg of uranium will undergo fission in an engine's lifetime of operation. The accumulation of fission products in the fuel elements will change the neutron-absorbing properties, and may in time reduce the efficiency of the fission process. All of this will depend on engine design, and the new development programme that will be required to prepare an engine for the Mars expedition.

The fuel savings are, of course, largest for the departure manoeuvre. As the mission proceeds, the payload decreases, and the fuel required for each manoeuvre also decreases. If there were a need to use the engine only once, for example, then this should be for the departure manoeuvre.

It is possible to calculate the total energy required for the four manoeuvres, and to compare this with the performance already achieved in the NERVA and earlier programmes. To do this we need to know the power output, calculated from the thrust and the exhaust velocity, and the thrusting time, calculated earlier. The relevant manoeuvres are **1**, **2**, **5**, and **6**. The power is represented by

$$P = \frac{1}{2} F v_e$$

Figure 6.11. The NRX engine on its test stand. This engine clocked up the longest continuous thrusting period of the NERVA programme: 3,270 seconds – longer than the longest Mars mission manoeuvre. (Courtesy NASA.)

which for our chosen engine is 1,549 MW. The times for manoeuvres **1** and **6** have already been calculated to be 3,296 seconds and 1,310 seconds respectively; and similar calculations for manoeuvres **2** and **5** yield 2,000 seconds and 769 seconds respectively. The total thrusting time for the engine is then 7,375 seconds, excluding the disposal burn. The total energy output is then the product: 11,423,875 MW-seconds, or just over 3,000 MW-hours. The engine must operate for more than 123 minutes. This is a long time compared with the few-minute burn times of current engines such as the Space Shuttle main engine. The Aestus engine operates for 1,400 seconds, which is still a long way short of the time required for a nuclear engine.

The historical nuclear engine programme produced highest-energy outputs of 2.6 million MW-seconds from Phoebus, and 4.19 million MW-seconds from NRX-A6, compared with the requirement of 11.42 million MW-seconds – certainly in the right region. The NRX-A6 also produced the longest continuous operation time of 3,720 seconds. This exceeds the longest single manoeuvre time for the Mars mission, but the total operating life is still short by about a factor of 2 in time, and a factor 3 in total energy output. The NERVA programme produced sufficient confident to specify, for future engines, a 10-hour operating life and a capability for fifty restarts – which is ideal for a Mars mission. It may be concluded, therefore, that it is highly probable that nuclear propulsion will be capable of executing the entire mission requirement of four manoeuvres, using a single engine.

6.12 THE BIMODAL NUCLEAR THERMAL ROCKET ENGINE

Many engineers and mission designers consider that it would be a waste of resources to take a nuclear rocket engine all the way to Mars and back, and to not use it to generate the electrical power required to support the mission. A nuclear-powered spacecraft with solar panels to provide electricity is an anomaly. It is therefore planned to make the reactor in the engine do double duty, both as a source of propulsion power at more than 1,000 MW, and to deliver electrical power at, say, 100 kW. This requires more hardware on board, and a more complicated reactor, but it is efficient in terms of the design of the whole spacecraft. The plan is to have part of the fissile core fitted with heat-extracting pipes containing liquid lithium, like those in an electrical power reactor, or to use sodium vapour in a heat pipe. The reactor will be run it at about 1,300 K, and a power level of a few hundred kW. This can be achieved by controlling the neutron flux to a much lower level than for propulsion. The production and extraction of the heat is not much of a problem, but there must be, of course, a means of converting the heat to electricity – either thermoelectric or mechanical – using a Stirling or a Brayton engine to drive a generator. Again there will need to be a significant area of radiator to dump the waste heat from the generation process. All of this, for nuclear power generation, was discussed in the previous chapter.

It may be necessary to have, for this purpose, separate fuel elements within the main engine reactor, because the conditions will be different for power generation and propulsion. The reactor operates at high temperature and high hydrogen pressure for propulsion, and at a lower temperature, with only the lithium cooling pipes, for power generation. It is improbable that these requirements will be compatible, as lithium boils at 1,615 K, and hydrogen cannot flow through the channels that contain the lithium pipes to cool the fuel elements during propulsion. It still makes sense to combine the functions, because the shielding and reactor casing, the safety features and the main control systems will all be common, and the fissile core is only a small fraction of the total mass. One approach is to use the central support tube of a fuel rod to facilitate the low-power mode. By placing the heat pipe inside this tube, perhaps separated by a vacuum layer, as shown in Figure 6.12, it will be partially insulated from the hot fuel element, thereby maintaining the heat pipe temperature below the boiling point of lithium. If the two requirements can be combined, then the cooling of the core, after the reactor is shut down – nominally carried out by blowing some of the hydrogen propellant through the core until the fission products have quietened down – can be done by the lithium pipes and the radiators. This would save some propellant mass, and propulsive manoeuvres would therefore be simpler. At present, the bimodal nuclear thermal rocket engine seems to be the most popular choice for a Mars expedition, and development of the bimodal fuel elements is in progress.

The main technical power consumption will be the electrical refrigeration of the liquid hydrogen propellant to prevent boil-off. Ample power will also be useful to the crew in their habitat, and to allow high-bandwidth radio communication with Earth; such systems are power-hungry, and are not normally used on deep

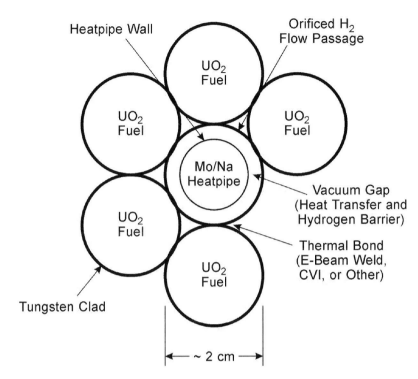

Figure 6.12. A possible design for a fuel element in a bimodal nuclear rocket engine. The heat pipe is vacuum-insulated from the rest of the fuel rod to protect it from the very high temperatures during rocket engine operations, but can extract low-level heat for electrical power generation. (Courtesy NASA.)

space mission because of power shortage. So, a mission powered by a bimodal nuclear rocket engine will be much more comfortable, for many reasons. While there exist designs and mission scenarios based on the bimodal engine, no such device has yet been tested, and it remains to be seen whether bimodality can be included in the first practical space-borne nuclear rocket engine without compromising the performance as a propulsion unit.

6.13 CONCLUSION

It may seem strange, after so many years have passed, that the space community is reviving ideas for nuclear propulsion; and the 1950s, when nuclear fission for peaceful purposes was seen as promising huge benefits to mankind – on looking back – seems a fantasy. But the evidence is there, and many developments were carried through; and there were even experimental nuclear aircraft engines. The horror of nuclear fission that pervades society today would have seemed equally fantastic, but it does appear to be a permanent mind-set. However, the physics of nuclear fission

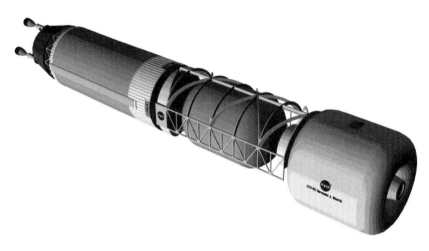

Figure 6.13. A proposed Earth–Mars transfer vehicle using a bimodal engine. The central tank holds propellant for Mars capture and return, and protects the crew from the reactor during electrical power production. The crew quarters use the 'transhab' concept – an inflatable cabin. (Courtesy NASA.)

and the physics of a Mars expedition have not changed. Chemical rocket technology is essentially very old. We could trace it back to ancient China; it is at least possible to call it technology of the first half of the twentieth century; all the inventions were produced before 1950 – by Goddard and Tsiolkovsky, and during the German, Russian and US rocket programmes.

If we attempt to use chemical rocket technology to go to Mars, then, saving some miraculous new development, the process will be cumbersome, and expensive in the extreme. If, on the other hand, we embrace the new technology and use nuclear fission, then we can be reasonably certain of mounting a successful expedition, within a reasonable level of cost and complexity. Of course, the lessons of the last fifty years have to be taken to heart. The safety of the crew, and of the population at large, has to be paramount; and we cannot afford even the smallest radiation accident with nuclear rockets. Unfortunately, the track record for accidents using chemical rockets is not good; and even the latest designs have experienced major and very public failures, causing the loss and deliberate destruction of the launcher. We have to ensure that no-one is put at risk if a launcher fails while carrying an inert nuclear reactor. Nuclear reactors and rocket engines have to be fail-safe, and all risk of a major accident needs to be eliminated. But this *can* be achieved – and, indeed, has already been achieved with RTGs – and similar ideas and technology can be applied to nuclear thermal rocket engines, and to reactors for power. If proper design rules are applied, and proper safety management is enabled, then there is no reason to expect any release of radiation to occur on Earth or on Mars, or for the crew to be exposed to unacceptable levels. The journey to Mars, by human beings, can be accomplished, and we shall for the first time stand on the surface of another planet, perhaps fifty or sixty years after the first human being stepped onto the Moon.

7

The return from Mars

No expedition to Mars can proceed without a guarantee that the members can return safely to Earth under all predictable contingencies. In the previous few chapters we examined how the different types of rocket technology can contribute to a safe and achievable mission, including a safe return; and as we have seen, a number of scenarios are possible. The basic mission described so far, and based approximately on Apollo, ensures a safe return by leaving the return vessel, fully fuelled, in orbit around Mars. Like all the elements of the basic Mars expedition, the return vessel has to be reliable; it has to be safe to leave unattended in orbit; and the rocket engines must also start reliably after a long period. Chemical engines – using storable propellants such as monomethyl hydrazine and nitrogen tetroxide – are generally regarded as the most reliable for human spaceflight, and have been used for all return manoeuvres, from the Apollo lift-off from the Moon, to the Space Shuttle deorbit manoeuvres. These chemicals ignite on contact, so to develop thrust it is sufficient to open the valves and let them enter the combustion chamber. They are also liquid at normal temperature and pressure, and therefore easy and safe to store for long periods.

It will not have escaped the reader's notice that these 'safe' propellants generate the smallest exhaust velocity, and that enormous quantities are required to carry out the mission. The total mass of propellant required for our model mission was just less than 5,000 tonnes, with the return journey from Mars' surface to Earth requiring nearly 400 tonnes. Cryogenic propellants could reduce this requirement by a large amount, but they are neither self-igniting nor easily storable. These disadvantages can be overcome, but the vessel becomes more complicated and, by definition, less reliable. The more complex a system, the higher the risk that the failure of even one component will lead to disaster. So, is there an alternative to carrying large quantities of propellant to Mars, to be stored for the return journey? Some people think so, and schemes for manufacturing propellant *on Mars* have been proposed and elaborated. Such schemes are so attractive that they have been adopted by NASA in what is called the Reference Mission – a baseline design against which all the necessary enabling technologies can be developed. In this chapter we shall discuss this and other techniques that can be used to reduce the quantity of propellant required, and at the same time make the mission more reliable.

7.1   PROPELLANT FOR THE RETURN JOURNEY

The simplest approach, outlined above, is to take the propellant to Mars, and store it. There are three return manoeuvres – the lift-off from Mars, the departure from Mars orbit, and the Earth capture manoeuvre – and there are therefore three sets of storage conditions. The propellant for lift-off from Mars is stored on the surface, in the fuel tanks of the lander–ascent vehicle complex; the propellant for Mars–Earth transfer is stored in orbit, in the return vessel; and the propellant for Earth capture is stored in the same vessel, during the journey to Earth. We first consider the problem of storing the propellants in the three locations, assuming, for the moment, that they are the room-temperature liquids monomethyl hydrazine (MMH) and nitrogen tetroxide. The liquids should neither freeze nor boil: MMH freezes at −52° C and boils at 87° C, and nitrogen tetroxide freezes at only −11° C and boils at 21° C. Obviously, the definition of 'room temperature liquids' is somewhat marginal. Both materials are stable and relatively safe to handle, although poisonous. MMH is a clear liquid that smells characteristically of rotten fish or ammonia, while nitrogen tetroxide is a reddish-brown liquid that smells strongly acidic – the characteristic 'chemistry lab' smell. Neither is a chemical that one would wish to spend any time with, on terms of intimacy; but for the explorers they guarantee a safe return.

To keep both propellants liquid they must be held in a narrow temperature range between −11° and +21°, which must be the conditions immediately before and during their use as rocket fuel. Allowing them to freeze during storage is not a big problem, provided expansion is properly allowed for in the tank design. In fact, the propellants *do* freeze during long storage in space, and heaters are provided for the tanks to reliquefy the propellant before use. (It has been considered that a possible cause of the loss of Mars Polar Lander was that the landing rockets did not fire because the tanks were too cold.) The temperature range on the surface of Mars is rather wide, but generally no higher than −31° C, and certainly the propellant would need to be melted before use, after unheated storage on Mars. In space, at Earth orbit, the temperature can be kept reasonable by rotating the spacecraft. The natural mean temperature of a spacecraft in Earth orbit is about 5° C, but higher and lower temperatures can be achieved by special surface finishes. Mars is one-and-a-half times as far away from the Sun than is Earth, and so the natural temperature is lower: −47° C. After storage in Mars orbit, therefore, the tanks would need to be heated before the firing of the rockets.

By and large, the use of these storable propellants is not very difficult. Heaters have to be provided, but in general the materials can be kept indefinitely and brought into use rather simply, and they remain the safest option for guaranteed Earth return. The big drawback is the quantity required, especially if it is to be used for the whole mission.

Cryogenic propellants produce much higher exhaust velocities, and enable the mission with much lower propellant quantities. They do, however, require much more aggressive storage, and starting capability is not so straightforward. Liquid hydrogen–liquid oxygen engines are restarted during launch sequences. The Apollo J1 engine was, in fact, restarted during injection into Earth–Moon transfer orbit. In

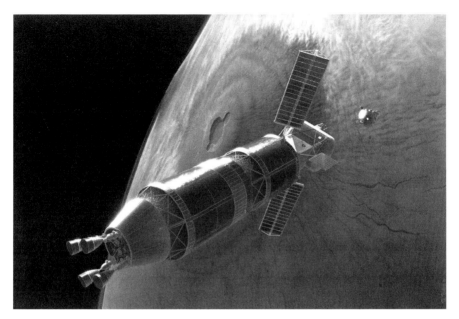

Figure 7.1. A nuclear thermal Earth return vehicle in Mars orbit awaits the approaching ascent vehicle. Liquid hydrogen for Earth return has been stored using refrigeration, powered by one of the four nuclear engines in idle mode. The truncated cone between the engines and the vehicle proper is the radiator of the electrical power system. (Courtesy NASA.)

general, however, the restarting of these engines is regarded as less reliable than the storable propellant option. The question of refrigerated storage for these propellants was discussed in Chapter 4. Considerable electric power is required, and the refrigerator cannot be allowed to fail. Once the temperature of hydrogen and oxygen begins to rise, boil-off will occur, and nothing can be done to stop it. This could be both immediately dangerous in the release of inflammable gases, and, of course, ultimately fatal to the mission and the lives of the expedition members, if the remaining fuel is insufficient for the return journey. The safe return then depends on a reliable refrigerator for the propellant tanks, which in turn requires a reliable source of electrical power. (The means of generating this power was described in Chapters 5 and 6.)

For a nuclear thermal rocket, liquid hydrogen is still required as the propellant, and it must be kept refrigerated. The dual mode reactor provides a useful source of electrical power for this, with the engine refrigerating its own propellant in idle mode. There is no ignition problem with a nuclear thermal rocket, as the reactor can be powered up, and the hydrogen admitted to the chamber, to develop thrust. In this sense it is probably more reliable than the purely chemical rocket, using hydrogen and oxygen. The refrigeration power supply is also less marginal than, say, a solar panel powered system that might be close to its maximum capability; a nuclear reactor always has a good margin of power production.

These options imply much reduced propellant requirements, but are more complicated and, moreover, require additional technology, like refrigerated storage, which is not yet qualified for space use. Clearly, if an expedition were to be mounted today there would be a strong inclination to use storable propellants for at least some of the manoeuvres, and so the quantity required would be significant. We have not yet considered the problem of placing this amount of propellant in Earth orbit; but it will be recalled that, using present-day launchers, the maximum payload into low Earth orbit is about 24 tonnes – which reveals that propellant quantity is still the major issue for Mars exploration.

7.2   INDIGENOUS RESOURCES: MAKING PROPELLANT ON MARS

We are accustomed to space expeditions being entirely self-contained. This was certainly true for Apollo, and is true for any mission to Earth orbit. On the other hand, most human exploration on Earth has relied heavily on available resources; land expeditions use food and water gathered locally. The great Victorian explorers, for example, used local transport and labour to carry their baggage, on foot, through Africa and Asia. The voyages of discovery during the seventeenth and eighteenth centuries were similar in duration to a Mars expedition; here again, water and food were taken on board whenever the opportunity arose. Propulsion used another indigenous resource: the wind. In terms of isolation and absence of local resources, exploration of the Arctic and Antarctic is closest to Mars exploration. The explorers have to haul all their fuel and food with them, with water available locally, provided fuel is available to melt the ice. If humans land on a planet, then planetary resources are available for use by the expedition. The resources that can be used depend on their nature, and on the ingenuity of the explorers.

The most desirable indigenous resource for any human expedition is water. Apart from the possibility of using it for human consumption, it is easy to convert it into hydrogen – providing fuel – and breathable oxygen – another essential for human life. The presence of water on Mars now seems certain, but whether or not it can be easily extracted is another question. And there is increasing evidence that water might be found on the Moon – particularly in craters near the poles, where sunlight never falls. This opens the possibility of making fuel and oxygen there, and even of refuelling spacecraft on the way to Mars, or elsewhere in the Solar System. The difficulty, in the assumption that enough water can be electrolysed to produce tonnes of oxygen and hydrogen, lies in the manual work required to extract the water. If open lakes or streams were to be available, then it would, of course, be a very simple matter. But if the water is in the form of permafrost – a mixture of ice and powdered rock – then the material must be mined, and the ice melted and separated from the rock; it is improbable that the use of space suits and shovels will result in the production of tonnes of water on Mars or the Moon. This kind of process could be mounted later in the history of Mars exploration, when large machines become available on the planet. Amongst those interested in a return to the Moon, there are many plans for setting up such mining operations there.

The early exploration of Mars requires something simpler, and requiring fewer problematic mining processes. Ideally, the raw material should be ubiquitous, and a fluid. In this way it can be found anywhere on the planet, and it can be ingested into whatever machinery will process it, without mining or any other extrinsic activity. The only fluid known to be ubiquitous on Mars is its atmosphere; and it is to the atmosphere that we must look to produce propellant on Mars.

7.3   THE MARTIAN ATMOSPHERE

Mars is much smaller than the Earth. Its radius is 53% of the Earth's radius, but its density – 71% – is lower than Earth's, and so its mass is only 15% of the mass of Earth, and its surface gravity is 38% of Earth's. This results in a much more tenuous atmosphere, but one which paradoxically extends further out into space. The surface pressure is much lower than on Earth, and is also much more variable, ranging by 25% from 6 to 8 millibars, depending on the season, compared with a mean range of less than 10% on Earth. The main reason for this variability is the composition of the atmosphere, which is mainly carbon dioxide, and the mean temperature, which is

Figure 7.2. A Viking 2 lander image of water ice on Utopia Planitia. The ice remains for about 100 days in winter. (Courtesy NASA/JPL.)

about –63° C. Like Earth, Mars has a tilted axis, and so when winter comes to one or other pole, the temperature drops sufficiently to condense the carbon dioxide and thereby reduce the atmospheric pressure. At other times the poles are warmer and the carbon dioxide evaporates, thus increasing the pressure. The daily temperature can range from –89° C to –31° C, and there are surface winds with a similar range of speeds to those on Earth: 2–10 m/s in normal conditions, and up to 30 m/s – equivalent to a force 7–8 gale – during dust storms.

The composition of the atmosphere is now well known: 95.5% carbon dioxide, 2.7% nitrogen, 1.6% argon, 0.15% oxygen, and 0.07% carbon monoxide. Water is present at 210 parts per million – compared with 1,000–10,000 parts per million in the Earth's atmosphere – and the atmosphere is permanently contaminated with dust. With a density of 0.02 kg/m^3, 50 m^3 of atmosphere are required to provide 1 kg of raw substance. Nevertheless, pumping martian atmosphere through a chemical processor is far easier than mining permafrost.

7.4   MAKING NUCLEAR PROPELLANTS FROM THE MARTIAN ATMOSPHERE

Compressing, purifying and liquefying carbon dioxide would produce a working fluid for a rocket engine; and with a nuclear thermal engine or an electro-thermal engine this is all that is required. So, it would be possible to refill the propellant tanks on Mars with quite simple apparatus: basically, a compressor, powered electrically, that would liquefy the carbon dioxide. It would take weeks to months to collect enough atmospheric carbon dioxide to fill the tanks, especially as many tonnes of propellant would be required; but it would be simple to achieve, and could be left to proceed while the rest of the exploration continues. The main disadvantage of this process is that a nuclear thermal rocket has to be used; and it is not possible to generate enough thrust to lift off from the surface of Mars using an electro-thermal rocket, as high thrust requires either the nuclear engine or a chemical reaction. The high molecular weight of carbon dioxide limits the exhaust velocity, even of the nuclear rocket; but the escape velocity on Mars is much smaller than that on Earth, so it would certainly be possible to use the carbon dioxide in a nuclear thermal rocket, to bring the expedition, and spare propellant, generated on the surface, into Mars orbit. This process is entirely feasible for a return journey, using propellant generated on Mars. The big objection would be the nuclear contamination of the martian surface and atmosphere that would be incurred during the launch from the surface. There are presumably no martians to complain, but there is a general presumption that we, as humans, ought not to contaminate other planets with radioactive material.

So, how much carbon dioxide would be required for the lift-off from Mars and the Earth return manoeuvres? This is relatively easy to estimate, because we can scale the exhaust velocity for the NTR, using the rule that the exhaust velocity is inversely proportional to the molecular weight. Carbon dioxide has a molecular weight of 44, and hydrogen a molecular weight of 2. The exhaust velocity of the NTR is 9,250 m/s, with hydrogen as the working fluid. The exhaust velocity would be

$$v_e = 9{,}250 \times \sqrt{\frac{2}{44}} = 1{,}972 \; ms^{-1}$$

with carbon dioxide as the working fluid. This velocity is about half that of the storable propellants. To calculate the quantity of propellant required, we need the fuel multiplier from Chapter 4, given by

$$\frac{M_f}{M} = e^{\frac{V}{v_e}} - 1$$

where V is the required velocity change for the journey from Mars' surface to Mars orbit, equal to 3.42 km/s. The multiplier is 4.67 – a huge value compared with even the 1.87 required for the storable propellants – the reason being that the multiplier depends on the *exponential* of the ratio of velocity change to exhaust velocity. Therefore, the mass of fuel required increases rapidly as the exhaust velocity drops. The mass of the payload, the ascent vehicle, is 5 tonnes in our basic mission, so the mass of propellant which must be made on Mars for the ascent vehicle is 23.4 tonnes – a much larger amount than the 9.4 tonnes required using storable propellants. But there is a major advantage, in that this propellant does not need to be brought to Mars, the propellant necessary to bring it to Mars can be saved, and the overall mission fuel requirements drop. Making the propellant for the return journey from Mars orbit to Earth is also a possibility. This has to be lifted into Mars orbit, and, with the inefficiency of the low exhaust velocity, quite a large quantity of manufactured propellant will be required.

The compression and liquefaction of carbon dioxide – essentially the martian atmosphere – is a simple process which could be made very reliable. Since it is required by – and indeed can only be used with – a nuclear thermal engine, the engine can also generate the electrical power to operate the compressor and effectively produce its own propellant. If we could overcome the objection to using a nuclear engine for lift-off, then this would be a simple and safe option for reducing the propellant brought from Earth.

7.5   MAKING CHEMICAL PROPELLANTS FROM THE MARTIAN ATMOSPHERE

If we cannot use a nuclear engine – and it is quite probable that contamination of the martian atmosphere will prove a significant objection – then we need to consider the production of chemical rocket fuel on Mars, still using the atmosphere as raw material. Carbon dioxide contains useful elements – carbon and oxygen – in combination; but is there a way of separating them with electrical power, and then using them as rocket fuel? If we could do this, then we would have a chemical rocket to lift off from Mars, and possibly to accomplish the whole return trip, using fuel generated on Mars. It would, in fact, be a conversion of electrical energy into chemical energy in the processing plant, and then the release of that chemical energy in the rocket engine. The energy required to take the expedition back to Earth would all be derived from the electrical power plant on Mars. It would be very difficult to do this using solar panels, and so a nuclear reactor again appears as the only viable power source. Note that the contamination issue would be subtly different from the nuclear rocket issue, as here a spent nuclear reactor would be left behind. Of itself it would be a local radiation hazard, but it would be isolated and contained, and the atmosphere and the open surface of Mars would not be contaminated.

7.6   CARBON MONOXIDE AND OXYGEN

It is of little use to search for an indigenous raw material other than atmospheric carbon dioxide, as for early expeditions the raw material must be ubiquitous and easily accessible. The oxygen in carbon dioxide is obviously a key chemical rocket propellant, and we must break down the carbon dioxide to release the oxygen. At the same time we cannot use carbon itself as a fuel – this would result in a coal-powered rocket! – and must find a liquid or gaseous compound of carbon to use. It will be clear to those with a little chemical knowledge that the only possibility, with carbon dioxide alone as the raw material, is carbon monoxide, which burns in oxygen and releases heat. It is a component of 'producer gas', made on Earth by restricting the air supply to burning carbonaceous fuels. During the Second World War, producer gas was used to power vehicles to save oil, and is still used in remote locations and by 'green' enthusiasts. Carbon monoxide has a well-known disadvantage: it is poisonous. Every year, thousands of lives are lost due to faulty domestic heating and car exhausts, both of which release it. It was a component of the 'town' gas used

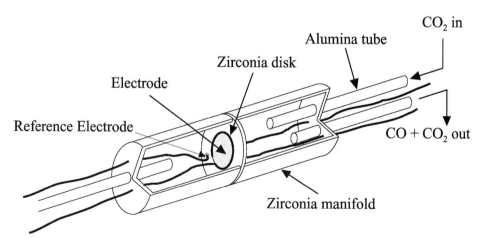

Figure 7.3. The zirconia solid electrolysis principle for extracting oxygen and carbon monoxide from martian carbon dioxide. An electric field is maintained across the zirconia membrane, causing the oxygen to migrate through it. Carbon monoxide remains on the inlet side. (Courtesy NASA/JPL.)

to heat and light homes before natural gas became dominant, and was also a favourite means of suicide. The astronauts would have to be careful.

The best way to produce carbon monoxide on Mars is to heat the atmospheric carbon dioxide to about 950° C, in contact with one surface of a special ceramic material, zirconia, which has platinum electrodes evaporated onto it. The reaction is partly catalytic, and partly thermal dissociation. Oxygen ions are released from the carbon dioxide, and pass through the ceramic under the influence of the electric field provided by the electrodes. Only oxygen passes through the ceramic, and so on the far side, oxygen accumulates, while on the near side, carbon monoxide accumulates. The pure oxygen can be liquefied by refrigeration, and stored. The carbon monoxide can be purified because it has a much lower condensation temperature than carbon dioxide. In this way, rocket fuel and oxidant can be made on Mars, using only heat, electricity, and the abundant atmospheric carbon dioxide.

This process is not science fiction. It has been developed for a different purpose: the conversion of carbon dioxide, breathed out by astronauts, into oxygen for life support – the carbon monoxide in this case being a dangerous waste product. A test of the process for producing oxygen from carbon dioxide was included on the lander

Figure 7.4. The In Situ Resource Utilisation (ISRU) experiment for the Mars Surveyor 2 lander. This was qualified, but was not flown, due to the loss of Mars Surveyor 1. The dome protects the zirconia cell which completes the conversion. (Courtesy NASA/JPL.)

of Mars Surveyor, so that it could be qualified on Mars. But the loss of the first Surveyor spacecraft forced NASA to simplify the second mission, and the lander was scrapped. The chemical processor languishes in store, waiting for another opportunity to fly; but the process works, and the apparatus has been space-engineered. Whether or not carbon monoxide is used as a rocket fuel, this means of producing oxygen from atmospheric carbon dioxide is a valuable way of increasing the oxygen supply on Mars, for use by the explorers. It would reduce the amount of oxygen necessary to be transported to Mars, or would provide a safety margin.

Carbon monoxide is a rather inconvenient chemical rocket fuel, as its boiling point is almost the same as that of liquid hydrogen at $-254°$ C. The rocket exhaust would, of course, be composed of carbon dioxide, with the accompanying depression of exhaust velocity caused by a high molecular weight, although this could be offset a little by the high flame temperature of carbon monoxide–oxygen. The combustion chamber temperature could be lowered by admitting excess carbon monoxide. Assuming that the combustion chamber temperature could be as high as for a liquid hydrogen engine, then it might be possible to achieve an exhaust velocity some 40% higher than the nuclear rocket, or about 2,700 m/s. This is quite respectable, but as yet it has not been seriously considered – partly because carbon monoxide has not been used as a rocket fuel on Earth, so there is no engineering experience, and partly because of its low boiling point, which would require refrigerated storage similar to that which would be required for liquid hydrogen. It can be stored under pressure, below its critical temperature of $-140°$ C, but the whole business of making it, liquefying it, storing it, transferring it to the rocket, and using it, is complicated.

Figure 7.5. The internal components of the ISRU experiment. The zirconia cell can be seen in the upper centre, with the pipes carrying carbon dioxide in from the atmosphere and out to the oxygen reservoir. The pumps appear in the background. (Courtesy NASA/JPL.)

Some scientists have expressed doubts about the ability to produce large quantities of oxygen and carbon monoxide using this process, as tens of tonnes are required. The special ceramic used, in what is essentially a solid electrolysis process, is yttria-stabilised zirconia: $ZrO_2 + 8\% Y_2O_3$ – fancy stuff. The material is said to be rather brittle, and thousands of cells, like that planned for the Surveyor lander, would be required to produce sufficiently large quantities. Nevertheless, we have to remember that it is established technology – at least on the small scale – and, but for an accident, would now be working on the surface of Mars.

7.7   METHANE AND OXYGEN

The more popular alternative is to take hydrogen to Mars in order to make methane and oxygen, using the martian carbon dioxide. At first this seems absurd; but it is, in fact, quite sensible. It is the means of producing rocket fuel on Mars favoured by NASA, and Robert Zubrin has made this process his own in his 'Mars Direct' mission scenarios. It was first proposed during the 1970s, by Robert Ash of JPL, who also devised the carbon monoxide–oxygen approach; he proposed the use of indigenous hydrogen produced by water electrolysis. This assumed, of course, that there would be usable quantities of water on Mars. The twist that Zubrin applied was to take the hydrogen to Mars from Earth, rather than hope to find water on Mars.

Before examining the potential of this idea, however, we should review the chemical process that produces methane and oxygen from carbon dioxide and hydrogen. This is a simple and robust process, and has been in use on Earth for more than a century. It is called the Sabatier process, after the inventor, and it has many advantages. The chemical equation is

$$4H_2 + CO_2 = CH_4 + 2H_2O$$

The reaction is *exothermic* (it produces heat), and is therefore self-sustaining once started. It requires a catalyst – nickel or ruthenium – and runs best at about 400° C. The reaction produces methane and water. The methane is a useful rocket fuel, in that it contains a considerable proportion of hydrogen, so that the molecular weight of the exhaust products will be low. No oxygen is produced directly by this reaction, but the water produced can be electrolysed to make hydrogen and oxygen. The hydrogen is then recycled with more carbon dioxide from the atmosphere, through the chemical reaction again, to produce more methane, while the oxygen is liquefied and stored, along with the methane.

Methane boils at –162° C, and oxygen boils at –183° C, so they require about the same degree of refrigeration – very much less than liquid hydrogen or carbon monoxide. Because the reaction that produces methane and water is exothermic, no electrical heating is required to keep it going, and so the power demands are tiny compared with those of the oxygen–carbon monoxide process, which is endothermic and requires the gas to be heated to 950° C in order to work. The other advantage of the Sabatier process is that it produces more or less pure methane and water, with one passage through the catalyst–reactor; the oxygen–carbon monoxide process

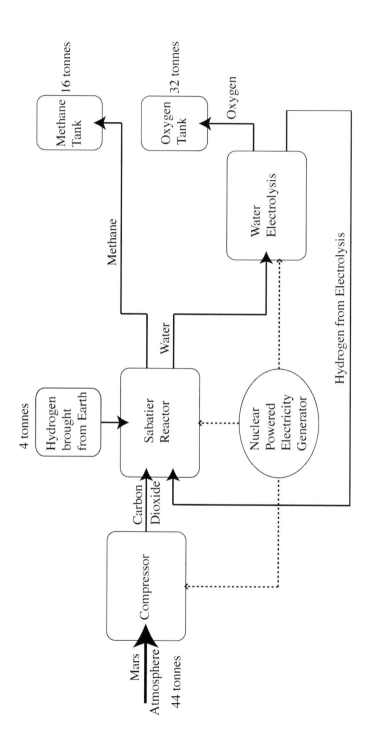

Figure 7.6. The process of producing methane and oxygen from the martian atmosphere, using hydrogen brought from Earth.

produces pure oxygen, but the fraction of carbon dioxide converted into carbon monoxide is quite small, so the gas emerging from the cell after a single passage is mostly carbon dioxide, and a further purification process is required to produce pure carbon monoxide. Making oxygen from the water produced by electrolysis consumes electrical power, and so the power saved by using the Sabatier process is somewhat offset by the electrolysis costs. On the other hand, the apparatus for electrolysis of water has been brought to a high state of engineering, for use in nuclear submarines, to produce breathable oxygen.

Now that we understand the chemistry, let us consider why it is worth taking hydrogen all the way from Earth, in order to make rocket fuel on Mars. It might parenthetically be noted that if quantities of liquid or easily obtainable water were to be found on Mars, then electrolysis could be used to produce the necessary hydrogen and to save the transportation; but this will not happen during a first exploration mission. The key to the advantage of taking hydrogen all the way from Earth lies in the mass of propellant created by the process. Remembering the rocket equation

$$V = v_e \frac{M_0}{M}; \qquad M_0 = M + M_P$$

we see that the velocity necessary to lift off from the surface of Mars and to gain Mars orbit or return to Earth, depends on both the exhaust velocity of the rocket engine and on the mass ratio. Methane has a molecular weight of 16, and hydrogen a molecular weight of 2. Two hydrogen molecules are required for each molecule of methane, so 4 tonnes of hydrogen will produce 16 tonnes of methane. At the same time, those 4 tonnes of hydrogen enable the extraction of 32 tonnes of oxygen from the carbon dioxide of the martian atmosphere. Therefore, 4 tonnes of hydrogen produce 48 tonnes of rocket propellant – an increase by a factor of 12. This is very significant, as it effectively reduces the propellant that has to be taken to the surface of Mars by a factor of 12.

For a proper comparison, the performance of the methane propellant should be compared with that of hydrogen as a propellant, both in a chemical rocket engine. The exhaust velocity for hydrogen is 4,550 m/s, so an engine burning pure hydrogen, with oxygen, would be more effective than an engine burning methane. If we assume the most effective methane–oxygen engine, the exhaust velocity would be 3,780 m/s. So, a crude comparison would be obtained by calculating the velocity change achievable with 4 tonnes of mixed hydrogen–oxygen propellant, as might be taken to the surface of Mars for a conventional approach, and one using 48 tonnes of methane and oxygen, made on Mars from four tonnes of hydrogen. This can easily be calculated using the rocket equation and assuming the usual 5-tonne payload. The result is 8.46 km/s with methane, and 2.67 km/s with hydrogen. We see that the mass advantage, using martian carbon and oxygen, has a much bigger effect than the exhaust velocity disadvantage, compared with using pure hydrogen and oxygen brought from Earth. The actual requirement is 3.42 km/s; so, by using hydrogen and oxygen more than 4 tonnes of propellant would be required to reach Mars orbit; but by using locally produced oxygen and methane produced from the 4 tonnes of delivered hydrogen, much less would be required to reach Mars orbit than was produced. The scheme obviously has merits.

The disadvantages have to be examined. In the first place, hydrogen has to be taken to Mars, safely landed, and stored on the surface; and the methane and oxygen produced on the surface have to be refrigerated and stored. This would also be true for a chemical rocket engine using cryogenic propellants. If we were to mount an all-cryogenic Mars expedition then we would have to solve the problem of long-term refrigerated storage for the liquid oxygen and hydrogen; so, the problem of storage of cryogenic hydrogen is not specific to the generation of propellant on Mars. Storage on the surface is somewhat different from storage in orbit, particularly if solar power is used. In orbit it is reasonably easy to ensure that the propellant tanks are always in shadow, while the solar panels are exposed to the Sun except for short passages round the night side of Mars. Solar-powered refrigeration is therefore a possibility – a challenging possibility – in Mars orbit. On the surface it is more difficult to shade the tanks. Mars is much warmer than deep space, and sunlight to power the refrigerator is available only during the day. Here it is almost essential to use a nuclear-powered electricity generator to run the refrigeration and propellant production, at least for a human mission. Electrical power for atmospheric pumping and propellant manufacture will be required for any *in situ* production of propellant, and this adds to the burden on the expedition power source. If it is solar, then the extra burden may well be of significance; but if the power source is nuclear, then there should be no problems. The saving in propellant mass has to be offset against the mass of the electrical power supply. This would be essential for any concept using cryogenic propellants, independent of where the propellants are produced. So, in comparison with an all-cryogenic mission, the advantage is clearly in the making of propellant on Mars.

If we compare this 'in situ resource utilisation' – or ISRU, as NASA calls it – with an all-storable propellant mission, then issues of safety and reliability have to be weighed against the huge saving in propellant mass. In the one case the explorers rely on a well-proven technology, used by all previous manned space missions, and on a comforting store of 'return home propellant' in their spacecraft when they land. In the other case they will have no propellant with them when they land on Mars, and if the automated propellant 'factory' fails to work, then they are marooned! To be fair, Zubrin and others have devised strategies to ensure that the arriving astronauts will find propellant already prepared for their return, as the automated factory is sent to Mars in advance, and operated before the travellers depart from Earth. But there must be some extra risk associated with ISRU compared with having a 'return ticket' in the form of fuel. This and other issues will be highlighted in the discussion on the NASA Reference Mission in the next chapter.

7.8   THE METHANE ROCKET, AND MORE CHEMISTRY ON MARS

Methane is not a regular rocket fuel, but it has been used experimentally – especially in the Russian space programme – and engineering experience exists, although much more experience has been accrued with other hydrocarbon fuels and liquid oxygen. The mighty first stage of the Saturn V used liquid oxygen and kerosene (paraffin), and the first successful liquid-fuelled rocket vehicle, the German A4, used liquid oxygen and alcohol. Methane is similar in performance to these fuels, except that it has to be liquefied and stored cryogenically, like the oxygen. The ratio of 4:1 hydrogen to carbon atoms ensures a relatively low molecular weight in the exhaust gases, and in general there is little difference between it and the other hydrocarbons in performance. Cryogenic storage in the rocket propellant tanks is found to be less of a disadvantage for upper stages of launch vehicles, where the quantity of propellant is smaller; and because of the low external pressure, many 'room temperature' liquids boil. For this reason, methane has appeared mostly in upper-stage engines. An optimal performance, based on that of a Russian RD-185 upper-stage engine, is an exhaust velocity of 3,785 m/s – a very respectable value, suitable for use in the return journey from Mars. The interesting factor here is that an exhaust velocity, not very much less than can be attained using liquid hydrogen as the fuel, is achieved with a much more tractable propellant, and, moreover, one that can be produced on Mars from a twelfth of its mass of transported hydrogen. It is not only easier to store than liquid hydrogen, but it has a much higher density, so that the same mass of propellant can be stored in much smaller tanks, both on the surface of Mars and in the rocket vehicle itself. Methane has a density of 0.425, compared with liquid hydrogen at 0.071, and is much closer to the density of kerosene: about 0.8. The tanks for the ascent vehicle can therefore be much smaller if methane, rather than hydrogen, is used; and it will therefore be much easier to transport the methane, made on Mars, into Mars orbit and beyond.

In the NASA programme, methane has not been used, but the RL-10 engine, originally developed for the Centaur upper stage at the very beginning of the Space Age, is still in use nearly forty years later. It has evolved over time, and is one of the most versatile cryogenic engines, with a very long history of successful use. Adapting this engine to use methane instead of hydrogen, making use of the Russian experience, is thought to be the best way to produce a Mars engine. The basic design can be retained, but everything will be easier than with hydrogen because of the higher temperature of the liquid methane, and its low permeation compared with hydrogen (the latter is notorious for leaking through seals, because of its small molecules). The Space Shuttle main engine is not suitable, because it is intended for reuse and, compared with the RL-10, is over-engineered for a single-use engine. This does not mean, however, that the RL-10 cannot be fired several times within its lifetime.

More martian chemistry was promised; and we now encounter a minor difficulty with the proposal to use the Sabatier process to peoduce martian methane and oxygen. It would be a miraculous coincidence if the only sensible way to make rocket propellant on Mars happened to produce the fuel and oxidant in exactly the right proportions for the rocket engine's best performance. It will be recalled that the performance of any chemical rocket engine depends very much on the exact ratio of fuel to oxidant; the Space Shuttle main engine, for example, uses a very fuel-rich mixture to produce a lower mean molecular weight in the exhaust. The general effect

Figure 7.7. An RL-10 engine in a test rig. This engine – adapted to use methane instead of hydrogen – is the baseline for a Mars engine. The cooling channels in the nozzle can be seen; and for operation in space, an extension can be added to make the nozzle longer. The RL-10 has a very long history of successful use; and – as the hairstyle and the oscilloscope indicate – this is an early version. (Courtesy NASA.)

is that the maximum chamber *temperature* is attained when the fuel and oxidant are used at the stoichiometric ratio; that is, when there is just enough fuel to use up all of the oxidant provided. The equation for the chemical reaction in the methane rocket is

$$CH_4 + 2O_2 = CO_2 + 2H_2O$$

or, inserting the molecular weights,

$$(12 + 4) + 2(32) = (12 + 32) + 2(2 + 16)$$
$$16 + 64 = 44 + 36$$

The stoichiometric ratio for a methane rocket is therefore 4:1 oxygen:methane mass. The Sabatier process, using electrolysis to convert the resultant water into oxygen and hydrogen, results in a ratio of 2:1 oxygen production, remembering that 4 tonnes of hydrogen produces 16 tonnes of methane and 32 tonnes of oxygen. A somewhat fuel-rich mixture is beneficial in the methane rocket. With water at a molecular weight of 18 and carbon dioxide at 44, having as much water in the exhaust as possible will lower the molecular weight and increase the exhaust velocity. However, a strong departure from the stoichiometric ratio will lower the combustion temperature, which will ultimately lead to a decrease in the exhaust velocity. The chemistry is also complicated by dissociation, which leads to a more complex exhaust stream composition; but having as little as half as much oxygen as would be necessary for complete combustion, will drastically reduce the combustion temperature and hence the exhaust velocity. With a 2:1 oxygen:methane ratio, only about 1,700 m/s could be achieved – about the same as the nuclear thermal engine with carbon dioxide propellant. Theoretical and experimental work in the Russian space programme suggests that an oxygen:methane ratio of 3.5:1 is about optimal to generate – conservatively – an exhaust velocity of 3,480 m/s. Thus it would be completely wrong to use the oxygen and methane in the proportions as generated by the Sabatier electrolysis process. There are then two options – one of which is to throw away some of the methane produced, and load propellant into the ascent vehicle tanks at the 3.5:1 ratio. Unfortunately, this will considerably reduce the advantage of generating the propellant on Mars. To arrive at a ratio of 3.5:1 requires that, of the 16 tonnes of methane prepared from the initial 4 tonnes of hydrogen, 6.9 tonnes have to be thrown away and only 9.1 tonnes flown. Then the leverage in generating propellant drops from 12:1 to 10.3:1, and more hydrogen would need to be carried to Mars in order to make up this difference. The saving compared with an all Earth-generated propellant mission is still, of course, very large.

Efforts have been made to solve the problem by using more martian-based chemistry: essentially, to produce more oxygen from the same amount of hydrogen, in order to improve the oxidant to fuel ratio of the production process. The solid electrolysis process, described above, could be used in parallel to produce more oxygen from the martian carbon dioxide. It would require no more hydrogen to be transported, as it uses only the atmospheric carbon dioxide and electrical power. But the above-mentioned disadvantages of this method remain: the process requires a large amount of power, and relatively sophisticated equipment. An alternative is to look for a simple chemical process that might be used. The most popular suggestion again takes us back to before the Second World War, and another method of producing propellant for petrol-engined vehicles, from solid fuel: the water gas process. If steam is blown through a bed of hot coals, then a mixture of carbon monoxide and hydrogen is produced. The reaction is represented by the following equations:

$$2H_2O + C = CO_2 + 2H_2$$
$$CO_2 + C = 2CO$$

The reverse reaction is also possible if a catalyst – preferably copper – is used, and amounts to

$$CO_2 + H_2 = CO + H_2O$$

It is essentially a reduction reaction, with hydrogen acting as a reducing agent to convert carbon dioxide into water and carbon monoxide. The water, as we know,

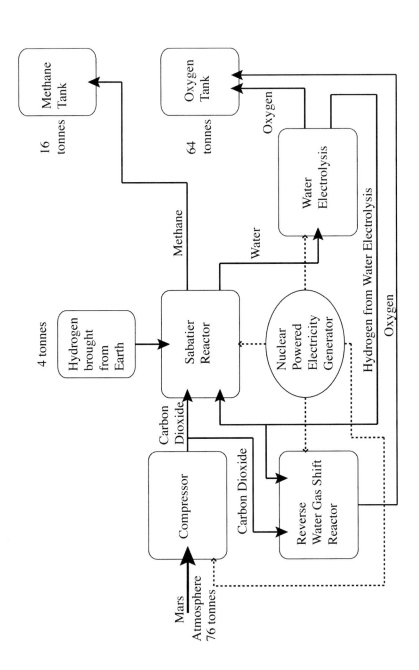

Figure 7.8. The production of oxygen and methane on Mars, using the Sabatier reaction combined with the reverse water gas shift reaction. This produces oxygen and methane in the correct proportions for rocket fuel, producing twenty times the mass of rocket propellant over the hydrogen brought from Earth.

can be electrolysed to produce oxygen – the required material – and hydrogen, which can be recycled to generate more water. A process using this 'reverse water gas' reaction can therefore extract oxygen from the martian atmosphere, with no net consumption of hydrogen. Electrical power is required to heat the catalyst to about 400° C to start the reaction, but as it is only mildly endothermic, very little heat is required to maintain it. It has, in fact, been proposed to combine the reverse water gas reactor with the Sabatier reactor in the same apparatus. The heat generated by the exothermic Sabatier process could be used to maintain the endothermic reverse water gas reaction, so that no net heat need be introduced to produce the two materials, methane and water. The water has to be electrolysed to separate out the oxygen and hydrogen, and this consumes power. The reverse water gas process is not so chemically efficient as the Sabatier process, and reacts only about 30% of the carbon dioxide; and to achieve even this, the reactor must be overloaded with one or other of the reactants, carbon dioxide or hydrogen. The emerging gas will therefore still contain carbon dioxide and hydrogen. The hydrogen must be recycled to minimise the wastage of this precious material, and it must be separated from the carbon monoxide, water, and carbon dioxide. This is a matter of chemical engineering, and is not too difficult to achieve, particularly as hydrogen has such a low boiling point. A combined reaction equation, as given by Zubrin, is

$$2CO_2 + 6H_2 = CH_4 + 2CO + 4H_2O$$
$$4H_2O = 4H_2 + 2O_2$$

which provides the correct production ratio for rocket propellant, with 4 tonnes of hydrogen producing 16 tonnes of methane and 64 tonnes of oxygen. The 4 tonnes of hydrogen now produces 80 tonnes of rocket propellant – a leverage of a factor of 20 – in the ratio appropriate for the most efficient use, and it should generate an exhaust velocity of at least 3,480 m/s. The extra mass of propellant has all been extracted from the martian atmosphere, either as the carbon atoms in the methane, or as oxygen.

The production of rocket fuel on Mars is not at all sophisticated or exotic. All the processes involved are used in the chemical industry on Earth, and are very well established. Some, indeed, are more or less amateur processes, and could be set up by anyone with glass vessels, red rubber tubing, and iron pipes for the reactors, with a few nickel and copper filings for the catalysts. Of course, forming this into a minimum-mass space-engineered rocket fuel factory, that will run unattended on Mars after a 258-day journey through space and, hopefully, a soft landing, is another matter. But much thought is being put into it now; and much effort, in prototyping and testing, will be required before the Mars explorers will trust their lives to it. Nevertheless it is a very serious proposition, and will almost certainly play its part in the first Mars expedition.

7.9   AN EXPEDITION USING MARTIAN ROCKET FUEL

Because propellant can be produced on Mars, with a factor of 20 saving in material mass delivered to the martian surface, it is imperative that we consider providing the whole of the propellant for the return journey in this way. We can calculate the propellant saving that would arise, and determine whether it results in a significant difference, making estimates for the mass of the power supply and of the automated rocket fuel factory itself. To simplify matters, and to anticipate the approach taken by NASA in its Reference Mission, we will assume that power is provided by a nuclear reactor, delivered to the surface, along with the propellant factory. It is necessary to choose a baseline mission with which to compare the martian rocket-fuel mission, so that the savings can be identified. Of the many possible scenarios it is easiest to choose an all-cryogenic outward journey, using liquid hydrogen and liquid oxygen. Then we should see how much the martian methane and oxygen could save us in best to propellant mass. The reason for this approach is that liquid hydrogen has, in any case, to be transported to Mars to make the methane and oxygen, and the engine used for hydrogen and oxygen will more probably be adaptable to a dual propellant, using hydrogen for the descent and methane for the ascent. To carry out the calculation we need the results of Table 4.2 (which for convenience is also included above as Table 7.1).

The delta-V requirements will be the same, but the fuel multipliers will be different. The delta-V requirements are approximately the same whether or not there is a period spent in Mars orbit before Mars–Earth transfer. The basic mission we have so far adopted assumes that the vessel leaving the martian surface would not be the vessel used to travel to Earth, because of the length of the journey and the crew requirements. If rocket propellant made on Mars is to be used for the whole return journey, then this might need to be reconsidered, as issues of propellant transfer from the ascent vehicle to the Earth return vehicle in zero-g arise, and would certainly add to the complexity. For the moment we shall assume that this can be handled.

It will be assuned that the manoeuvres taking the mission to Mars, and the landing on the surface, will use liquid hydrogen and liquid oxygen; while those for the ascent, the departure from Mars, and Earth capture, will be carried out using liquid methane and liquid oxygen. Each manoeuvre will be made at the exhaust

Table 7.1. Fuel requirements using cryogenic propellants

Manoeuvre	Delta-V (km/s)	Multiplier	Vehicle	Fuel mass (tonnes)
1	3.55	1.18	ABCD	828.1
2	2.45	0.72	BCD	249.6
3	3.42	1.12	CD	90.3
4	3.42	1.12	D	5.6
5	2.52	0.73	B	113.4
6	3.55	1.18	B	82.6

velocity and mass ratio appropriate to the propellant used. As in Chapter 4, the propellant requirements will be calculated 'backwards', with the later manoeuvres determined first.

The lift-off from Mars – manoeuvre 4 – uses liquid methane, and has a conservative exhaust velocity of 3,480 m/s. Applying the rocket equation, and calculating the fuel multiplier, we proceed as follows:

$$V = v_e \log_e \frac{M + M_f}{M}$$

$$\frac{M_f}{M} = e^{\frac{v}{v_e}} - 1$$

where $\frac{M_f}{M}$ is the fuel multiplier – the factor by which the payload mass is to be multiplied in order to calculate the required propellant mass. For manoeuvre **4**, the ratio V/V_e is 3.42/3.48 = 0.98. Substituting this in the equation gives, for the fuel multiplier, 1.67. This is larger than the multiplier for liquid hydrogen, reflecting the smaller exhaust velocity with the methane propellant. Using the same process, the other multipliers can be computed, noting that manoeuvres **4**, **5** and **6** only need be recalculated. For a proper comparison, the payload for the ascent vehicle should now include the propellant for the other two manoeuvres, **5** and **6**, and so these have to be calculated first. The payload for Earth capture is the Earth–Mars transfer vehicle, weighing 70 tonnes. The methane-powered deceleration then requires 127.4 tonnes of propellant, calculated in the same way. This, plus the 70-tonne Earth–Mars transfer vehicle, forms the payload for the departure from Mars orbit. The propellant required for this manoeuvre is 209.2 tonnes. The total propellant for these two manoeuvres must be manufactured on Mars, and, together with the ascent vehicle, forms the payload for the ascent. The sum of these three propellant requirements is 907.1 tonnes – the total required to be produced on Mars – and from this we can calculate the amount of hydrogen that has to be brought to Mars and landed on its surface. This is just one twentieth of the total propellant required, or 45.4 tonnes of hydrogen. This has to be landed on Mars, so the lander payload is this quantity plus 75 tonnes for the habitat and the ascent vehicle. The lander will use liquid hydrogen as fuel, and the multiplier is unchanged from Chapter 4. With the new, heavier payload, 139.4 tonnes of propellant are required to land on Mars. We can then calculate the amount of liquid hydrogen–liquid oxygen propellant required for Mars capture and Earth departure as before, using as payload the vehicle plus propellant for the subsequent manoeuvres. The results are presented in Table 7.2.

Comparing the results using martian propellant with those using Earth-produced propellant, we might feel disappointed. Some of the figures are larger, and those that are smaller are not *much* smaller. For example, we now need 570.5 tonnes of methane–oxygen propellant to ascend from the surface of Mars, whereas before, using hydrogen–oxygen we required only 5.6 tonnes. The saving on Earth departure propellant is only 168 tonnes of 828 tonnes – which is useful, but not outstanding. The total saving is clear when we add up how much propellant has to be taken from

Table 7.2. Fuel requirements using propellant production on Mars

Manoeuvre	Delta-V (km/s)	Multiplier	Vehicle	Payload (tonnes)	Fuel mass (tonnes)
1	3.55	1.18	ABCD	559.5	660.2 (H_2)
2	2.45	0.72	BCD	352.3	234.2 (H_2)
3	3.42	1.12	CD	120.4	134.9 (H_2)
4	3.42	1.67	D	342	570.5 (CH_4)
5	2.52	1.06	B	197.4	209.2 (CH_4)
6	3.55	1.82	B	70	127.4 (CH_4)

Earth for the entire mission, which for the martian production mission is simply the sum of the first three figures in the last column of Table 7.2: 1,034 tonnes. This is to be compared with the all Earth production mission, where we have to sum the whole column in Table 7.1: 1,370 tonnes. The total saving is therefore 336 tonnes – which is still not dramatic, but certainly useful. The poorer performance of the methane fuel greatly effects the amount of propellant required for the later manoeuvres; but these do not impact directly on the total, because the propellant is made on Mars. The 45 tonnes of hydrogen that has to be landed on Mars does have an impact, however, as it raises the payload for landing from 75 tonnes of vehicle plus 5.6 tonnes of lift-off propellant, to 75 tonnes of vehicle plus 45 tonnes of hydrogen – which in turn raises the descent propellant requirement from 90 tonnes to 134 tonnes. The first two manoeuvres show a net saving, because the fully fuelled return vessel, left in orbit, is now empty, and waiting for propellant from Mars.

But this is not the complete story, as we have allowed nothing for the power supply, for the propellant factory, or for the tanks to contain the hydrogen on the lander and the large amount of methane for the ascent vehicle. Both will have a significant effect on the bare vehicle masses. It should also be noted that this is *not* a fair assessment of the Mars Direct mission approach as proposed by Zubrin and his collaborators, who have in mind a multi-mission scheme in which the hydrogen, power supply and propellant factory are first sent to Mars before the human expedition leaves Earth. This completely alters the economics of the scheme, rendering it much more favourable. However, our object here is to compare each technological advance in turn, using the same basic mission scenario. In this way we can make a direct assessment of the potential contribution of each technology to the human Mars expedition.

In reviewing this calculation and its results, the somewhat disappointing performance gain from the production of propellant on Mars should be taken in context. If the message of the previous chapters is anything, it is that there is no magical solution to the technical problems of mounting an expedition to Mars. Each advance in technology or concept pares a little off the fuel bill, and it is only by a combination of appropriate new technology and an intelligent approach to mission design that we can make the mission achievable. In this context, bringing in our crude basic mission at about 1,000 tonnes with all-chemical propulsion is a major

Figure 7.9. The propellant factory on Mars, preparing fuel for the explorers' return journey. The factory is on the ground at lower right, and the nuclear power truck is at lower left. The insulated fuel tanks are on the lander-ascent vehicle structure. (Courtesy NASA.)

achievement for ISRU. We obviously now need to consider other methods and determine what else we can do to make savings. We know that nuclear propulsion, where it can be used, will reduce the overall fuel requirements even more; and it is clear that the intelligent use of *all* available resources will be necessary in order to achieve the maximum fuel savings. This will require a reassessment of our basic mission scheme, which is rapidly becoming seen as useful only as a baseline, to compare the advantages of the different technical advances discussed so far.

7.10 USING THE ATMOSPHERE AS A BRAKE

The atmosphere of Mars, and indeed of Earth, can also be used as a brake. We have seen, in the above, that no single rocket technology will enable the expedition, and that we shall have to use every resource available to reduce the fuel cost. Nuclear propulsion, electric propulsion, martian propellant, refrigerated storage of cryogenic propellants; all these will play their part. But there are other resources, mentioned in Chapter 3: the use of gravity assist – stealing momentum from the planets themselves – and the use of aerodynamic braking. The Apollo flights made use of the atmosphere to return the astronauts to Earth, and the Space Shuttle also uses it. In both cases, rocket engines were used to position the spacecraft correctly, but not to slow it significantly; and, in fact, every astronaut brought back safely to Earth has returned using aerodynamic braking. Sadly, there has now been an accident during re-entry with the Space Shuttle, involving the loss of seven astronauts; whatever the cause of this accident,[1] aerodynamic braking remains the best method of returning astronauts to Earth. The basic Mars mission implicitly requires aerodynamic braking, if only because the Shuttle or a similar vehicle was assumed as the means of returning the expedition from Earth orbit, to its home on Earth. The landing on Mars is also capable of using aerodynamic effects, as mentioned in Chapter 3.

It is now time to seriously consider atmospheric braking as a means of saving propellant on both the outward and the return journeys; it is one of the elements of the NASA Reference Mission. In general terms, using the atmosphere to slow down a vehicle – aerodynamic braking – is a feature of a number of different spacecraft manoeuvres. The nomenclature adopted by NASA defines *aerobraking* as the process of converting a highly elliptical orbit into a circular orbit by using the atmosphere of a planet, This was used to place an orbiting planetary observatory, Mars Global Surveyor, into its final circular orbit, with a major fuel saving over a purely propulsive orbital correction. *Aerocapture*, on the other hand, is defined as the manoeuvre that slows a spacecraft, travelling in a hyperbolic orbit, into an elliptical or circular orbit with *one passage* through the upper atmosphere of the planet. This could be used to provide the whole of the capture delta-V, with no expenditure of propellant. It is a much more dramatic use of aerodynamic braking than aerobraking. In the former case the spacecraft may make hundreds of passages through the thin upper parts of the atmosphere, losing a little velocity with each passage, until the elliptical orbit has been circularised. It will be recalled that to circularise an elliptical orbit to the smaller radius requires velocity loss at periapsis. So gentle is the process of aerobraking that the spacecraft uses no protective shield, and, indeed, the extended solar panels are used to increase the area presented to the atmosphere and to increase the efficiency of the braking. The configuration of the panels is sometimes changed to maintain the stability of the spacecraft during atmospheric passage.

Aerocapture requires a major conversion of kinetic energy into heat, and a protective shield is essential. Furthermore, it also requires very careful control of the trajectory through the atmosphere, to avoid either deeper passage leading to excessive heat generation, or shallow passage leading to a bounce off the atmosphere and failure to capture. The main concern for us is the third application of aerodynamic braking, entry into the Mars atmosphere, deceleration, and landing on the martian surface. For NASA this is *entry* or, in the case of Earth, *re-entry*. Here the vehicle is slowed quickly, converting all the kinetic energy into heat, and then lands using parachutes, followed by rockets, to slow the final descent, or a combination of parachutes with airbags to cushion the fall. Re-entry is also a military space term, referring to warheads re-entering the atmosphere from ICBMs. The thermal protection technology is the same.

---

1   The final report on this accident identifies a failure of the ablative cooling on the Shuttle port wing as the cause. The impact of a detached block of insulation from the main tank damaged the insulating tiles during launch.

7.11   EARTH CAPTURE AND RE-ENTRY

The Apollo return used a small capsule, in which the three crew-members could lie, side by side, with just enough room. Below them were storage compartments for oxygen, fuel for the attitude-control rocket engines, and the heat shield that protected them from the intense heat of re-entry; above them, in the 'pointed' end, were stored the parachutes that would later be deployed to slow them down for the eventual drop into the ocean. Very similar arrangements were used in the Russian space programme, with the single difference of a landing on the ground. The sea, in fact, does not help much with the landing loads; but it covers a much larger area and there is a greater choice of sites for splash-down, which rescue and recovery teams can easily reach. There is a big difference, in the re-entry requirement, between the Space Shuttle and the Apollo capsule. The Shuttle has to be reusable, and it must land in a designated location: a particular airstrip. With respect to re-entry and aerodynamic braking, therefore, the entire design is different. In the one case the whole capsule can be discarded after launch, as the heat shield is a single-use device and can be a unitary construction with no joins or gaps. Moreover, the shape of the capsule can be optimised for re-entry. In the case of Apollo it was a reversed blunt cone, which maintained its stability during re-entry; it is a shape still favoured for atmospheric entry of robotic missions to Mars. In the case of the Space Shuttle, the vehicle must be used both as a re-entry capsule and as an aircraft, to allow it to be flown and landed at a particular place. It is improbable that this requirement would be included on a return capsule from Mars, as the tried and tested, stable, reverse cone could be used.

There is no reason why, for the Mars expedition, return, re-entry and landing cannot be carried out using only aerodynamic braking, in a return capsule such as that used for Apollo. The attraction of this idea is immediate, having seen how much propellant, made on Mars, is required for Earth capture. A direct saving of 127 tonnes, together with the consequent savings in not having to bring up this amount from the martian surface, is well worth considering. The direct fuel saving is really in the one-twentieth of hydrogen required to make the methane and oxygen, but the tanks and other ancillaries associated with so much propellant will be significant in a real mission. If we intend to use the atmosphere to enter and land on the Earth, it is important to establish how well it can work, for return from Mars. We know it worked well many times for the Apollo missions, and it even worked for the Apollo 13 return, in which the trajectory was not optimum. But return from Mars is somewhat different.

The simplest way to examine this is to compare the return velocity of the Apollo capsule with the return velocity of the Mars capsule. What we are doing is to conceive a more sophisticated return for the Mars expedition, where expedition members and their goods are transferred to a small re-entry capsule, as the whole complex approaches Earth at transfer velocity. Once the expedition and its samples are in the re-entry capsule, the large Mars–Earth transfer vehicle is then shed. The transfer vehicle is not slowed down, and passes Earth in a hyperbolic orbit, while the re-entry capsule is guided into the atmosphere, for aero-braking. Its initial velocity will be the same as before: the perihelion velocity of the Mars–Earth transfer orbit.

Reference to Chapter 3 shows this to be a velocity, relative to Earth, of 3.55 km/s, at closest approach – relative to Earth because in our basic mission the expedition ended when the spacecraft was captured into a circular orbit about the Earth. Now we have to consider the full velocity, including the 7.6 km/s circular orbit velocity, and all this, and more, has to be lost in the return to Earth's surface. The same calculation as was performed for the landing on Mars shows that the necessary velocity change, to land from the 7.6 km/s circular orbit, is the sum of 7,619 – 7,473 = 145.7 m/s, to enter the descent orbit from the circular orbit, plus 8,060 m/s, which is the velocity of the capsule, in the descent elliptical orbit when grazing the Earth's surface. This produces a total of 8,206 m/s for transferring from the circular 500-km orbit, and to this must be added the 3.55 km/s that has to be removed from the spacecraft for it to enter the circular orbit around the Earth. It is not necessary for the spacecraft to first enter the circular orbit; this is just a convenient fiction to allow the calculation to be carried using available data. It is correct to add these velocities in order to calculate the total velocity change necessary to land safely on Earth. This total – 11,760 m/s – represents the total kinetic energy that has to be removed from the spacecraft by aerodynamic braking.

We can now compare this figure with the Apollo return velocity. The Apollo circular orbit velocity was 7.79 km/s, the velocity for escape to the Moon was 10.83, and the circular Earth orbit was at 100 nautical miles (185.3 km). The descent ellipse was therefore different. Applying the same calculation, the result is 10,604 m/s. Thus the Mars mission would arrive on the Earth's surface 1,156 m/s faster than the Moon mission, if it were not for the atmosphere. The atmosphere has to take this additional velocity away from the spacecraft – just 11% more than the Apollo capsule had to cope with. There should therefore be no real problem in using aerodynamic braking to save the return and landing propellant; and if it could be achieved in 1969, it can certainly be achieved in the twenty-first century. It would save the 3.55 km/s accounted for in the basic mission, and, in addition, the full landing velocity change that we had assumed would be taken care of outside the scope of the basic mission. The success of the Apollo return method guarantees the success of aerodynamic braking for return to Earth from Mars. Apart from manoeuvring, and separation from the Mars–Earth transfer vehicle, which use only a small amount of propellant, reactive propulsion is not required for Earth capture and landing. This will save a considerable amount of propellant being lifted from Mars surface, or taken all the way from Earth to Mars and back again.

7.12 MARS CAPTURE, ENTRY AND LANDING

This chapter is mainly concerned with the use of local resources on Mars. The martian atmosphere is not only a source of raw materials, but can also be used for aerodynamic braking. But the conditions are markedly different from those on Earth. Mars has only 10% of the mass of Earth, and gravity about 38% of Earth's, and the velocities associated with a Mars landing are therefore much smaller. The escape velocity is only 5 km/s, but at the same time, the atmosphere is also much thinner and less able to slow down the spacecraft. At 7 millibars, the surface pressure is, on average, about 1% of the Earth's, but the scale height – the measure of how rapidly the pressure drops with altitude – is 44% smaller than Earth's. Consequently, above a certain altitude the atmosphere of Mars is denser than the atmosphere of Earth, at the same altitude. In evaluating the efficiency of aerodynamic braking it will be useful to determine this altitude. The variation of pressure with altitude is represented by

$$p = p_0 e^{\frac{h}{s}}$$

where S is the scale height – 7.5 for Earth, and 10.8 for Mars – and h is the actual height. By setting the required pressure to be equal for Mars and Earth, and then solving for h using the two scale heights and the two surface pressures, the cross-over altitude can be determined: 122 km. Above this altitude, the atmosphere of Earth is thinner than the atmosphere of Mars.

If we now consider the Apollo return we find that aerodynamic braking began at 400,000 feet – coincidentally, just 122 km. It was this point at which the Apollo capsule was aimed to begin the entry manoeuvre; braking was complete at about 180,000 feet, or 55 km. It is obvious that aerodynamic braking on Earth all took place in the region where the atmosphere was denser than the atmosphere of Mars; and aerodynamic braking on Mars will be significantly different from aerodynamic braking on Earth. The scale height is larger, so that the atmospheric density increases more slowly as the spacecraft penetrates; and as the gravity of Mars is much less than that of Earth, the velocities are all smaller. Approaching from deep space, a vehicle reaches about 7.3 km/s by the time it encounters significant atmospheric braking. The Apollo craft began braking at 11 km/s – the actual Apollo velocity. The value calculated above makes certain simplifying assumptions in order to be consistent with our other calculations.

Figure 7.10. The entry of the lander into the martian atmosphere. (Courtesy NASA.)

The fact that Apollo braking occurred at higher atmospheric densities, and at higher velocities, implies that the martian braking will be gentler, with both the approach velocity and the atmospheric density being smaller. To attempt to calculate the atmospheric braking for Mars missions is beyond the scope of this book; but there are several successful examples of the landing of robotic probes on the surface of Mars using aerodynamic braking, from Viking in the 1970s through to Mars Surveyor in 2001. A typical entry and landing profile begins with the spacecraft approaching at interplanetary fly-by velocity – about 7.3 km/s – and aimed at a point about 130 km above the planetary surface. The angle of entry is about 14° from the local horizontal (although Apollo entered the Earth's atmosphere at a flight path angle of about half this, around 5–7°). The martian atmospheric density at the entry altitude is about 1.2×10^{-7} kg/m^3. The important parameter for aerodynamic braking is the *dynamic pressure*, which represents the effective pressure exerted on the spacecraft due to its motion through the atmosphere. It is a well-known parameter for vehicle flight in an atmosphere, and is defined by the equation

$$q = \frac{1}{2} \rho V^2$$

where ρ is the atmospheric density, and V is the vehicle velocity. In the case of an entering spacecraft, the velocity decreases with altitude as the vehicle slows, while the density increases. It is obvious, therefore, that the dynamic pressure passes through a maximum value, somewhere between the point of entry into the atmosphere and landing. If it can survive this maximum pressure, then the rest of the journey, before and after the maximum, can also be survived. The maximum dynamic pressure is referred to by engineers as 'maximum q', or 'max q'. The forces acting on the spacecraft during the manoeuvre comprise *drag* and *lift*. Drag is of most interest to us, but lift is also important in that it is an upward force countering the vertical force of gravity. Too much lift will result in a bounce off the atmosphere, and it has to be taken into account when determining the trajectory of an entering spacecraft. It falls to zero if the *angle of attack* – the angle which the primary axis of the spacecraft makes with the flight path – is zero. Drag and lift both depend on the dynamic pressure, together with the cross-sectional area of the spacecraft. The general equation for the drag force acting on the spacecraft is

$$F_D = \frac{1}{2} \rho V^2 C_D A$$

The dynamic pressure is modified by the coefficient of drag, C_D. For hypersonic flight, which this certainly is, it has a value of about 0.15. The cross-sectional area of the spacecraft is represented by A.

At the point of entry, 130 km above the surface, the dynamic pressure is small – only 3.16 N/m^2 – which is a very gentle beginning. During the next minute, the spacecraft slows to about 6 km/s, while descending to an altitude of about 40 km above the surface, where it reaches the point of maximum dynamic pressure, 8,867 N/m^2. The density has increased by a large amount, while the thin atmosphere has not yet had time to slow the spacecraft appreciably. This is also the point of maximum heating. The greatly increased dynamic pressure now quickly decelerates the spacecraft, and 90 seconds later it is moving sufficiently slowly – only 400 m/s – to release the parachute. This takes place at an altitude of 8.6 km at which point the dynamic pressure is down to 722 N/m². With the greatly increased area of the parachute, the velocity lessens to 100 m/s, 20 seconds later. The dynamic pressure is now more or less constant at about 30 N/m², and the spacecraft is falling at a constant speed of about 60 m/s. The atmosphere on Mars is not sufficiently dense for the velocity to drop below about 60 m/s, while using a parachute, and other means have to be used to slow the vehicle for landing. This can be a fully controlled landing, using rockets that enable the vehicle to hover, or it can be a combination of a rocket burst to slow the velocity to zero about 50 m above the surface, followed by a drop to the ground, cushioned by airbags. The latter would be very uncomfortable, if not fatal, for the explorers, because it involves a number of high bounces accompanied by random rotation of the spacecraft. It will not be used for manned landings.

7.13 PROTECTING THE SPACECRAFT

Some idea of the forces involved, and of the energy dissipation, can be gained by applying the drag equation given above to the point of maximum dynamic pressure. Substituting the maximum value in the equation and applying the drag coefficient produces, per square metre of cross-section, 1,330 N for the retarding force. The work done by this force, per second, is $1,330 \times 6,000$ for a speed of 6,000 m/s, or 8 MW. This power has to go somewhere, and it goes into heating the atmosphere. A crude idea of the temperature can be gained by considering the cylinder of air of cross-section 1 m², through which the spacecraft passes in 1 second. The mass is $6,000 \times 4.9 \times 10^{-4}$, or 3 kg. Taking an approximate value for the specific heat as 240 calories per kg per degree, and 4.2 joules per calorie, produces, for the temperature rise associated with 8 MW power dissipation, 2,645 K. Note that this cannot in any way be considered as an accurate figure, because of the simplification used here.

   The spacecraft has to be protected from this very high temperature plasma, but fortunately the period of high temperature does not last long: about 100 seconds. Techniques long used for the protection of solid rocket nozzles from hot gases at a similar temperature can therefore be used. This technique is called *ablative* cooling. The spacecraft is protected by a heat shield consisting of an insulating layer, forward of which is a thick layer of composite material designed to burn away during the entry. As the composite material burns away it absorbs heat by being vapourised, and at the same time the products of vapourisation form a layer of cool gas that flows across the surface of the heat shield and round the rear of the spacecraft. This keeps the surface of the vehicle relatively cool by preventing the really hot gases from reaching the surface. This is a very well established technology that relies on the fact that the heating is soon over – and it works. Obviously, if the ablative material all burns away then the spacecraft has no protection and will burn up, and so the period

Figure 7.11. The Mars transfer and landing vehicle on its way to Mars. The rear portion – abaft the barred ring in the centre of the image – is the nuclear stage. Forward of this is the biconic aeroshell containing the lander. The dark area has more ablative shielding, and is presented to the martian atmosphere at entry, with the nose well up. (Courtesy NASA/JPL.)

of heating has to be limited, for ablative cooling to work. Fortunately, this is the case. The materials used are generally fibrous, and contain silica and phenolic compounds. More advanced composite materials such as carbon–carbon are now being used, and all have the property that they do not melt at high temperatures, but, rather, sublime away, keeping their strength, up to moment of sublimation. The fibrous nature ensures that cracks, which might prove fatal, do not develop.

For these robotic probes it is possible to enclose the spacecraft in a blunt cone, or back-shell, with most of the ablative material in the form of a heat shield on the base. The surface of the back-shell, which trails the base during entry, needs less protection, and is covered with a thinner layer of ablative material. This type of shield and conical back-shell is appropriate for a robotic lander, or for the return capsule used during Earth re-entry, but it is not adaptable for bringing the large, cylindrical vehicles, required a Mars expedition, to the surface. That other geometries are possible is exemplified by the Space Shuttle, which has no back-shell, and instead relies on the hot plasma, produced by the heat shield, flowing round the unprotected parts of the vehicle without contacting them – the heat shield in this case being the tile-covered bottom and leading edge surfaces of the wings and Space Shuttle body. The tiles are made of a composite silica material[2] similar to that

---

2   Carbon–carbon material is now used to replace silica tiles on some areas of the Shuttle.

used in the integral heat shields discussed above. The general expectation for the Mars entry of an expedition vehicle is to use a biconic cylindrical heat shield, combined with a steep angle of attack, so that the plasma is flung wide of the rest of the vehicle. This is the equivalent of the Space Shuttle body without wings.

7.14 LANDING ON MARS

There are some issues that make the landing of expedition vehicles less easy than robotic probes. In the latter case it is not too important where the probe lands, provided it is within the designated area, as the parachutes and air bags are allowed a reasonable margin of error in the final resting place of the lander. For the expedition, precise landing will be required, for a number of reasons. The expedition site will have been selected carefully to meet the requirements of scientific investigation, crew safety, and easy access to neighbouring regions; explorers will want to explore. The return journey must also be secured. For a single vehicle approach, such as in our basic mission, the lander has to be the launch pad for the return vehicle. The landing must therefore take place on a firm site, with the appropriate down-range clearances for the take-off. For a multiple vehicle approach, such as the Mars Direct, or NASA Reference Mission, then several vehicles have to arrive at different times, but at the same location on Mars. In practice they will need to be within 100 m of one other, so that the crew can access them easily. This precludes the airbag solution, even for cargo, and requires a certain amount of manoeuvrability during the last few hundred metres of descent. This would allow selection of the landing site, depending on the small-scale terrain, and would allow vehicles to be positioned close to each another – but not *too* close, as landing one vehicle on top of another would be embarrassing! All this means is that after the parachutes have slowed the vehicle to about 60 m/s, a fully controllable rocket engine will have to be used, to hover, traverse if necessary, and soft-land the vehicle. This will involve propellant and engine dead weight, but it is essential for expedition safety. The engines will have to be reasonably large to land a 70-tonne payload – weighing about 27 tonnes on Mars – but propellant use will be quite modest, given a velocity change requirement of only 60 m/s.

7.15 PROPELLANT SAVINGS

It seems from the above that the use of the martian atmosphere to capture and land the expedition spacecraft is perfectly reasonable, and that the technology and experience is available now. The many tonnes of saved propellant have to be offset against the extra mass of the biconic shield, but in general, most of the propellant mass in the basic mission, intended for Mars- and Earth-capture and landing, can be dispensed with. This is shown in Table 7.3.

Comparing Table 7.3 with the previous table, we find a very significant saving, because of the fuel cost effect. The Mars capture, entry and landing, and Earth capture, re-entry and landing, all have a high fuel cost, being late in the mission. If

Table 7.3. The basic mission modified to include maximum
use of aerodynamic braking

Manoeuvre	Delta-V (km/s)	Multiplier	Vehicle	Payload (tonnes)	Fuel mass (tonnes)
1	3.55	1.18	ABCD	155.3	183.2 (H_2)
2	2.45	0.72	BCD	155.3	0.0 (H_2)
3	3.42	1.12	CD	85.3	0.0 (H_2)
4	3.42	1.67	D	79.2	132.3 (CH_4)
5	2.52	1.06	B	70	74.2 (CH_4)
6	3.55	1.82	B	70	0.0 (CH_4)

we dispense with these propellant requirements and replace them with aerodynamic braking, then, in our approximation, propellant is required only for ascent from Mars, departure from Mars, and departure from Earth, and we save the high fuel cost of transporting those propellants to the place of use. Counting the 10 tonnes or so of hydrogen required to make the methane and oxygen propellant on Mars, together with the lander payload, we require a total of only 183 tonnes of hydrogen and oxygen propellant from Earth, for the entire mission. We use also 206.5 tonnes of methane and oxygen produced on Mars from the 10 tonnes of Earth-hydrogen included in the lander payload.

This very large reduction in propellant requirement, which arises from the use of indigenous resources, atmospheric braking, and the production of propellant on Mars, is very significant in bringing the Mars expedition within reach. We have, in the interest of simplicity, avoided the detailed accounting associated with these novel means of enabling the expedition. Among these are the mass of the nuclear power supply and the propellant factory, both to be landed on Mars, the mass of the biconic heat shield, and the small amount of propellant required for the entry and re-entry manoeuvres and a fully controlled landing on Mars. The tanks to take down the hydrogen and bring the methane back have also not been taken in account – and all of these could add perhaps 15 tonnes to the total. It is, however, not worth the effort of strictly accounting for these complicated scenarios, because a whole new mission strategy is opened up by these possibilities, and the number of permutations grows. There could be further propellant savings using nuclear stages for certain manoeuvres, but not for all of them. We have already alluded to contamination aspects of using nuclear rockets on Mars itself, and it therefore does not make sense to use a nuclear engine with indigenous propellant, even for the departure from Mars orbit. A nuclear stage would in this case be used only for departure from Earth orbit, as capture at Mars would be enabled by aerodynamic braking, and departure from Mars by chemical rocket engine. We have glossed over the return stage, left in Mars orbit, as this would need to be placed in Mars orbit by aerobraking or aerocapture, and refuelled in orbit with martian propellants.

All of these detailed considerations will have to be taken into account in devising an optimal Mars expedition scenario. Since these considerations have been applied, in a much more rigorous way, in defining the NASA Reference Mission, Mars Direct, and Mars semi-Direct, and many other potential means of realising the Mars expedition, it would be nugatory to consider further possible permutations here. In this and the preceding chapters, the basic technological issues connected with the transportation requirements of a Mars expedition have been explained, and a rough comparison of their relative performance demonstrated. Simple ways of combining them to reduce the propellant cost of the expedition have shown how each propulsion technology can contribute to a basic Mars expedition. The next chapter will show how NASA, and others, have made use of these technologies to produce fully self-consistent Mars expedition scenarios. These in turn are being used to define technology-development 'road maps', so that the required engines, reactors, propellant factories, heat shields, and the many other devices that will be required for a Mars expedition, can be developed.

8

The expedition to Mars

Before the beginning of the Space Age – which we can date from the flight of the first Sputnik in 1957 – most people connected with the drive to space had in mind human exploration of the planets. From Kepler through to Tsiolkovsky, and beyond to Oberth and the early German pioneers, as well as for Goddard, the object was to enable humans to travel in space. If satellites were considered, they were manned satellites – forerunners of the International Space Station. It would be a surprise to these early thinkers to observe, today, the proliferation of unmanned satellites in Earth orbit, robotic probes sent to the planets, and so little effort expended in manned space flight. Reality is so often different from the dream. Seen from our perspective, it seems natural that robotic satellites and probes should be used, because we are familiar with the huge advances in large-scale integrated circuits and microcomputers, of which the early pioneers could have had no idea. Computers did not impact, even on specialists, until the early 1950s, and nobody could have predicted that computing power would reach current levels. It would have been even more incredible that computing power, as we know it today, should have been available, effectively, in a matchbox. It is only necessary to read the science fiction of the 1950s to see vast buildings and power stations, holding and supplying the supercomputers of the authors' imaginations.

History has taken a turn, and the early predictions and dreams of manned spaceflight have not come to pass – *pace* Apollo. Manned spaceflight is seen as unnecessary amongst a large community of space users who are quite happy to make money, and obtain information – military, strategic or scientific – from robotic spacecraft.

Nevertheless, there still exists the dream of human space exploration. It persists amongst space enthusiasts of the present day; it persists in the popular imagination; and (if they are honest) it persists in the hearts and minds of many space professionals – managers, engineers and scientists. This dream – the subject of this book – is on the verge of becoming reality. Professional space engineers and scientists, of the world's space agencies, are working on it, and in this chapter we shall examine the best ideas of those who are seriously engaged in preparing for a human expedition to Mars.

The dream of a Mars expedition has been endemic in NASA since its foundation. As we know, von Braun planned for it to follow Apollo, and NASA was fully prepared to continue on from the Apollo missions to manned exploration of the Solar System. But after the first few successful Moon landings, the sudden budget cut put paid to these plans, and a long deviation, through the Space Shuttle and the International Space Station, ensued. Then, in an address presented on 20 July 1989 – the twentieth anniversary of the Apollo 11 landing – President George Bush again opened the prospect of bringing the dream to reality by announcing a 30-year plan leading to a Mars landing in 2019. This led to renewed vigour in NASA, among those enthusiastic for human exploration, and to new hope for those who had been resigned to waiting, yet another generation, for the fulfilment of the Apollo legacy. In the event, there has been slower progress than might have been hoped, although there has been substantial progress towards defining how a human expedition to Mars can be mounted. In this chapter we shall review the outcome of this process: the best guess today of the form of the Mars expedition.

8.1   A REALISTIC MARS EXPEDITION FOR THE 2020s

The basic Mars mission, as outlined in Chapter 4, is prohibitively costly in propellant terms. Huge masses of propellant would be required, and the freighting of such masses into low Earth orbit leads to a huge financial outlay. Even so, the basic Mars mission has shown us where savings in propellant might be made, and which technologies should be developed to enable the expedition. The NASA engineers and scientists who worked on the Mars Reference Mission will have taken similar issues into account in producing their findings; and as they have access to sophisticated computer modelling tools, based on forty years of experience, their findings will be better founded. Our basic mission used a single vehicle complex, launched from Earth orbit, and made up of smaller vessels that could be used for the individual stages of the journey there and back. The result of the NASA deliberations was to separate the functions into several separate vehicles, and launches from Earth, with more use of robotic vehicles to transport elements of the expedition to Mars and to the martian surface. Separating the functions in this way led to savings in vehicle complexity and ultimately in propellant, but also significantly increased the safety margins for the crew.

In moving from our basic mission to a real mission – or its equivalent, the NASA Reference Mission – there are two primary problems that have to be solved: placing a sufficient mass of materiel for the mission into Earth orbit, and building in a sufficient margin of safety for the crew. Each of these challenges imposes significant constraints on the basic Mars mission. We do not possess a heavy launcher that can carry hundreds of tonnes into orbit, and we have no positive experience of protecting human beings from the long-term environmental effects of deep space, or, indeed, the martian surface. We need to understand and address these problems before we can mount a realistic expedition to Mars.

8.2   THE EARTH-TO-ORBIT CARGO VEHICLE

We, on the Earth's surface, lie in at the bottom of a very deep gravitational potential well. So deep is this well, that it is only marginally possible for us to escape from it and explore space. The frequency of launches of satellites for a multitude of purposes, and of astronauts for the International Space Station and Space Shuttle flights, obscures the desperate nature of escape from the Earth's surface. Rockets were invented more than a millennium ago; Man has clawed his way into space only during the last fifty years or so. If we had lived on a smaller planet, with a smaller gravitational field, we might have been in space much earlier; and perhaps ancient Chinese bamboo rockets, powered by gunpowder, would have succeeded in reaching Earth orbit. In truth, however, only the most advanced engineering can overcome the huge difficulties of escaping from Earth's surface – which is why we had to wait until 1957 for the first success. The huge advances in engineering and metallurgy during the twentieth century, and the advances in measurement and control systems, finally enabled us to break the bonds of Earth's strong gravitational field; but fifty years on we are still unable to lift, into space, the payload of a single jumbo jet, let alone that of a small sailing ship. Our exploration of space is much more difficult than Columbus' mission to find the New World.

The R-7 rocket that put Sputnik into space in 1957 had a payload capability of 1.4 tonnes; in 1961, Gagarin's Vostok could lift 4.7 tonnes. Today, an Ariane V can put 18 tonnes into low Earth orbit, the new Titan IV is capable of putting 22 tonnes into low Earth orbit, and the Space Shuttle can place about 24 tonnes in orbit – about a factor of 5 greater than Gagarin's rocket. Part of the slow-down in the development of heavy launchers has been caused by the reduction in mass of the main payloads, due to improved electronics and lightweight construction techniques. Before large-scale integrated circuits, space electronics was heavy and unreliable. This state of affairs quickly improved for NASA because of Apollo, and satellites could increase their capability without increasing their mass and power requirements. But it was not so in the Russian space programme, in which up to sevenfold redundancy was sometimes used to counter the unreliability, with resultant large power supply requirements and cooling problems. Heavy launchers remained a strong feature of the Russian space fleet, while in the US they gradually became less important. Neither should it be forgotten that nuclear warheads also became much smaller – and this had an impact on both US and Russian rocket development.

The main drive for large payload capability was, of course, the manned spaceflight programme. In 1969 the Saturn V could place 47 tonnes into the translunar orbit necessary for the Apollo programme; and it had a payload capacity, into low Earth orbit, of 118 tonnes. After Apollo, the heavy launcher programme in the US stagnated, while the Space Shuttle was developed as the next stage in space technology. The Shuttle's design, however, was somewhat compromised by military requirements – particularly the need (associated with military payloads) for it to be manoeuvrable in order to land at specific landing fields. The Russian Proton rocket – the second workhorse for the International Space Station assembly – can place 22 tonnes into orbit. Indeed, it was putting 20 tonnes into low Earth orbit in 1968 –

Figure 8.1. The Titan IV launch vehicle. (Courtesy NASA.)

although it is, of course, a much simpler vehicle because it is not reusable. This brings us to the fundamental dilemma of heavy launch capability in the twenty-first century: reusable or expendable?

During hard times for the space programme – and they do not come much harder than at present – it is tempting to argue that this or that new approach will save money. Invest *now* in new technology, and access to space will become cheaper: a useful argument to present to managers, and to the holders of purse strings; but it is rarely completely honest, or born out by events. It is not completely honest because the promoters of a particular technology stand to gain if it is selected; and it is often not successful in saving money because new technology has hidden problems which do not emerge until late in the programme, when there is no turning back. This tendency led to new technology such as the Space Shuttle, and away from the Saturn V, and the expendable launcher programme. Something like the Space Shuttle was clearly needed, but the compromises in its design made it much more expensive per tonne than even existing heavy launchers. On the other hand, given the refusal by Congress to fund Apollo follow-on, NASA would have had *no* human spaceflight programme without the Space Shuttle. Taking astronauts into orbit and bringing them back can be achieved by using expendable launchers and return capsules with aerodynamic braking. This is what the Russians do, and they have significant achievements in manned spaceflight. Nevertheless, we are where we are. The two methods of placing humans in orbit remain the Space Shuttle and the Soyuz vehicles; while cargo, in appreciable quantities, again requires the Shuttle, the Titan IV, the Ariane V or the Proton.

The archetypal reusable vehicle is the single-stage-to-orbit, or SSTO – another idea that could reduce the cost of access to space. Here it is hoped that technology will emerge that will allow access to space to be like an aeroplane trip. The vehicle would be launched with its payload on board, discharge its payload in orbit, and soft-land, back on the launch pad, ready to be refuelled for the next launch. It would be cheap, because nothing is thrown away after just one launch: the cost of hardware can be amortised over many flights. This seductive idea has attracted significant effort in space agencies and industry, and many enabling technologies have been reviewed; and some, indeed, are promising. But there are two major difficulties that have not been overcome. Gravity is so strong that it is virtually impossible to reach orbit by using only a single stage. Tsiolkovsky realised this more than 100 years ago, and the physics has not changed. This is the first difficulty. The second difficulty is that, to make it work at all, every aspect of the engineering will have to be stretched to the limits of material properties and ingenuity – which does not make for easy turn-round on the launch pad. Easy maintainability depends on systems being rather 'detuned', so that margins are reasonable; but this is not the case with present SSTO concepts, even should they prove achievable in practice. On the positive side, however, several technologies developed for SSTO could be used to improve the engineering of Mars expedition vehicles. Among these are aluminium–lithium propellant tanks, lighter than anything we have used before, and precision control systems for rocket-powered landing. So, even if the dream of SSTO does not become reality, some of the technology developed for it will find a home in the Mars expedition.

8.3   A MODERN HEAVY LAUNCHER

At present there is no heavy launcher available, although the main commercial launchers are evolving towards bigger payloads. The Ariane V has recently been equipped with a cryogenic upper stage to increase the payload to 10 tonnes in geostationary orbit; which scales to about 24 tonnes in low Earth orbit, almost equivalent to the Shuttle payload. The Titan IV can place 22 tonnes in low Earth orbit, although so far it has apparently not been used for that purpose. For a Mars expedition, and perhaps for bigger robotic missions and space stations, a 200-tonne payload in low Earth orbit may be required. This would have been a modest extension to the capabilities of the Saturn V, but it is a major leap in capacity for modern launchers. Nevertheless, work has been carried out – at least on paper designs – for such a launcher. It is an expendable launcher, because to make it in any way reusable would downgrade its payload capability per unit cost.

It is useful here to examine the present-day heavy launchers, compare them with the Saturn V, and to consider how a new launcher could be conceived, based on the experience of the world's space agencies and industries. Table 8.1 shows the heavy

Table 8.1. Comparison of modern and past heavy launchers

Launcher	Boosters	Stage 1	Stage 2	Stage 3	Payload to LEO
Ariane V EU	Two solid MPS; 13.4 MN	Single Vulcain engine; LO2/LH2; 1,096 kN	Single Aestus engine; N2O4/UDMH; 28 kN	None	19 tonnes
Proton (Russia)	None	Six RD 253 engines; N2O4/UDMH; 10.74 MN	Four RD 0210 engines; N2O4/UDMH; 2.5 MN	Single RD 0212 engine; N2O4/UDMH 642 kN	22 tonnes
Titan IV	Two solid USRM; 14.2 MN	Two LR-87-11 engines; N2O4/Aerozine; 2.44 MN	One LR-91-11 engine; N2O4/Aerozine 467 kN	Centaur; two RL-10A-3A engines; LO2/LH2; 147 kN	22 tonnes
Delta IV EELV-heavy (US)	Two liquid RS-68 engines; LO2/LH2; 6.8 MN	Single RS-68 engine; LO2/LH2; 3.4 MN	Single RL-10B-2 engine; LO2/LH2; 112 kN	None	23 tonnes
Space Shuttle (US)	Two solid RSRM; 23 MN	Three SSME engines; LO2/LH2; 6.9 MN	Two OMS engines; N2O4/MMH; 106 kN	None	24 tonnes
Energia (Russia) no longer in production	Four liquid: RD-170 engine; LO2/kerosene; 31.6 MN	Four RD-0120 engines; LO2/LH2; 7.84 MN	One RD-0120 engine; LO2/LH2; 1.96 MN		88 tonnes
Saturn V (US) no longer in production	None	Five F-1 engines; LO2/kerosene; 34 MN	Five J-2 engines; LO2/LH2; 5.2 MN	Single J-2 engine; LO2/LH2; 902 kN	118 tonnes

launchers. The two with the largest payload capability – the Saturn V and the Energia – are no longer in production; but they indicate the direction in which present-day launchers should evolve, in order to make one suitable for the Mars expedition.

Launchers have two elements which are both essential for high performance into orbit. The first is a high-thrust lower stage to lift the payload and the large quantity of propellant from sea level to high altitude, and the second is one or more upper stages with the highest achievable mass ratio and the highest exhaust velocity to accelerate the maximum payload to orbital velocity. A high exhaust velocity, it will be remembered, is essential in achieving a high vehicle-velocity, and thrust is less important for an upper stage.

It is sensible to base the new launcher on existing technology, whenever possible; and there is no lack of existing, or at least past, technology, which meets the requirements. For high thrust we need look no further than the F-1 engines used on the Saturn V. These used liquid oxygen and kerosene – or RP1 (Rocket Propellant 1), as it was known – the first 'rocket fuel'. These engines were very big, and developed a very high thrust, on the Saturn V first stage. Each of the five engines produced 7.7 MN of thrust, with an exhaust velocity of 2.65 km/s at sea level – equivalent to 3.1 km/s in vacuum. The determinants of a rocket engine's thrust are the combustion chamber pressure, and the diameter of the *throat* – the narrowest part of the nozzle, before it expands to the well-known bell shape. For the F-1, the throat diameter was 0.9 m, the chamber pressure was 70 bar, and the outer end of the nozzle was 3.6 m in diameter. To produce its huge thrust, the engine burned about 3 tonnes of propellant per second. It was developed in the 1950s as a 'million-pound thrust' engine; and between 1967 and 1973, sixty-five were flown on Saturns, with no failures. It is now out of production, but an improved design exists on paper: the F-1A, with a thrust of 9.2 MN. It is possible that for a heavy launcher of the 200-tonne class, liquid boosters, powered by something like the F-1 engine, fuelled with liquid oxygen and RP1, will be used. The Russian programme also has something to offer: the RD-170 engine. This has quadruple combustion chambers and nozzles, fed by a single propellant distribution system. Four of them, on the Energia boosters, delivered 31.6 MN of thrust, not much different from the 34 MN delivered by five F-1 engines on the Saturn V. For these very large engines, the major design problem was combustion instability; but once this was solved, the engines were very reliable.

The SSME is the most efficient rocket engine in use, using liquid hydrogen and liquid oxygen to produce an exhaust velocity of 4.55 km/s and a thrust of 2.4 MN. It is the obvious choice for the upper stages of the new launcher, but it is very expensive to produce, and its potential for reuse would be unnecessary in an expendable launcher. A development for the NASA Advanced Launch Vehicle and the National Launch Vehicle is the Space Transportation Main Engine (STME). This programme was never completed, but elements of it will probably find their way into any new 200-tonne launcher. The STME is a cheaper, expendable version of the SSME, but with a somewhat lower exhaust velocity of 4.3 km/s and a thrust of 2.9 MN.

With the engines for a heavy launcher defined, we can now consider the possible configurations of launchers that would provide us with the desired 200-tonne capability.

8.4   HEAVY LAUNCHER CONFIGURATIONS

Most Mars expedition scenarios have, in the past, called for a heavy launcher of the '200-tonne class', in order to place the components of the mission in orbit. In the last few paragraphs we have seen that our present portfolio of launchers is inadequate for the Mars expedition. If we had an active Saturn V launcher, then there would be no problem. The Saturn V could put 47 tonnes into lunar transfer orbit; but its payload for low Earth orbit was 118 tonnes. If the Mars expedition had followed on directly from Apollo, then the placing of sufficient cargo into orbit would not have been the major problem. Forty years on, however, it *is* the major problem.

Better ideas and improved engineering have evolved since the Saturn V was designed. The use of strap-on boosters reduces the complexity of the main rocket vehicle, and it is possible, as in the case of the modern Delta IV, to vary the number of boosters with the size of the payload. This makes for a more flexible and cost effective approach for general commercial launching. Liquid-fuelled engines are more efficient because of the higher exhaust velocity, and they can therefore be lighter and more effective than solid-fuelled launchers; although, of course, they are much more complicated and costly. If we compare the Proton and the Delta IV, which have similar payload capabilities, we see that the launch thrust is about the same, because the boosters on the Delta IV fire at the same time as the main engine. But the more modern Delta IV requires only a single low-thrust upper stage, while the Proton – a traditional, sturdy, Russian design – uses two more high-thrust stages to put the payload into orbit. The higher efficiency of the Delta IV upper stage is due to the hydrogen–oxygen propellant, with its much higher exhaust velocity. Note that thrust is most important for the early part of the launch, during which the rocket is moving near-vertically against gravity. Later in the launch, when its trajectory is much flatter, exhaust velocity is the most important parameter. This can be seen most clearly in the Saturn V, in which the first stage uses liquid oxygen and kerosene. This produces a relatively low exhaust velocity of 2.65 km/s, but a high thrust, because of the relatively high density of the exhaust. The upper stages use liquid hydrogen and liquid oxygen, producing an exhaust velocity of 4.21 km/s, for this early design. Five engines are still required to provide adequate thrust to accelerate the huge payload, but the efficiency is much greater. The third stage, with a very flat trajectory, can afford to use only one J-2 engine to complete the injection into low Earth orbit.

In designing a 200-tonne launcher for the Mars expedition, there are a number of guidelines to be followed. The technology should, if possible, be taken from existing production launchers, so that there is no development cost and the reliability is well established; this applies particularly to engines. For the high-thrust lower stages, propellants such as liquid oxygen and kerosene, or even solid propellant, can be used; high exhaust velocity is not vital here. For the upper stages, the use of liquid hydrogen and liquid oxygen is mandatory, because here, high exhaust velocity is essential. The launcher will probably be used both for cargo and for manned flights. In the latter case, additional reliability is demanded, but this is not necessary for the cargo flights. For manned flight, *engine-out* capability is necessary; in other words, the vehicle still needs to be able to convey the crew to a safe place – to orbit, or back to Earth – even if one engine fails. With its three engines, the Space Shuttle has this capability, and the Saturn V was able to fly with just four out of the five engines on the first and second stages. It should be possible to arrange this for the Mars expedition launcher, but it will not be necessary for cargo flights, and one or more engines can be removed, or more cargo carried with the same number of engines.

In what follows we shall examine a number of configurations of existing technology that could provide the necessary heavy lift capability for a Mars expedition. In some cases, launcher configurations have been designed and their performance determined; and in others, only generalised statements are made as to what is required. For simplicity we here take the configurations discussed in the NASA Reference Mission, and the Ares launcher proposed by Zubrin and his co-workers. As already mentioned, none of these launchers exists at the present time, and one or more of these ideas would need to be developed and qualified as part of a programme to put human beings on Mars. Each makes use of existing components, in order to minimise development costs.

8.5   THE NLS-BASED VEHICLE

This is largely a concept derived from the National Launch System studies carried out by NASA contractors up to 1992. The core vehicle is based on the Shuttle external tank, stretched by 1.5 m to accommodate more propellant – 727 tonnes extended to 768 tonnes of liquid hydrogen and liquid oxygen, and strengthened to take the larger payload. Four STME engines – the cheaper version of the SSME – are mounted behind the tank, on a skirt attached to it. Four strap-on boosters are mounted round the core, and these use an engine derived from the Saturn V F-1: the F-1A. It burns liquid oxygen and RP1, and develops a somewhat higher thrust than the F-1: 9.2 MN compared with 7.7 MN. Each of the four newly-developed 8.4-metre-diameter, boosters, strapped round the core, contains 1,636 tonnes of liquid oxygen and RP1, and each has three F-1A engines. The initial thrust of the first stage, plus boosters, is 104 MN – about three times the thrust of the Saturn V. The upper stage, for orbit circularisation, is based on the Space Shuttle Orbital Manoeuvring System, has a thrust of 106 kN, and uses storable propellants, amounting to ten tonnes. This truly enormous vehicle was proposed by the Synthesis Group in 1991, in response to the initial proposal from President Bush, to develop a Mars expedition. It had a payload capacity to Earth orbit of 250 tonnes. A smaller version, with only two boosters, was intended for manned lunar expeditions. The demise of the NLS programme, and a major rethink about the Mars expedition, prevented this design from being carried forward to the Mars Reference Mission. However, the next four options described here *do* form part of the Reference Mission.

8.6   THE ENERGIA-BASED VEHICLE

This uses a modified Energia core with eight strap-on boosters; the original Energia

Figure 8.2. An artist's impression of a Mars Reference Mission payload, side-mounted on the Russian Energia-based launch vehicle. (Courtesy NASA/JPL.)

had four boosters. This is coupled to an upper stage based on the Space Shuttle external tank with a single SSME engine, with a payload capacity of 179 tonnes to low Earth orbit. It requires no new engines: the RD-170 liquid oxygen and kerosine engines were developed and qualified in the Russian programme, as were the core stage and boosters of the Energia, and the SSME is in current use. The element of new development is in the structure. Early in its development, the Energia was modified to carry the Russian version of the Space Shuttle, Buran, which was mounted on the side of the Energia. It was originally assumed that the Mars payload would also be mounted like this, as shown in Figure 8.2. The new design requires the upper stage to be mounted on top of the Energia core, with corresponding changes to the structure. Redesigning structures is nowhere near as costly as developing new engines, and the rest of the vehicle is based on designs that have already flown. The main disadvantage is that the payload capacity – 179 tonnes – is rather low compared with the 200-tonne initial requirement.

8.7 SHUTTLE-BASED VEHICLE 1

In this design the core is based on the NLS modifications to the Space Shuttle external tank with three SSMEs burning liquid hydrogen and liquid oxygen to provide the thrust. There are seven liquid oxygen and RP1 strap-on boosters, each using a single RD-170 engine. The upper stage is the same as before, with a single SSME burning liquid hydrogen and liquid oxygen. This vehicle has a payload capacity of 209 tonnes, and again makes use of developed engines. The liquid boosters use a Russian-developed engine, but the tanks and other requirements would be newly developed.

8.8   SHUTTLE-BASED VEHICLE 2

Here the core stage is the same as above, with the longer version of the Space Shuttle external tank; but it is equipped with four of the more powerful STME engines, which have yet to be developed and qualified. This core stage is the same as that of the NLS-derived launcher. The programme also requires the four boosters, newly developed, each using two F-1A engines. The upper stage is the same as previously, using a single SSME. This vehicle has a payload capacity of 226 tonnes, but uses almost no currently available engines and components, the only exception being the SSME.

8.9   THE SATURN V-BASED VEHICLE

This is the evolved version of the Saturn V vehicle, designed to meet the requirements of a Mars expedition It might have been developed directly in the 1970s, had the Apollo programme not been cut. The core stage is virtually the same as the Saturn first stage, with five F-1 derivative engines, and is surrounded with four newly developed strap-on boosters, each with two F-1-derivative engines burning about a third of the propellant carried by the core stage. The upper stage uses six Saturn V J-2 engines, burning liquid hydrogen and liquid oxygen. This stage is much larger than the equivalent Saturn V stage, and carries the payload into orbit. It has a payload capacity of 289 tonnes.

   In the above sequence we see how the payload capacity into orbit can be increased by progressing from the use of existing technology through the use of newly developed technology and finally the resurrection of the technology of the 1960s.

8.10 THE MAGNUM VEHICLE

The vehicle finally selected for the Mars Reference Mission, as updated in 1998, was called Magnum. Ideas, as we shall see, had gradually moved round to a smaller payload capacity into low Earth orbit. This was driven by the high cost of developing a new launcher of the 200-tonne class, based on one of the above concepts. The Magnum still uses Space Shuttle technology, being based on a core using the extended Space Shuttle main fuel tank. However, because of the smaller payload it uses two RS-68 engines rather than Space Shuttle-derived engines such as the SSME or STME. The RS-68 is a newly developed engine, with simplified construction, so that costs are reduced. It is currently in use on the Delta IV launcher, and develops a thrust of 3.4 MN. The two newly developed *liquid flyback boosters*, attached to the central core, are a concept emerging from SSTO technology. The idea is that the boosters can use expensive liquid hydrogen–liquid oxygen engines, and that the entire booster can be recovered and refuelled. Being reusable, the cost of the boosters is amortised over many launches. The boosters work right up to the edge of space, and when empty, detach and return to the landing field automatically, powered by

Figure 8.3. An artist's impression of the Magnum launch vehicle. (Courtesy NASA.)

jet engines. It remains to be seen whether or not this is really a cheap solution to a heavy launcher, but the Magnum concept uses these as its baseline. The proponents also claim that the advanced SRB, as mentioned above, could also be used, at a quote 'higher cost'. Again, it is a question of whether new 'fancy' technology is the correct way to proceed, or whether rather old technology is cheaper in the long run. The Magnum vehicle has a payload capacity of 89 tonnes into low Earth orbit, but is expected to be cheaper per tonne than other designs. The lower capability is more in line with recent views on the 'right' way to go to Mars.

8.11 THE ARES VEHICLE

The final version of the heavy launcher to be considered is that proposed by Zubrin and colleagues for the Mars Direct mission. It has an apt name – Ares – and uses as much existing technology as possible. Because it is designed for a specific Mars expedition scenario, it has a smaller payload capacity than some of the other vehicles described. The core stage is again based on the Shuttle external tank, with three SSME engines, which is the same as the first Shuttle-derived vehicle above. The two boosters, however, are an improved version of the Shuttle boosters – the ASRM (Advanced Solid Rocket Motor), designed, after the *Challenger* disaster, as a replacement for the SRM, although in the event it was not used. It has a thrust of

15.7 MN. The upper stage uses a single, newly designed, high-thrust liquid hydrogen–liquid oxygen engine to inject the payload directly into Mars transfer orbit – an essential feature of Mars Direct. An alternative is to use seven RL-10-derived engines, burning the same propellants. The RL-10 type has a long history. It was part of the early Saturn programme, and is the same engine as was used in the Centaur upper stage of the Titan IV. Seven of them would produce a thrust of 517 kN.

With the Ares vehicle we conclude our examination of the available concepts for a heavy launcher. The task of development is not particularly challenging, as all the necessary elements already exist, or have existed in the past. Once a decision is taken to send humans to Mars, the resources, already available, can be brought together to provide the essential first step. It is clear that without the heavy launcher there can be no expedition to Mars, although there is much attractive technology associated with the other aspects of the expedition. And without this mundane application of existing technology, mankind will be doomed to remain inside the orbit of the Moon.

8.12 THE SAFETY OF THE CREW

The whole ethos of the Mars expedition lies in the exploration of Mars by human beings, and their safe return. The propulsion technology required to accomplish the interplanetary journeys has been set out in the previous chapters. Here we need to look at the hazards associated with the expedition, and at how these hazards will be dealt with. Of all the eventualities that would horrify us, were it to happen, the possibility of the expedition being stranded on Mars, with no means of returning to Earth, and no means of being rescued, would be the worst. It takes little imagination to recognise the anguish of those on Earth, and the bravery of those on Mars, as the supplies run out. We have to ensure that a safe return is built into the mission design, and that sufficient back-up exists to bring the members back under all conceivable eventualities. We already know, for instance, that it is not possible to return to Earth along the continuation of the Earth–Mars transfer orbit, should something go wrong on the way to Mars, or Mars capture prove impossible. The Apollo programme had this built-in safety margin; and, indeed, Apollo 13 had to make use of it. The LM simply carried on round the Moon and returned to Earth, with no action necessary by the astronauts until they began to approach Earth. We know that if this were to be attempted during the Mars mission, the Earth would not be in the right place when the astronauts returned. So, if anything goes wrong on the way to Mars, the crew will either have to wait in Mars orbit or land on Mars, to await a suitable moment to return. This presupposes that the capture manoeuvre can be accomplished, and that whatever has gone wrong has not damaged the ability to make the return trajectory manoeuvre. Waiting for the right moment, at Mars, is the only way to return safely to Earth. Given that everything will have to work properly to enable the crew to do this, it would be wise to ensure that if they succeed in reaching Mars, the return to Earth can be secure. This leads to the suggestion that

Figure 8.4. The fully-fuelled Earth return vehicle awaiting the ascent of the explorers from the surface of Mars. It has been in Mars orbit for four years, having been sent, unmanned, before the explorers left Earth. The large solar array provides power to refrigerate the cryogenic propellants in order to minimise boil-off during the four-year wait. (Courtesy NASA/JPL.)

there should be a *fully fuelled return vehicle* in orbit round Mars, waiting for the crew to arrive there. It would have to be sent, unmanned, to Mars, before the expedition members leave Earth.

Once the concept has been established, of using unmanned vehicles to send advance resources to Mars, in order to enhance the crew's safety when they arrive, then other examples spring to mind. A spare habitat on the martian surface, fully equipped with life support consumables and so on, would provide additional safety for the expedition. They could perhaps wait for a rescue mission if they had sufficient spare supplies, or all would not be lost if the main habitat became damaged. Making rocket fuel on Mars, as described in Chapter 7, would be less risky if the factory were to be sent in an unmanned vehicle and set in operation remotely, so that the fuel would already be prepared before the astronauts left Earth. Such ideas are, of course, not new. It is exactly the planning involved in polar expeditions and mountain climbing. Forward and return base camps are prepared before the final attempt on the pole or the summit. These considerations suggest that there should be *two* types of vehicle going to Mars: the *crewed expedition transport vehicles*, and the *unmanned cargo vehicles*. The cargo vehicles can be simpler than the crewed vehicles, and perhaps carry larger payloads. The crew safety issues will only impact on the design of the crewed vehicles.

8.13 CREW HAZARDS

There are logistical hazards which basically arise from malfunctions of the equipment. Some of these have been mentioned above, and the approach is to provide back-up systems and plans. The other hazards are environmental. Once the surface of the Earth has been left behind, the members of the expedition are exposed to an environment for which humans were not designed. Some elements of this are relatively easy to combat: the sealed spacecraft provides a safe atmosphere to breathe; and on Mars the habitation, the pressurised surface vehicles and the Mars suits provide the same safe atmosphere. Others, however, are less easy to combat; and the two most important are space radiation and the zero-g environment.

8.14 RADIATION

We are all scared of radiation. From 1945, the horrible effects of massive exposure to radiation loom in our folk memory. As people exposed to minor doses of radiation grow older, more subtle but nonetheless deadly effects have emerged. Away from man-made sources of radiation we feel relatively safe; but in fact, living on the Earth's surface we are exposed to radiation from a number of sources. The Earth itself contains radioactive minerals, and from space, *cosmic rays* bombard us continuously. Exposure varies with location. North-west Scotland and Cornwall, for example, have higher levels because of the granite bedrock, which is also used for buildings in these areas. And people are radioactive: marriage is said to double a person's exposure to radiation! Human life evolved on Earth amongst all these hazards, and we have to accept these levels of radiation as normal.

The Earth's surface is, in fact, rather a safe place in terms of radiation. The rocks are old, and there is not much radioactive material left in them; the atmosphere partially protects us from cosmic rays; and, perhaps less well known, the Earth's magnetic field has a major protective effect against them. Once outside these protective shields, however, human beings are unprotected. This is one of the major hazards for a Mars expedition. The atmosphere of the Earth provides about 1 kg/cm^2 of material between space and ourselves. This acts as a shield, which would be very difficult to reproduce on an interplanetary transfer vehicle because it would be too heavy. The Earth's magnetic field extends out into space, and protects astronauts, in low Earth orbit, rather well from cosmic rays. But a Mars expedition must journey beyond the protection of the Earth's magnetic field.

Cosmic rays generally come from far away. It is possible that the highest-energy particles emanate from distant galaxies, while the lower-energy radiation emanates from our own galaxy, and the lowest energies come from the Sun. The majority of the cosmic radiation consists of protons travelling at high velocity; many travel at close to the velocity of light. They act on matter, and on human tissue in particular, by ionisation – the damage from which depends on the particle energy. Paradoxically, low-energy protons are much more damaging than high-energy protons because of relativistic effects (which will not be discussed here). Moreover, because the spectrum of cosmic rays falls off very quickly with energy, there are many more of these more damaging low-energy particles. Nature is, however, kind: the more damaging the radiation, the easier it is to shield against it. The very process that damages human tissue – the ionisation generated by the particles – also rapidly takes away their energy when they encounter a shield. This means that, in normal circumstances, a rather thin shield will protect the crew from excessive exposure. The real problem is (as mentioned above) that the lowest-energy particles – the most damaging kind – come from the Sun, and the Sun is not a steady companion with which to share a Solar System. It is highly erratic: sometimes it is quiet, emitting a low, steady flux of radiation; and at other times it is violently active, emitting great bursts of particles that absolutely dominate all other sources of radiation damaging to humans. These *solar flares* are caused by the twisting and untwisting of the magnetic fields associated with sunspots on the surface of the Sun, and like the sunspots, their frequency and intensity varies over an 11-year cycle. The present cycle reached its maximum around the year 2000, and the Sun is still very active as I write; but there will ensue a quiet period of up to seven years, when the frequency and intensity of solar flares will be low. Unfortunately, the voyages to Mars and back have to take place at the proper time, defined by the relative positions of the planets. These will, in general, not coincide with periods of low solar activity, and as a result, the expedition members will need to be provided with special protection from solar flares.

Shielding against radiation is a complex business. As stated above, the lower-energy particles are both the most damaging and the easiest to shield against. But shielding, like radiation damage, is a stochastic process: particles have a certain chance of being stopped in a given thickness of shield. As the thickness increases there is a decrease in the probability that a particle will survive with enough energy to damage a human being, but it never falls to zero. During a solar flare, the intensity of particles rises up to a million-fold. So, even if a shield only allows one in 100,000 dangerous particles through, ten in a million will survive to do damage. This strong variability shows that when the Sun in quiet, the normal aluminium pressure shell, having a thickness of about 5 mm, will provide all the protection that is required. During solar flares, however, much thicker shielding will be required, although this need not be incorporated in the pressure shell. Solar flares last from a few hours up to a few days, so if a *small* but thick radiation shelter can be provided for the crew to live in for up to, say, three days, much mass can be saved in protecting the expedition members from radiation. Since any material will suffice, it makes sense to put the shelter in the centre of the spacecraft, so that all the surrounding material will add to the protection provided by the special shield.

It is notoriously difficult to establish a safe level of human exposure to radiation. Ideally, it should be zero; but we have seen that on Earth there is already a certain level of radiation that cannot be avoided. The effects of low-level radiation on health are again stochastic. It is thought that the probability of developing a cancer is related to the level and duration of exposure to radiation; and for the low levels experienced on Earth, this probability is very low. Since the expedition members will be exposed to a higher level of radiation, they will have a higher probability of developing cancer than the general population, but this should be kept small.

Figure 8.5. A possible configuration for the Earth–Mars transfer habitat for the expedition. The central region is where the crew can shelter during a solar flare. Notice that *meetings* are provided for, even in zero-g. (Courtesy NASA.)

Because astronauts have been operating for long periods in low Earth orbit, since the 1960s, it is not surprising that a safe radiation exposure level has already been defined for them. This is the equivalent total radiation dose that would increase the risk of any cancer developing over the astronaut's lifetime by 3%; in other words, an astronaut who has been exposed to the maximum allowed dose would be 3% more likely to develop cancer than the general population.[1] Since such a risk increases with the total dose, it is important to monitor the dose, and to retire astronauts who have reached that level during their missions in space. For the expedition to Mars it seems

[1] 3% *more* likely, not a 3% chance of developing cancer.

likely that the total dose will be kept to that level for the whole expedition, including the period of residence on Mars as well as the journey there and back.

The magnetic field of the Earth has been described as acting as a shield that protects us from space radiation. It does this by trapping the electrically charged solar and cosmic ray particles in two belts, high above the Earth. The highest belt extends out to about 40,000 km, and the lowest belt begins at about 600 km above the surface. The intensity of radiation in these belts can be more than a million times higher than on Earth. Most of the current activities of astronauts take place below this altitude, where the radiation level is small. If they were to spend any appreciable time within the radiation belts themselves, then their dose would be very high indeed. This must obviously be avoided; but to do so requires that the departing and arriving spacecraft pass rapidly through the Earth's radiation belts, in order to minimise the crew exposure. This poses a significant difficulty for electric propulsion. Despite the major propellant savings that would arise from the use of this technology, the slow acceleration away from Earth, in a tight spiral path, would expose the crew to an unacceptably high total dose of radiation from the belts.

If the expedition members can be carried safely through the Earth's radiation belts, and can safely make their voyage to Mars, sheltering from solar flares, from time to time, inside the radiation shield, there is one more place where radiation exposure is a risk: the surface of Mars. Unlike the Earth, Mars is not a particularly safe place, in respect of radiation from space. The atmosphere, as we know, is very thin, and offers much less protection than the Earth's atmosphere; and, moreover, Mars has no radiation belts. While around the Earth there is a strong magnetic field generated by circulation in the Earth's interior – the 'dynamo' effect – Mars has nothing at all of this nature. Instead, localised regions of Mars' crust have high surface magnetic fields – a little less than the field on the Earth's surface – but outside the atmosphere there is no coherent field to act as a shield against the solar wind, solar flares, or cosmic rays. As far as radiation is concerned, the surface of Mars is not as safe as the surface of the Earth.

Because the details of the mission depend heavily on these radiation safety conditions, it is important to eaxmine, numerically, the relative levels of radiation. The unit used when evaluating human exposure to radiation is the *rem* – although purists will want to use the *sievert*, 100 of which equal 1 rem. For the above-mentioned risk to the crew – a 3% increase in the probability of cancer – the maximum dose is 100–400 rem – a broad range which reflects a risk dependent on age and gender. The annual dose, calculated for interplanetary space, is 57 rem, while for the surface of Mars the annual dose is only 12 rem. The martian atmosphere attenuates the interplanetary flux by about 75%; moreover, in space, the radiation can arrive from any direction, while on the surface, the planet itself acts as a shield, cutting off radiation from half of the sky. For comparison, the natural dose on the Earth's surface, for the general population of the United States, is 0.36 rem. Up to 10 rem, however, is allowed for workers in occupations involving radiation, although their actual mean level is less than 5 rem. Note that these are all annual figures; the actual dose will depend on how long the exposure lasts. For the astronauts – away from home for 2 years 8 months – the mean annual dose should be less than 38 rem, in order to keep their total dose below 100 rem. This, then, points to an important conclusion. The expedition members are *very much safer on Mars* than in space. Short journey times and long stay times are beneficial in reducing the overall radiation risk to the expedition. As we shall see later, this will have an impact on the strategy for the mission.

8.15 THE ZERO-G ENVIRONMENT

Perhaps more serious, in its deleterious effects on the human body, is long-term exposure to zero-g. Despite the vast database built up on astronauts in weightless conditions, since Gagarin's flight, no antidote against the effects of weightlessness has emerged. Immediate results include space sickness, which affects quite a high proportion of astronauts; but the most dangerous are the loss of muscle and bone mass, and the degradation of the cardiovascular and immune systems, which arise from exposures longer than about a month. Our muscles, bones and hearts have evolved to work in a 1-g environment; and once that force is removed they begin to lose condition – which is why medical advice, following surgery, is not to stay in bed, but to get up and move around as soon as possible. In space there is no equivalent – no continuous force to keep the bones and muscles under tension, and the heart has quite a different job to do. Most astronauts prefer zero-g because of the sense of bodily freedom it induces; and it is also tremendous fun. But adaptation from zero-g to any level of planetary gravity takes a long time, from days to weeks. This might not matter on return to Earth, but time spend adapting to the 0.38-g gravity of Mars will be lost to the expedition. Venturing on strenuous activity, before full acclimatisation has been achieved, invites accidents: at best, muscle strains, but at worst, broken bones. This worry suggests that long journey times are to be avoided, if at all possible, and that sufficient time on the martian surface, to allow full acclimatisation *and* the full exploration programme, must be allowed.

Exercise programmes can help to keep the expedition members fit during the voyages. One can imagine a gym in the interplanetary transfer vehicle like that in the Shuttle orbiter, but more extensive, as shown in Figure 8.5. It is, however, more or less impossible to do enough work against artificial restraints, to counteract the effects of zero-g. Nevertheless, the crew will need to be assiduous in exercise during the voyage if they are quickly to adapt to work on the martian surface. Experience of long-duration spaceflight is limited, at present, to just over a year, in the Russian space programme. The International Space Station has, at its heart, a programme to investigate the effects of weightlessness on human beings; and there are several institutes of space medicine in Europe, Russia and the US where basic research on techniques to minimise the effects of zero-g exposure is carried out, both on the ground and using the astronauts in the ISS. Countermeasures such as body fluid management, and time spent with the lower body in a reduced pressure suit, to pull fluid down from the upper body, are being used. So far, however, no fully effective procedure has emerged. The chief result seems to be that exposure to 1 g *eventually* restores all the damage suffered in zero-g. In most cases this happens within a period of several days to a few weeks. No-one knows if long-term exposure to 0.38-g martian conditions is of positive benefit, or whether it will lead to further deterioration. At present, the best approach seems to be to shorten the time spent in zero-g – which again implies short journey times.

There is, of course, the alternative of providing artificial gravity. From the very beginning of the space dream, the idea of spinning the spacecraft to generate artificial gravity has been ubiquitous in literature. Unfortunately it is much more difficult to achieve in practice than in the imagination of science fiction writers. Spinning the spacecraft is not a problem, and it is routinely done for a variety of reasons, mostly thermal. It is the astronauts that are the problem. They become sick. On Earth we have the same gravity acting all over our bodies, and our heads feel the same force as do our feet. But in a relatively small, spinning spacecraft, this is not the case. To generate sufficient force, the spacecraft needs to spin rapidly; the force depends on the radius and the speed of rotation. The parts of the body closest to the outer rim experience the full force, and those near the centre of rotation experience very little. Moreover, our organs of balance malfunction in this rapidly rotating system, just as much as in zero-g, and so space sickness again ensues. The production of an acceptable form of artificial gravity requires slow rotation over a large radius; and for a spacecraft with a diameter of less than 10 m, this cannot be achieved.

Ingenuity has been applied to this problem, and two suggestions have been made. The crew compartment could be separated from a heavy mass – the engine, or a nuclear reactor – by a cable or tether, and the two could rotate about their mutual centre of mass. This would produce a sufficiently large radius to remove the above-mentioned problems, but the control of the cable is an added complication; if it were to break or become tangled, the expedition could be lost. A different approach is to design a long vehicle, with the engines and most of the dry mass at one end, and the crew compartment at the other. This works best with a nuclear rocket that has significant mass in the engine and its shield, like the bimodal nuclear vehicle shown in Figure 6.13. The vehicle could then rotate about its short axis to produce the same effect. Both ideas have been considered, in detail, for the Mars Reference Mission. The extra mass associated with the design of an artificial-gravity spacecraft could be as much as 10%, and this has to be offset against, say, extra fuel, to provide a shorter transit time.

8.16 THE ELEMENTS OF THE MARS REFERENCE MISSION

We now have all the basic ideas that are necessary to understand the arguments that have led to NASA's Mars Reference Mission. The propulsion issues have been dealt with in previous chapters, and the basic problems of the launcher and the issue of crew safety and health in this chapter. A large body of work is incorporated in the results of the Reference Mission studies, and NASA publication SP-6107 provides a summary of the conclusions, in 275 pages. The reason for the Reference Mission was to set out the scheme for a near-term Mars expedition, and to identify the best technical solutions for each element of the scheme. The details were then published as a basis for further work on any or all of the elements, to identify optimal solutions. To some extent this has already occurred, and since publication in 1997, a

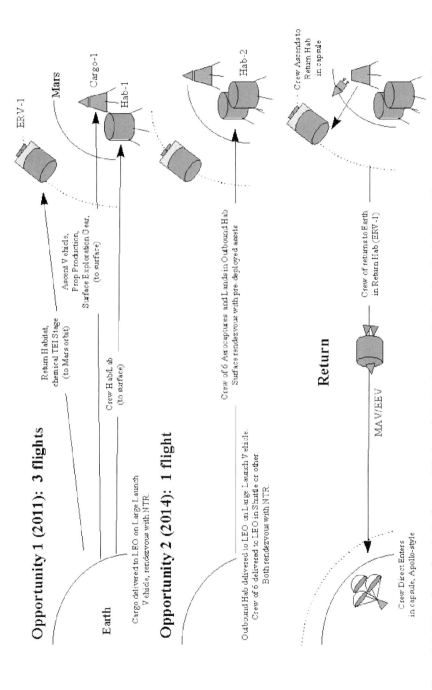

Figure 8.6. The scheme for the Reference Mission of the NASA Mars Exploration Study Team. Three unmanned cargo flights are sent at the first opportunity; and two years later, the crew are sent. After exploring the surface, the crew return using the previously placed Earth return vehicle. (Courtesy NASA.)

new concept, which puts less pressure on the launcher mass capability, has been identified. We shall deal with this update later; but first, the Mars Reference Mission of SP-6107 will be examined.

The basic scheme for the first four launches is set out in the diagram in Figure 8.6. The plan is to have a continuous series of launches, rather like the Apollo programme, as the opportunities arise, every two years or so.

8.17 FIRST LAUNCH OPPORTUNITY

We have already alluded to the idea of first sending unmanned cargo flights to Mars in order to improve the logistics of the mission and crew safety. In the Reference Mission, at the first flight opportunity – optimistically set for the conjunction in 2011 – three cargo flights are launched directly into Hohmann Earth–Mars transfer orbits. The first contains the fully fuelled Earth return vehicle, which is to be inserted into Mars orbit by aerocapture, to await the returning crew; the second contains the ascent vehicle, the propellant production equipment, and the exploration equipment; and the third launch carries the expedition habitat. The contents of both second and third launches are to be soft-landed – hopefully, in the same place – on the martian surface, using aerodynamic braking and rocket engines. The fuel production facility then begins operation, to provide the ascent propellant. It is an important safety feature of this plan that the return propellant manufactured on Mars is ready, and the ascent vehicle checked out, *before* the crew depart from Earth. The Earth return vehicle is also waiting in Mars orbit. To maximise the payload, these launches use the minimum-energy Hohmann orbits and a 258-day transfer time. No humans are involved, and so a long voyage presents no problems. The masses of these payloads are listed in Table 8.2. In all cases, this mission scenario requires a 200-tonne-class launcher.

Figure 8.7. The habitat on Mars, sent two years earlier on an unmanned cargo flight, before the explorers arrived. On arrival the explorers attached an inflatable extension to provide more space. (Courtesy NASA.)

Table 8.2. Manifests for the first three cargo launches to Mars*

Launch 1: Earth return vehicle		Launch 2: cargo lander 1		Launch 3: cargo lander 2	
Payload element	*Mass (tonnes)*	*Payload element*	*Mass (tonnes)*	*Payload Element*	*Mass (tonnes)*
EMTV habitat	56.4	Ascent vehicle (dry)	8.1	Expedition habitat	38.5
Crew and samples	(0.5)	Propellant factory	4.8		
		Hydrogen feedstock	4.5		
		Nuclear power system	14.7	Nuclear power system	14.7
Dry return stage	5.2	Exploration equipment	24.2	Exploration equipment	3.1
Earth return propellant	52	Lander engines and propellant	16.6	Lander engines and propellant	16.6
Mars capture aerobrake	17.3	Entry aeroshell	17.3	Entry aeroshell	17.3
Nuclear engine	28.9	Nuclear engine	28.9	Nuclear engine	28.9
Earth-Mars propellant	86	Earth-Mars propellant	86	Earth-Mars propellant	86
Total	246.3	Total**	205.1	Total**	205.1

* These all occur one synodic period (about 2 years) before the explorers depart for Mars.

** The nuclear stage used to propel the Earth-Mars transfer vehicle is sized and fuelled to transport the Earth return vehicle - a payload of 132 tonnes - on the Hohmann transfer orbit; with a reduced payload, a much faster transfer time can be achieved.

8.18 SECOND AND SUBSEQUENT LAUNCH OPPORTUNITIES

At the next opportunity just over two years later, the Earth–Mars transfer vehicle is launched into low Earth orbit, and the crew are sent up on the Space Shuttle. After rendezvous in low Earth orbit, the EMTV leaves for Mars, carrying the explorers and a habitation identical to that sent ahead at the previous opportunity. The whole is then aerocaptured, and landed on the surface close to the previously delivered hardware. Note here that the crew compartment of the EMTV and the Mars expedition habitat are combined. The crew lands on Mars in the habitat. This provides, on the surface, a complete habitat fully stocked with life support, next to the habitat already delivered. The expedition then has a complete back-up habitation, with stores. If necessary, the explorers can remain on Mars for one additional synodic cycle, to wait for a rescue mission. The manned Earth–Mars transfer vehicle does not travel on a minimum-energy Hohmann orbit, but instead uses the extra mass ratio, arising from the smaller payload, to enter a short transit orbit, which arrives at Mars within about 180 days – two and a half months earlier. Depending on the particular synodic cycle, this journey can be as short as 120 days. From the point of view of the health and safety of the crew, this short transit is a very important part of the scheme, because it reduces the total exposure to zero-g and space radiation by six months, over the whole mission, compared with our basic scheme using all-Hohmann orbits. Table 8.3 shows the payload masses for this launch, and for two subsequent launches intended to prepare for a second crewed mission.

Once the crew has arrived on Mars, there follows the exploration of the planet during the wait for the proper departure time. When that moment arrives, the crew enter the fully fuelled ascent vehicle and lift off, to rendezvous with the return vehicle. This has been waiting in orbit, ready to return to Earth, since the first series of launches. They then make the Mars–Earth transfer. On arrival near Earth they enter an Apollo-style re-entry capsule, which then aerobrakes and lands. The transfer vehicle is not braked, and is not captured into Earth orbit.

Note how this scenario utilises all the propellant-saving manoeuvres and devices that were described earlier. Wherever possible, planetary atmospheres are used to brake and capture vehicles. In situ propellant production on Mars is used, and nuclear engines are used, to minimise the propellant to be delivered into low Earth orbit. Note also that there are three rendezvous operations, if rendezvous on the surface of Mars is included. The crew meet up with the fuelled nuclear stage in Earth orbit, for transfer to Mars; they meet up with the fully fuelled ascent vehicle and the spare habitat on the surface of Mars; and they again meet up with the fuelled return vehicle, in Mars orbit, for transfer back to Earth.

With the general scheme outlined, we can now consider the individual elements, and why their particular design was chosen.

Table 8.3. Manifests for the first crewed voyage and two subsequent cargo voyages*

Launch 4: Crewed vehicle		Launch 5: cargo lander 4		Launch 6: cargo lander 5	
Payload element	Mass (tonnes)	Payload element	Mass (tonnes)	Payload element	Mass (tonnes)
EMTV habitat	56.0	Ascent vehicle (dry)	8.1	Expedition habitat	38.5
Crew	0.5	Propellant factory	4.8		
		Hydrogen feedstock	4.5		
		Nuclear power system	14.7	Nuclear power system	14.7
		Exploration equipment	24.2	Exploration equipment	3.1
Lander engines and propellant	16.6	Lander engines and propellant	16.6	Lander engines and propellant	16.6
Entry aeroshell	17.3	Entry aeroshell	17.3	Entry aeroshell	17.3
Crew shield	3.3				
Nuclear engine	28.9	Nuclear engine	28.9	Nuclear engine	28.9
Earth-Mars propellant	86	Earth-Mars propellant	86	Earth-Mars propellant	86
Total	208.2	Total	205.1	Total	205.1

* The two cargo voyages are to prepare for a second exploration mission.

8.19 THE EXPEDITION MEMBERS

The size of the expedition is clearly the major determinant of the propulsion requirements, and we know that the propellant dominates the mass to be placed in Earth orbit. A reduced number of explorers would save resources, but below a certain size the team would not function. The explorers are absent from Earth for nearly three years, and for this entire time they must work together and remain viable as a team. The long delay-time involved in radio communication will also place an extra burden on the crew members, compared with, say, the Apollo programme. In the latter, the support team on Earth was available to advise instantly on problems, and an arbitrarily large team of experts could be brought in to devise solutions. But the round trip for a radio message from Mars to Earth and back again can take up to 42 minutes. Instant support from Earth therefore cannot be provided, and so the crew will need to be more self-sufficient. Given the long time away from home, the possibility of death or an incapacitating injury to one or more crew-members cannot be discounted; and at the same time, the social and psychological dynamics of the team will improve with numbers.

The solution adopted for the Reference Mission was to identify essential specialists, and build the expedition team around them. A mechanical engineer will be required to maintain and repair the large number of mechanical devices on Mars; an electrical/ electronics engineer will also be needed, based on a similar argument; for exploration, a geologist and a life scientist are the essential specialists; and to deal with medical emergencies, a doctor will be needed. This adds up to a specialist crew of five; and to cover for incapacitation of one of the experts, each specialist will need to be cross-trained in another discipline. Having taken this as a basic crew size, all the operations of exploration can be modelled to determine whether they can be carried out with so few members. For example, expeditions, on foot or in vehicles, will require at least two members, for safety reasons; and at the same time there must be somebody back at the base. With a crew of five it is just possible to do this, assuming one vehicular and one pedestrian expedition, running concurrently. So, a crew of five is the minimum that can be taken. In order to cover for the possible loss or incapacity of a crew-member, another person, possibly trained in a different discipline, will be essential, and so the Reference Mission is sized for a crew of six. It is not clear that a specifically trained pilot will be needed, because the landings have to be automatic, given that most will be unmanned cargo missions; and docking in Mars orbit, for the return journey, should also be automatic. Nevertheless, one or more of the crew will probably be trained as back-up pilots, alongside their other specialities.

Given an expedition made up of six members, it is now possible to size the habitat and the resources for their support.

8.20 THE HABITAT

The habitat used for the Reference Mission is loosely based on the units designed for the International Space Station, but is of a larger diameter. They are aluminium structures with a 5-mm thick pressure shell; a significant number of these are now in orbit. The same idea is used for the habitat structure. The same basic habitat is used throughout the mission, so the expedition members travel to Mars in the habitat that will be their home when on Mars. A second identical habitat will have already been landed on Mars, as a back-up, before they arrive, and the return vehicle will use the same design of habitat. This commonality is an essential feature of the Reference Mission, and is intended to reduce development costs.

The habitat is a squat cylinder, 7.5 m in diameter and 4.6 m high, with domed end caps. Including the caps it has a total axial length of 7.5 m. In space, its division is largely irrelevant, but on Mars, under 0.38 g, it will rest with its cylindrical axis vertical, and be divided into two floors, each about 3 m high. It will be fitted out with windows, airlocks and hatches, and power systems, life support, environmental control, waste management and communications will also be provided. The stores of food, water and oxygen, sufficient to last for 800 days, will be incorporated in the crewed habitat, for transit to, and use on, Mars. The spare habitat will have 500 days' supply of consumables, to allow the crew to wait out a complete synodic cycle, should something go wrong. As the stores are consumed, space will be created to develop extra laboratory space and possibly a 'greenhouse' to attempt food production. The radiation shielding of the aluminium skin will already be significant, but the galley, in the centre of the habitat, will be used as the storm shelter. This will have extra shielding, and the crew will stay in the shelter during dangerous radiation storms, in space, or on the surface.

Figure 8.8. In an alternative scenario, after landing on Mars the explorers link the two habitats by joining the one in which they arrived to the one already there. This provides additional living space, and one of the habitats can be isolated and used should there be a failure of any of the equipment in the other. In order to be linked, the habitats–landers need to be mobile on the surface of Mars. (Courtesy NASA/JPL.)

In total, three of these habitats will be used: two on the surface of Mars – one having also been the crew cabin for the journey to Mars – and one delivered to Mars orbit as part of the return vehicle. They do not provide luxurious accommodation for six people. The floor area is about the same as that of a small house, and given that much of this space will have to be used for stores, and for laboratories and workshops, it is not much. Expedition members are thought to need private sleeping cabins, however small, for long-term psychological and social health, and these can be provided, together with common eating and recreation areas. Much of the basic concept is derived from, for example, submarines, in which crews are thrown together for long periods in a confined space.

Naturally, the environment will be 'shirt-sleeve', as to wear special clothing for the whole mission would clearly be impossible. The atmosphere will not be at Earth pressure, and this allows the pressure shell skin to be thinner and lighter, which thus reduces the mass of the habitat. A probable atmosphere for the habitat would be at a pressure of 350 millibars, with 75% oxygen concentration. The oxygen partial pressure is approximately the same as in the terrestrial atmosphere, so that physiological processes function normally. The reduction in nitrogen pressure is unimportant for these, but helps greatly with the pressure shell. There is a choice of buffer gas for the atmosphere: nitrogen is natural. of course, but argon or helium could be used. Helium is sometimes used to reduce the mass of buffer gas carried, but on Mars there is nitrogen and argon, so either could be used. Argon and nitrogen can be derived from the martian atmosphere, as by-products of propellant production. The life support equipment removes carbon dioxide from the atmosphere, and replaces the oxygen. The buffer gas is recirculated, but some is lost each time the airlock is used, and must be replaced.

For maximum flexibility the two habitats should, ideally, be joined together, and this requires that they should have some mobility on the surface. They are delivered to the surface on the lander, and it therefore requires some clever design work to achieve this mobility. An automated landing on the rough surface of Mars requires the use of stabilising legs to adjust for any unevenness, just as on the surface of the Moon. These are folded during the journey, but are deployed, together with the wheels, once the atmospheric entry aeroshell has been released. The rocket engines used for landing are mounted between the wheels, and the propellant tanks are mounted on the roof of the habitat to fit within the tapered shape of the aeroshell. All of this complicates the design of the lander, and requires considerable developmental effort. This does not apply, however, to the habitat used in the Earth return vehicle, which is simply attached to the propulsion stage. On the other hand, this habitat has to be provided with space docking equipment not required for the surface habitats.

8.21 ENTRY AND LANDING

The martian atmosphere is used as much as possible to slow down the landing spacecraft, which involves the use of aerodynamic braking to capture the spacecraft on arrival at Mars orbit and to accomplish atmospheric entry. Once the spacecraft has slowed so much that aerodynamic braking is ineffective, parachutes are deployed to drop the velocity to a few hundred m/s. Rockets are then employed for a soft landing. They are also used to circularise the aerocapture orbit and for the re-entry burn. For simplicity and commonality, the engines are based on the RL-10 engine adapted for, and fuelled with, liquid oxygen and liquid methane. This is not necessarily the optimum propellant combination for the purpose, but since a methane–oxygen engine is required for the ascent from Mars, it is sensible to use the same engine for other purposes. In particular, the development costs are reduced if only one chemical engine is used for the Mars expedition.

The general issues relating to atmospheric braking have been dealt with in Chapter 7. The landing vehicle will be protected by a complex aeroshell from the heat of entry. It is not possible to make the entry vehicle flyable, like the Space Shuttle, but, on the other hand, precision landing is required, and so some manoeuvrability is necessary both during aerobraking and in the final stages. This is vital, to enable the crew to meet up with the ascent vehicle and the spare habitat on the surface. The design of the aeroshell is based on this necessity. It is called *biconic* because it is in the form of a blunt cone with two different cone angles. Depending on the payload–crew lander–habitat, or ascent vehicle–propellant factory, the full aeroshell is either 18 m long or 15 m long. The forward section is a 25-degree half-angle cone topped by a spherical end cap, and is 6 m high. The centre section, which varies in length, depending on the payload, is conical, with a 4-degree half-angle. The aft section is also conical with a 4-degree half-angle. This rather improbable vehicle enters the atmosphere with an angle of attack of 25°; that is, in a nose-up attitude, with the axis of the vehicle making an angle of 25° to the flight path direction. The odd design ensures that the vehicle remains stable under aerodynamic forces while

Figure 8.9. The biconic aeroshell slows down the spacecraft as it flies past Olympus Mons on the way to the landing site. (Courtesy NASA.)

entering the atmosphere, so that the aeroshell is properly presented to the direction of motion, and the parts that are heated by atmospheric friction are those that are intended to be heated. As described before, the shield ablates away to produce a protective layer of cool gas around the vehicle. The lift:drag ratio for this vehicle is about 0.6.

The thin atmosphere of Mars eventually ceases to be effective as the aeroshell slows down (remember that the dynamic pressure, which controls the drag and lift, depends on the square of the velocity). At this point – at an altitude of about 8 km and with a vehicle speed of 700 m/s – parachutes are deployed. These then slow it down to about 200 m/s at an altitude of 3 km. The aeroshell is discarded once the parachutes are deployed, to reduce the mass to be slowed. Finally, the parachutes are detached, and the rocket engines take over to bring the vehicle to rest on the surface; and they also enable some fine guidance on the trajectory, to ensure that it lands in the right place.

The four rocket engines used for landing are RL-10-derivative engines burning liquid oxygen and methane. It has already been mentioned that development costs can be cut by using the same engine and fuel throughout the mission. These engines develop a thrust of 66.7 kN with an exhaust velocity of 3.79 km/s, have a dry mass of 4.9 tonnes and use 11 tonnes of propellant, and are sized to land a payload of 40 tonnes on the surface of Mars.

8.22 IMMEDIATE CREW ACTIVITIES ON MARS

The crew are, of course, on Mars to explore it; but much has to be done before exploration can begin. Once the immediate safety of the expedition has been assured by checking out the functioning of the life-support system and the integrity of the habitat–lander, the most important task will be the adaptation of the crew to martian gravity. They may well be in poor condition immediately on arrival, because the entry and landing procedure will expose them, for short periods, to up to 5-g deceleration, after months of zero g. Medical assistance can only be provided by the medical specialist, but he or she will also be suffering. A period of recovery and rest will be needed before any real activities commence.

Once the crew has recovered from the landing, the habitat has to be properly checked out, and the long-term stability of the life-support systems ensured. This is to establish that the expedition can remain safely in the habitat, and to begin the necessary routine maintenance of the habitat systems. The crew will not venture outside.

The next phase will be to establish the safety of the immediate environment of the habitat, and the security of egress and ingress. Assuming the habitat is securely standing on the surface, and that there are no rocks waiting to fall on it, the main dangers will be related to martian dust. On Apollo 11, dust was a major concern, and the crew remained in space suits to avoid contact with it; there was also the probablility of the clogging of dust in the suit seals. On Mars the dust is certainly different, and may contain reactive chemicals harmful to humans. There is little oxygen in the martian atmosphere, but rather a lot in the red soil and rocks, so oxidation reactions could easily take place between the martian dust and the atmosphere in the habitat. There is, perhaps, little possibility of a martian 'bug' waiting to infect the expedition. During this period the atmosphere will be chemically investigated, and the immediate locality will be explored by robot rovers, carrying analysis experiments, cameras, and other equipment.

Once the environment is declared safe, the explorers can emerge from the habitat and begin their tasks. It is tempting to think that they may be impatient and emerge earlier, as they will already have spent six months in the habitat 'can', and be eager to explore outside. All exploration on foot will involve 'Mars suits'. These will be little different from the space suits worn by lunar explorers; 7 millibars is as good, or bad, as a vacuum, when the human body is intended to operate at 1,000 millibars. Life support supplies will have to be carried on the backs of the Mars explorers. This limits the scope and duration of any expedition on foot; the 'walk-back' time has to be maintained, just as on the Moon; explorers must always have sufficient life support to be able walk back to the habitat; and they must not stray too far. The expedition will also be provided with a non-pressurised, powered rover or truck to widen the region that can be explored beyond the walk-back radius. There may possibly be a pressurised truck, which can carry exploration teams much further afield, perhaps for several days at a time. The truck provides a safe, shirt-sleeve environment, and carries several days' supply of life support consumables, as well as providing powered transport for the explorers. The Moon rover was electrically powered, but on Mars there is rocket fuel, produced for the ascent vehicle. This could be used to power the truck via an internal combustion engine adapted for methane and oxygen – not much different from the gas-fuelled vehicles being driven around our cities.

8.23 JOINING THE TWO HABITATS

At some point, the two habitats need to be joined to maximise the workspace. Once the spare has been checked to ensure that it is not leaking, and that the life support system is functional, they can be rolled together on their powered wheels and the linking tunnel connected. This will provide much-needed space for the crew to carry out their work and take their recreation. It should not be forgotten that it is only in the habitat that they can live without a Mars suit; the wide open spaces of Mars can be seen only through the suit visor.

8.24 EXPLORATION

It is not the object of this book to discuss the details of the exploration of Mars, or the scientific experiments that will be carried out. The aims of the expedition are, of course, to search for life, or evidence of past life; to understand the planetology of Mars, and how it is different from that of Earth; and to understand how human beings can live on Mars. The first two aims are related to pure science, with a flavour of 'Are we alone?' The last aim is unashamedly connected with the Great Dream: can humans colonise other planets, and eventually move out into the Galaxy? We should not deny the first Mars explorers their dreams.

8.25 PROPELLANT PRODUCTION

Leaving them with these dreams, we now need to secure the safe return of the expedition. A nuclear-powered autonomous propellant factory will have been working on Mars since its delivery to the surface, together with the ascent vehicle, more than two years previously. The nuclear reactor cannot be left close to the factory and the ascent vehicle, because the crew will need to work around the latter, and, of course, enter it for the return journey. When this particular payload is landed on Mars, and once everything has been checked out, the reactor will have to be removed to a distant place so that the local radiation hazard around the ascent vehicle can be reduced to an acceptable level while the reactor is working. It will therefore be provided with powered wheels, to enable it to trundle off into the distance. At the same time, the electrical power has to be brought back to the factory, and to the ascent vehicle, and so the reactor will have to trail behind it about 2.5 km of power cable. The design of this system is no mean task, and it will be rather difficult for the wheeled robotic reactor to successfully execute the manoeuvres in negotiating rocks and other terrain hazards, while ensuring that the cable does not become tangled. Nevertheless, it must be done, and considerable development will be required to ensure that it *can* be done. The reactor, once powered up, will provide power for the propellant production; but once the expedition arrives, it can also be used to provide power for the habitat and for many other purposes. Simply moving the reactor more than a kilometre away from the 'base camp' will provide adequate radiation protection for the explorers. The reactor will convert heat into electricity using Stirling or Brayton engines to drive generators, and it will generate up to 80 kW. The line voltage will be as high as possible, to minimise the thickness of the conductors in the long power cable, and so reduce its mass. Up to 5,000 V will be used for transferring the power back to the factory and the base camp. It is worth mentioning here that the martian atmosphere, being at such a low pressure, is not a good insulator. On Earth we often rely on air gaps to isolate high voltages – the overhead power cables used in the UK to transmit power at up to 33,000 V are a good example – but this cannot be done on Mars, because the power would arc through the low-pressure atmosphere across the gap. All conductors will therefore need to be well insulated and sealed to prevent arcing or coronal discharge.

Propellant production (as explained in Chapter 7) will begin as soon as the reactor is ready and powered up. In addition to the propellants oxygen and methane, the plant will also make water and breathable oxygen, and extract the buffer gases nitrogen and argon from the atmosphere for life support. To fuel the ascent vehicle, the plant must produce 5.8 tonnes of methane and 20.2 tonnes of oxygen. It uses liquid hydrogen, brought from Earth, to reduce carbon dioxide – the major constituent of the martian atmosphere – to make both methane and oxygen. For life support supplies it must also make 23.2 tonnes of drinking water, 4.5 tonnes of oxygen, and 3.9 tonnes of buffer gas. The propellant needs to be liquefied and stored in the tanks of the ascent vehicle, and the life support supplies have eventually to be stored in the habitat. Since this is to be completed before the expedition members leave Earth, the whole process must be automatic. Again, this process is vital to ensure human safety, so it must be made to work in this mode. The Sabatier process is used to make methane and water, and oxygen can be made by electrolysing the water, with the hydrogen being fed back into the process. If more oxygen is required than can be produced by this process, another system, using solid electrolysis of carbon dioxide, can be used to make the additional oxygen. However, it would be better to use the excess water produced, for human consumption. It is all saved mass, and it therefore does not matter whether the mass saving is in propellant or in consumables like water. It all has to be landed on the surface of Mars.

The lander carries the propellant factory, the seed hydrogen, and the empty tanks to hold the life support products. The propellants are sent directly to the tanks of the ascent vehicle. The life support consumables will need to be transferred to the habitat by the explorers, either using pipelines or by physically transferring the full tanks to the habitat.

The explorers will want to ensure that all is well with this system, and that the ascent vehicle is ready to be used, before beginning their main task.

8.26 THE ASCENT AND EARTH RETURN VEHICLES

The ascent vehicle is just big enough to transport the explorers and their samples into Mars orbit and allow for docking with the Earth return vehicle. It will carry life support consumables for this short flight, plus some margin for delay in docking. To ensure that nothing can interfere with the success of lift-off, it is landed on Mars on top of the lander, which also acts as a launch pad – a heritage of the Apollo programme. The ascent vehicle is just a small capsule atop two RL-10 engines and tanks full of liquid oxygen and liquid methane. A total velocity change of 5.7 km/s is needed to overcome the gravity of Mars and to enter orbit ready for docking with the Earth return vehicle. It is a single-stage-to-orbit vehicle – which can be used on Mars because of the lower gravity. The ascent requires 26 tonnes of propellant to be burned with an exhaust velocity of 3.79 km/s; and the dry mass of the propulsion unit – tanks, engines, pumps, and so on – is 2.6 tonnes. The crew capsule has a mass of 2.8 tonnes, and is in the form of a flat cylinder, 4 m in diameter and 2.5 m high – not a comfortable ride. It also has the manoeuvring jets and docking systems necessary to mate with the Earth return vehicle.

The Earth return vehicle incorporates the trans-Earth habitat, the Earth entry capsule, and the rocket engines and fuel tanks. It will have been delivered to Mars orbit four years previously, and will be in orbit around Mars, waiting for the crew to come up from the surface. With regard to propellant mass, the optimum rocket system would be nuclear. However, it was believed by the designers of the Reference

Figure 8.10. The ascent vehicle being prepared for launch. The lander and ascent vehicle use the same engines, but now those required for landing are being removed, leaving only those required for the ascent, which is much less demanding on thrust. The tanks are already filled with liquid oxygen and liquid methane from the propellant factory, which is also being removed. The door on the small conical ascent capsule is open. (Courtesy NASA.)

Mission that there would be significant cost savings if, again, it were powered by liquid oxygen and liquid methane, using the same chemical rocket engines. The development cost of these is, of course, shared with the engines for the lander and ascent vehicles. This is a cost saving. The other factor was the need to store the propellant for the 258-day voyage out, and for four years in Mars orbit. Compared with methane, hydrogen – which is essential for the nuclear stage – is much more difficult to store cryogenically for this long period. We have seen earlier that a nuclear stage could provide power, in stand-by mode, to refrigerate the hydrogen, and this is still an option; but the planners decided on the methane option. With adequate insulation, and a manoeuvre programme to keep the correct side of the spacecraft pointing towards Mars – the main source of heat – almost no refrigeration power is needed to store the oxygen and methane. A solar array, intended to provide power during the flight, can also provide some power for top-up cooling during the four-year wait, should events not quite take place according to plan. The habitat will be the same as used on the outward journey and on Mars. The Earth return vehicle is fitted with two RL-10 class engines – the same as those used for the other stages – and requires 52 tonnes of propellant to enter a fast transit 180-day transfer ellipse to Earth.

Having completed the fast transit to Earth, the crew enter another capsule, this time intended for Earth entry, and separate from the Earth return vehicle. We know that, without braking, the Earth return vehicle will pass by Earth, and not be captured. Using small rockets, the freed capsule will adjust its trajectory in order to enter the atmosphere at the correct altitude and angle. It weighs only 5.5 tonnes, including life support sufficient for several days. The entry technology is based on the Apollo system, using a large ablative heat shield on the base of the blunt cone, in which the crew lie. Once entry is completed, a proposed steerable parafoil (rather than the Apollo parachutes) is used to complete the landing. Note the differences compared with Mars entry and landing. The gravity is much stronger on Earth, so that the entry velocity is much higher and the period of heating is much longer, requiring a much thicker heat shield. On the other hand, the atmosphere is thicker, and parachutes are all that is necessary to bring the vehicle to a soft landing or splash-down.

8.27 THE EARTH–MARS STAGE

Having defined the Mars expedition, with its separate space vehicles, habitats and support equipment, the whole needs to be transported to Mars, using the Earth–Mars transfer stage. This is essentially the same whether the transfer involves a cargo vehicle or a manned vehicle. From earlier chapters it will be apparent that this stage has the toughest job, in that it has the largest payload to accelerate. The whole system has to be very efficient, because we have identified the need to transfer the humans quickly to Mars, and this cannot be achieved using a minimum-energy Hohmann orbit. A payload of about 74 tonnes has to be placed in this fast transfer orbit. It could be accomplished with chemical engines, but as we know, a very large amount of propellant would have to be lifted into Earth orbit. A nuclear thermal engine, on the other hand, requires much less propellant because of its higher exhaust velocity. This is what has been chosen for the Reference Mission. It requires about 86 tonnes of liquid hydrogen propellant, and incorporates four nuclear thermal engines similar to those discussed in Chapter 6. Each engine has a thrust of 67 kN and an exhaust velocity of 9 km/s. Four engines are required for the fast transit human mission, both to provide the necessary acceleration to the higher departure velocity, and to provide a margin of safety with an engine-out capability. The crew could still be brought to a safe place with only three engines, should one of them fail.

To reduce gravity losses, the final trans-Mars velocity is achieved in two perigee burns. The first burn places it in a highly elliptical orbit round the Earth, and when it returns to perigee the engines are fired up a second time, at which time the final departure orbit and velocity are achieved. This is not ideal, because the crew have to make three passages through the radiation belts instead of one; but it is not possible, with nuclear engines, to produce enough thrust to make Mars transfer in one firing of the rocket and have a reasonably small gravity loss. If a single firing were to be used, then the burn would last well into the climb portion of the ellipse, and energy would be wasted lifting unburnt fuel against Earth's gravity (see Chapter 5). More propellant would therefore need to be carried. The sum total firing time of the two burns is 35 minutes.

Once any mid-course corrections have been accomplished, and the mission is safely on its way to Mars, the nuclear stage is separated from the aeroshell containing the payload, whether crew or cargo. It then performs another manoeuvre to place itself into a stable interplanetary orbit that will not interfere with Earth,

Figure 8.11. The Earth-Mars transfer vehicle in orbit, ready for departure. The nuclear stage, with three engines and its large hydrogen tanks, is attached to the vehicle proper, which for Mars entry is protected by its biconic aeroshell. (Courtesy NASA.)

Mars or the spacecraft for a million years or so. We saw, in earlier chapters, that the natural path of a spacecraft on a trans-Mars orbit would take it past Mars on a hyperbolic orbit and into deep space. So, only a small manoeuvre is required to ensure that this takes place in a controlled manner. The aeroshell, with its contents, then travels on to Mars.

Note how this scheme is designed to avoid safety problems, and to minimise complexity. The hydrogen propellant does not need to be stored for long – only from launch until the final safety manoeuvre is completed. Super-insulation of the hydrogen tank can probably achieve this, without the need for active refrigeration. This saves a complete cooling system, and reduces the complexity and cost of the mission. If hydrogen were to be taken to Mars, a much more costly and complex system would be required.[2] Secondly, we see that the nuclear stage fires in such a way

---

[2] In the context of the large amount of propellant required for the nuclear engine manoeuvres, the few tonnes delivered to the surface of Mars to make propellant for the Mars ascent vehicle do not present a major storage problem. Super-insulation and a small refrigeration unit can be used with little penalty in mass or power.

that it transports itself and the payload away from Earth. Its final disposal in deep space is then simple and inevitable. Were we to use a nuclear stage to return to Earth from Mars, it would be moving towards the Earth. The disposal orbit would be equally safe, but the perception of safety would be much worse. One can imagine the headlines: 'Nuclear Rocket Heads for Earth'. There are still sensible arguments for using nuclear propulsion more widely in the mission, but we need to tread carefully in this area, and at the present time the approach used for the Reference Mission is probably best.

8.28 ARRIVAL AT MARS

The vehicle moving on to Mars is the biconic aeroshell containing the lander with the expedition habitat and its members, or one of the cargo landers, or the Earth return vehicle. It is intended that the whole business of Mars capture, or entry and landing, should be carried out using aerodynamic braking. As we know, this will save all the propellant that would be needed to carry out these manoeuvres propulsively. It is clearly also a fairly risky procedure. The spacecraft is swooping down on the martian atmosphere from interplanetary space, at a very high velocity between 7.6 km/s and 8.4 km/s, depending on the particular synodic period. For the manoeuvre to be accomplished safely, it has to enter a narrow corridor in the atmosphere. Too high, and the spacecraft will bounce off; too low, and it will plunge into the atmosphere and burn up. Thus very precise guidance and navigation will be required as the vehicle approaches Mars.

There are two different manoeuvres to be accomplished. For the Earth return vehicle, the requirement is to capture into a parking orbit round Mars, to wait for the expedition to arrive in the ascent vehicle, some four years later. It is sensible for this orbit to be highly elliptical, for several reasons. The velocity change required to enter this orbit, by aerocapture, is smaller for an elliptical orbit; in fact, it is doubtful if a single aerocapture passage could achieve anything other than capture into a highly elliptical orbit. Correspondingly, it requires less propellant to leave this highly elliptical orbit, to enter the Mars–Earth transfer ellipse, for the return journey; and any saving in propellant mass is, of course, very important. The other reason is thermal. It is relatively easy to shield the liquid oxygen and liquid methane tanks from the Sun in deep space, simply by keeping a deployed sunshade in the right orientation; but close to Mars, the planet itself becomes a major source of heat. It is much warmer than the liquid oxygen and liquid methane, and it is difficult to shield against the Sun and Mars at the same time, while the spacecraft is circling the planet. So, it is better to spend as much time as possible away from Mars, and in deep space – which happens naturally in an elliptical orbit, because of Kepler's law. The one disadvantage of this approach is that at periapsis the spacecraft will be moving more quickly than it would be moving in the equivalent circular orbit, and the docking of the ascent vehicle to the Earth return vehicle might therefore be more difficult. However, Mars' gravity is weaker than Earth's gravity, and so the orbital velocities are correspondingly smaller. The vehicle velocities for the docking manoeuvre at

Mars, with one in an elliptical orbit, are therefore about the same as at Earth, with circular orbits.

For the Earth return vehicle, the aerocapture is completed in one passage through the upper atmosphere, and some slight correction can be carried out using small chemical rockets, collectively known as the reaction control system, or RCS.

For the landers, a more complicated manoeuvre is required. The landing site will have been chosen based on exploration requirements. Natural choices might be areas where evidence of water is very strong, and there will be many other considerations that suggest the landing site. The arrival at Mars is governed by celestial mechanics, and the orbit that the spacecraft enters on aerocapture may not be ideal for the descent manoeuvre to begin, given that a particular landing site has to be reached. Aerocapture can change the altitude and shape of an orbit, but it cannot change the inclination with respect to the martian equator. It is also quite difficult to choose the position of closest approach, or periapsis, independently, as it is fixed by the interplanetary approach trajectory. Thus, landing on a particular spot is difficult to achieve by aerodynamic effects alone. It is very difficult to change the orbital inclination, even with the use of rocket engines, and it requires a great deal of propellant. The inclination will therefore be fixed by the approach trajectory, and this will in turn be defined by the instant of departure from Earth, and the chosen transit ellipse. There is little that can be done to change this once the spacecraft has arrived at Mars. What *can* be changed, however, is the position in the orbit at which the descent manoeuvre begins, and so the landing point can be chosen provided it is on a path directly below the final orbit. To do this, the spacecraft has to be placed in a circular orbit so that any point in the orbit can be chosen for the descent to begin. Changing the aerocapture ellipse to a circular orbit provides just that degree of flexibility in the spacecraft dynamics to make the arrival of the equipment, and the expedition, all at the same place, achievable. So, following aerocapture, the spacecraft must be further slowed until it is in a circular orbit. This could achieved by aerobraking, but it would take a long time – perhaps months of grazing passages through the atmosphere. Instead, a short burn by the lander engines is used to accomplish the circularisation with one manoeuvre.

The proper place to begin the descent manoeuvre can then be chosen. The spacecraft is slowed a little by using the lander engines, and then the aeroshell takes on its second task, of killing most of the orbital velocity, until the parachutes can take over. The lander engines then ignite for a third time, to guide the payload to the appropriate place, and to provide it with a soft landing.

8.29   ACHIEVING THE REFERENCE MISSION

The Reference Mission that we have examined above is about the best plan for getting humans to Mars in our lifetime. We have shared in the deliberations of the NASA engineers and scientists, and we have seen how the choices they have made have been based on the physics of the situation, and on the performance of the propulsion systems that have been discussed in previous chapters. Simplification and commonality have each played their part in reducing the complexity and cost of the proposed mission, and in increasing the safety of the human explorers: the sending of cargo vehicles ahead with vital equipment, and the return vehicle placed ready in Mars orbit; using the same engine, and fuel for many different applications; making use of planetary atmospheres for decelerating vehicles; and using advanced technology where it will have the biggest impact – for example, the nuclear engine for Earth–Mars transit, and making propellant on Mars. All these are examples of the careful thought, and profound knowledge of the discipline, which has formed this Reference Mission.

So, why is it not happening? There are, of course, political and financial reasons; but the main obstacle is the absence of a suitable heavy launcher – the 200-tonne-class vehicle which was postulated at the beginning. At present there are no other customers for such a large launcher, and although, technologically, it is perfectly feasible, no-one is prepared to shoulder the development costs. And without this vehicle, the Reference Mission, as detailed in NASA SP-6107, cannot even begin. The NASA team were, of course, aware of this. They could hardly have been unaware; and as the study reached its close, the absence of a heavy launcher loomed large in their minds as the single major obstacle to a human Mars expedition. To a scientist or an engineer, this is, in a sense, a stupid reason. No new or fancy technology is required to produce a heavy launcher – we had them in the 1960s – but politics and economics, both inexact sciences at best, dictated that this problem had to be solved. The ink was not dry on SP-6107 before the team had begun to examine how this difficulty could be obviated. Of course, there was no magic solution; but there was significant progress, and the result was made public in the so-called 'Addendum to the Human Exploration of Mars', published in June 1998, less than a year after SP-6107 was printed.

8.30 THE ADDENDUM

There was to be no 200-tonne launcher, and if Mars were to be accessible to humans, some other route had to be taken. Could it be done with a smaller launcher? The 200-tonne launcher would be significantly bigger in capacity than the old Saturn V. Could anything be done with something smaller than the Saturn V: somewhere between the present-day 25 tonnes and the Saturn V's capability of 118 tonnes? The Reference Mission would have to be slimmed down.

The team reviewed every item in the mission, to determine what changes could be made to reduce the mass. This was not only a case of looking to save a tonne here and there, in the whole manifest, but also of looking at the mission concept, to determine whether a different approach could reduce the demands on the launcher. In order to avoid confusion we shall, following NASA, refer to the original concept, as detailed in SP-6107 and above, as Reference Mission 1.0, and the updated mission as Reference Mission 3.0. In time-honoured tradition, there is no sign of Reference Mission 2.0.

8.31 EARTH ORBIT RENDEZVOUS

A glance at Table 8.3 shows that, as packaged for Reference Mission 1.0, each launch has to carry in excess of 200 tonnes into low Earth orbit. There is no in-orbit rendezvous, except for the crew launch, where the nuclear-powered Earth–Mars transfer vehicle and the crew are separately put into orbit. Since the nuclear stage, propelling the Earth–Mars transfer vehicle, weighs 118 tonnes, and for the crewed mission the payload to inject into trans-Mars orbit is 90 tonnes, it is clear that if the whole vehicle is to be launched intact, we are dealing with an all-up mass of 208 tonnes; hence the 200-tonne launcher. It is natural to examine whether this could be split between two smaller launchers, at around 100 tonnes each. This is within the capability of the Saturn V, and is more digestible. It would require two separate launches: the nuclear stage, fully fuelled; and the aeroshell, containing the habitat and lander, together with the crew. These two units then have to dock in Earth orbit before setting off for Mars, as previously described. In other words, we have introduced an Earth-orbit rendezvous into the mission, in order to enable the individual launcher payload to be halved. More launches will be required, but the launchers themselves can be smaller, and closer to available technology. Launching the nuclear stage, always on a separate launcher, has conveniences for security and safety; it could, for instance, be launched from a special, secure site.

This is the first improvement – and so far it is without pain, as far as descoping the mission is concerned. There then remains the matter of an Earth-orbit docking manoeuvre, which is not at all difficult. Such manoeuvres have been executed since the very early days of the space programme, and it is routinely carried out with the Space Shuttle and the International Space Station. And it will be recalled that the Apollo mission involved a docking manoeuvre, carried out on the way to the Moon. So, this is nothing new, and should have few cost or programmatic difficulties. However, we have doubled the number of launches. It is a matter of judgement whether twice as many smaller expendable launchers will cost more or less than the original plan. The hope is that when the development costs are taken into account, there will be a significant cost saving.

8.32 THE SPARE HABITAT

The next step was to determine whether the number of launches could be reduced. This, however, cannot be done without pain, because each payload is required for Reference Mission 1.0. Reducing the number of half-payload launches implies the omission of some payload elements. So, what could be excluded? There are four payload elements in mission 1.0: the Earth return vehicle; the crew habitat and lander; the ascent vehicle and propellant factory; and the spare habitat. The obvious payload to consider is the spare habitat. The mission could be accomplished without it, but with some increase in pain: less space for the expedition members on Mars, and less security, in terms of being able to wait-out a synodic cycle, if rescue were to become necessary.

Habitable space for the expedition base camp is a serious issue. The explorers have to remain on the surface of Mars for more than 450 days, which is much the longest-duration segment of any: on a fast transit, the journey to and from Mars takes only 180 days. Careful studies of past habitable volumes on Skylab, Mir, Apollo, and the ISS, confirmed that the requirement of 90 m^3 per crew-member should be retained. The habitation, as designed for the mission, provided about 44 m^3 per crew-member; and the link-up with the spare habitat was essential to raise this value closer to the required 90 m^3. However, a new development was proposed, initially by Boeing. Here there had been work on inflatable habitats, using composite materials that were strong enough to form the pressure shell on Mars, and flexible enough to be collapsed: the martian equivalent of an explorer's tent. Composite materials consist of layers of fibres – usually carbon fibre – bonded together with epoxy resin to form a very strong but light skin, and they have a much better strength:weight ratio than aluminium. They are ubiquitous in spacecraft structures. Even some solid rocket casings are made from them, and they are also present in racing cars and aircraft. They therefore have an extended heritage, and making a pressure shell from composites is perfectly feasible. Flexibility is a relatively new development, but is still current technology, although there are, of course, some disadvantages. Designing a habitat that can be inflated, but that has all the necessary services already plumbed in, is tricky; recall that it is not only the bathroom that has to be plumbed, but the whole life support system, for oxygen, water, and carbon dioxide removal. An airlock and windows also have to be provided. Nevertheless, this offers both a very significant weight saving and significant space saving within the launcher shroud.

If an inflatable habitat could be used to provide the necessary volume augmentation for the explorers, then there is a possibility of dispensing with the spare habitat launch. If the inflatable habitat can be made light enough and small enough, it could be accommodated in one of the other launches. A mass estimate for the inflatable habitat, without crew accommodation and life support, was only 3.1 tonnes, compared with 10 tonnes for the bare shell of the Reference Mission 1.0 spare habitat. Moreover, in its collapsed form the inflatable does not need to occupy most of the volume of a launcher shroud. The proposal, therefore, is to consider the inflatable as a pure extension to the living space provided by the crew lander habitat – the aluminium structure unchanged from the old concept. Deflated, the 'tent' can be placed within the launcher shroud on one of the cargo launches. When the expedition arrives on the surface of Mars, it can be extracted from the lander, attached to the habitat, and inflated, to provide an extension to the living quarters and laboratories. Thus, one launch is saved.

Much has been lost, however. The extra life support, food and water contained in the spare habitat will no longer be available, and neither will it be possible independently to occupy the inflatable should the main habitat become untenable. The consumables and life support amount to 13.5 tonnes, and it seems improbable that this could also be fitted in somewhere in another cargo launch. The back-up provided by the spare habitat will have to be abandoned. But what could the explorers do instead, should something go wrong? Simple leaks in the habitat, and

Figure 8.12. The lander with attached inflatable habitat, in the revised concept of the 'Addendum'. In the distance is a cargo lander. The lander is now a different shape because, to save on combined weight, its structure is now integrated with the aeroshell. (Courtesy NASA.)

mechanical malfunctions of the life support system, are assumed to be repairable, and each life critical system will in any case have a back-up that can be brought on-line if there is a failure. The main risk is that, due to some malfunction or accident, the stocks of food, water, or oxygen might be lost, or depleted to dangerous levels. It should be possible to solve the food problem, as it is non-perishable, and not the major component of the delivered life support consumables, and some extra could be provided. Water and oxygen, however, are much more serious, and a thoroughly reliable back-up solution has to be in place.

The first level proposed is to use the output of the propellant factory to provide oxygen and water. In principle, it could also be used to remove carbon dioxide from the habitat atmosphere. Some design changes would be required, although these would chiefly be alterations to plumbing, and could easily be incorporated in the basic design to provide this extra safety factor for the expedition. The second back-up would simply be to return to Mars orbit and live in the Earth return vehicle until the next trans-Earth slot becomes available. These contingencies assume that no more cargoes are arriving on the surface, from Earth. The Reference Mission actually constitutes a programme of Mars exploration with successive crews and cargoes arriving on Mars, at each synodic interval. In this plan there would be another ascent vehicle, with its propellant factory, arriving on Mars a few months after the first human landing; and this could be used in the same way. A second

Earth return vehicle would also be arriving in Mars orbit, and could also be used by the distressed explorers.

It is clear that this approach is less safe than the original plan. The return to Mars orbit assumes that the crew is sufficiently capable, and can enter the ascent capsule, using what remains of their life support, and that it is ready to go. It might become a condition of the mission that the ascent vehicle is always ready for launch.

8.33 MODIFYING THE VEHICLES

The next major change is to incorporate the aeroshell into the structure of the crewed vehicle habitat. This can be seen in Figure 8.12. The aeroshell then becomes both the structure of the vehicle and the habitat itself, and also the shroud for the launcher. It is not discarded before landing on Mars. An identically-shaped aeroshell is used for the cargo missions, but cannot be incorporated into the cargo elements, and instead is a normal cargo container. This shell forms the shroud of the launcher, acts during Mars capture and entry, and is discarded during the landing manoeuvre. The aeroshell of the Earth return vehicle is discarded after aerocapture into Mars orbit. This design change creates a new aeroshell that is smaller than the aeroshell used in Reference Mission 1.0. Because it is also the launcher shroud, the entire diameter of the launcher is reduced from 10 m to about 7.5 m. This is much closer to the natural diameter of existing components, like the Space Shuttle main propellant tank.

Having two independent sets of engines for the lander and the ascent vehicle seems excessive, and the new design of the lander for the ascent vehicle and the propellant factory retains the same engines and propellant delivery system for both landing and ascent. The empty propellant tanks of the lander are left behind, and some of the combustion chambers and nozzles are removed before ascent, and the system re-plumbed – which is no more than removing some of the propellant pipes and blanking off the orifices. The fuel tanks are refilled by the propellant factory. Some redesign of the lander/ascent vehicle complex is required so that the cargo – mainly the nuclear reactor – can be unloaded, and the ascent vehicle can separate from the factory, landing gear, and the empty lander tanks, for departure from the surface. At the same time, the ascent capsule itself is now also used for Earth entry and landing, and it is therefore docked with the Earth return vehicle, and undocked on arrival at Earth.

8.34 MASS REDUCTION

The remaining efforts involving Reference Mission 3.0 are devoted to the saving of mass on existing systems. To those outside the space business it may seem a strange occupation, but most users of space and most space engineers and scientists have to think about it more or less all the time. Mass is absolutely critical to the design of any space hardware, and long and difficult are the struggles between designers –

whose creations add to the mass – and managers – who need to reduce the mass. In the initial design phases, mass estimates are using past design experience, with the addition of a little extra to provide room for manoeuvre in the inevitable haggling once the full design is complete. The mass of sub-systems therefore tends to decrease as the programme proceeds – unless someone has forgotten something, and more mass has to be found to create it. The usual 'forgotten' items are nuts and bolts to join two pieces of hardware, and cables to connect two electronics boxes; but they all have to be taken into account. Reference Mission 1.0 is still in the 'Have we forgotten anything?' stage, and the design margins for each sub-system are still fairly wide. With Reference Mission 3.0, some serious mass-reduction exercises – referred to, by the team, as 'mass scrubbing' – were carried out to sweep up these margins. At the same time, new and more accurate methods of estimating the mass of sub-systems were incorporated.

This work has resulted in a practicable plan to accomplish the first human mission to Mars in six launches, with less than 80 tonnes of payload for each launch. The launch manifest is shown in Figure 8.13. The payload for the first launch is the Earth return vehicle, now pared down to a total mass of 74.1 tonnes. It is accompanied into orbit by a second launch, which takes up a fully fuelled nuclear Earth–Mars transit stage. This docks with the payload in its aeroshell, before injecting it into the Earth–Mars transit orbit. Because of the payload reduction, the mass of these nuclear stages is also reduced, and they now use only three nuclear rocket engines. For the first launch, the nuclear stage weighs 73.4 tonnes, of which 50 tonnes is liquid hydrogen propellant. The second payload to be launched is the cargo: the ascent vehicle, the propellant factory, the nuclear reactor to provide electrical power, and the inflatable habitat. This cargo vehicle docks with its nuclear stage, weighing 68.6 tonnes, and departs for Mars. The crewed vehicle, with a mass of 60.8 tonnes, is launched at the next opportunity, more than two years after the cargo and Earth return vehicles have been dispatched. It is sent on a fast transit to Mars by its nuclear stage, weighing 76.6 tonnes. If the programme is followed in full, then similar cargo and crewed launches will take place at each launch opportunity.

8.35   THE BOTTOM LINE

We have advanced a long way since the first tentative definition of a Mars expedition. The first proposals, cited in Chapter 5, estimated the mass to be placed in Earth orbit at 5,000 tonnes; but this figure has now been reduced – courtesy of the NASA Mars Exploration Study team – to below 480 tonnes on six 80-tonne-capacity launchers. Of course, we do not yet have the 80-tonne launcher; but such a vehicle is much less problematic than a 200-tonne launcher, and it may well be possible to develop such a launcher and also find other customers who would use it and so help with the amortisement of the development costs.

This is as far as we can go. With six 80-tonne launchers, three nuclear rockets, the habitats, the aeroshells, and the martian propellant factory, we humans can journey to Mars.

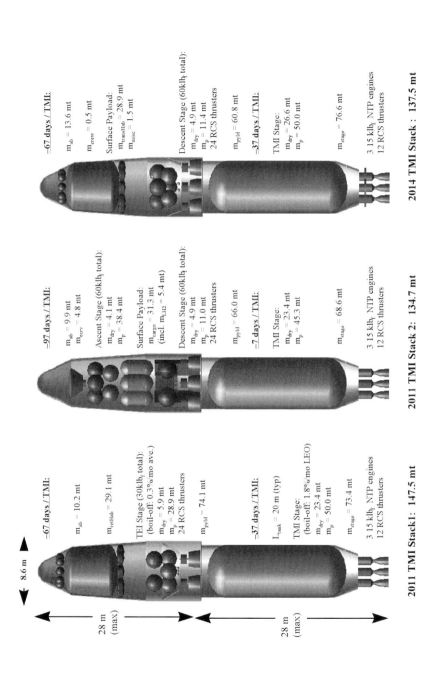

8.6 m

28 m
(max)

28 m

−67 days / TMI:

$m_{ab} = 10.2$ mt

$m_{retlab} = 29.1$ mt

TEI Stage (30klb$_f$ total):
(boil-off: 0.3%/mo ave.)
$m_{dry} = 5.9$ mt
$m_p = 28.9$ mt
24 RCS thrusters

$m_{pyld} = 74.1$ mt

−37 days / TMI:

$L_{tank} = 20$ m (typ)

TMI Stage:
(boil-off: 1.8%/mo LEO)
$m_{dry} = 23.4$ mt
$m_p = 50.0$ mt

$m_{stage} = 73.4$ mt

3 15 klb$_f$ NTP engines
12 RCS thrusters

2011 TMI Stack1: 147.5 mt

−97 days / TMI:

$m_{ab} = 9.9$ mt
$m_{ecrv} = 4.8$ mt

Ascent Stage (60klb$_f$ total):
$m_{dry} = 4.1$ mt
$m_p = 38.4$ mt

Surface Payload:
$m_{cargo} = 31.3$ mt
(incl. $m_{LH2} = 5.4$ mt)

Descent Stage (60klb$_f$ total):
$m_{dry} = 4.9$ mt
$m_p = 11.0$ mt
24 RCS thrusters

$m_{pyld} = 66.0$ mt

−7 days / TMI:

TMI Stage:
$m_{dry} = 23.4$ mt
$m_p = 45.3$ mt

$m_{stage} = 68.6$ mt

3 15 klb$_f$ NTP engines
12 RCS thrusters

2011 TMI Stack 2: 134.7 mt

−67 days / TMI:

$m_{ab} = 13.6$ mt

$m_{crew} = 0.5$ mt

Surface Payload:
$m_{transflab} = 28.9$ mt
$m_{misc} = 1.5$ mt

Descent Stage (60klb$_f$ total):
$m_{dry} = 4.9$ mt
$m_p = 11.4$ mt
24 RCS thrusters

$m_{pyld} = 60.8$ mt

−37 days / TMI:

TMI Stage:
$m_{dry} = 26.6$ mt
$m_p = 50.0$ mt

$m_{stage} = 76.6$ mt

3 15 klb$_f$ NTP engines
12 RCS thrusters

2014 TMI Stack : 137.5 mt

Figure 8.13. The three Mars–Earth vehicles from the 'Addendum' concept, as assembled by docking in Earth orbit. The lower half of each vehicle is the nuclear stage, and the upper half is the cargo or crewed vehicle. The first complex contains the Earth return vehicle, fuelled with methane and oxygen, and the second complex contains the ascent vehicle and the propellant factory. Both of these are sent ahead of the crew. The third complex contains the crew and the lander, plus cargo. (Courtesy NASA.)

9

How do we get there, from here?

The plan for a human expedition to Mars is prepared, and, as discussed in Chapter 8, has been published by NASA. The existing technology that can be used, and the (very little) new technology that needs to be developed, has been identified, and the state of technological readiness for a human expedition to Mars far exceeds the state of the US space programme at the time when President Kennedy made his famous commitment to go to the Moon. And yet the Mars expedition is not included in the approved programme of any space agency. We should ask why. We should ask what the space agencies are doing. We should ask how we could move from the present unsatisfactory position to the day when the Mars expedition begins.

9.1   THE TECHNOLOGICAL BARRIERS

Most of the technology required for a human expedition to Mars is ready: it is 'off the shelf'. But there are some key technologies that are not ready, and that require development programmes before they can be used. But this is no reason not to decide to go to Mars today. When the Moon programme began, the Saturn V was not in any sense 'ready', and neither were most of the other elements that would be required. They were developed specifically for the Apollo programme, once it had begun. This is precisely what should happen for the Mars expedition. Which are the 'off the shelf' technologies; and which are those that need development? Those requiring development are the 80-tonne launcher, the nuclear stage, the aeroshell, and the nuclear power generator. All the others exist, and are either already in use in space, or have been demonstrated in trials on the ground. We should first examine the issues surrounding the development of these new technologies, which can only begin once the decision to explore Mars is taken.

9.2   THE 80-TONNE LAUNCHER

The bare bones of this proposal were included in the previous chapter. The vehicle

Figure 9.1. The launch of one of the Earth-Mars transfer vehicles on a Magnum rocket. It will be united in Earth orbit with a separately-launched nuclear stage, before departure to Mars. In this picture it is side-mounted on the central propellant tank, rather like the Space Shuttle. (Courtesy NASA.)

was called Magnum by its proponents, and a version of it is illustrated in Figure 9.1. In this picture it carries, as its 80-tonne payload, the Earth–Mars transfer vehicle, without its nuclear stage, and has the advanced solid booster, or ASRM, rather than the fly-back booster.

In one sense, the production of this launcher is just a matter of assembly – the putting together of existing elements. Currently, there are two strands of launcher development that could lead to an 80-tonne launcher: the launchers based on the Space Shuttle and its components, and the launchers based on expendable vehicles, mainly the Atlas and Titan in the US, and the Energia in Russia. The Ariane V family of launchers could also evolve towards a launcher of this capability. The concept depicted above is an attempt to reduce potential development and recurrent costs by using existing designs and existing NASA facilities. The Space Shuttle-like configuration is chosen simply to allow the Shuttle boosters to be used, and to allow the use of Shuttle-handling facilities. It is often not appreciated that a considerable fraction of the cost of a launcher is in the launch site and the rocket-handling facilities. For safety and security, launch sites cover many square kilometres of territory; and apparatus for safely handling large rockets, and for producing, storing and delivering rocket fuel, are vastly expensive to create, and to maintain and operate. The cost of a launch can therefore be minimised by utilising existing facilities, and by adhering as closely as possible to existing rocket designs.

The use of Space Shuttle facilities and boosters would lead to a cost saving; but using Shuttle engine technology would be very expensive. The hope here is that the propellant tank – which must be somewhat longer than the Shuttle tank – can be developed reasonably cheaply, using existing designs from the new expendable launcher programmes such as the Extended Expendable Launch Vehicle (EELV) and the Reusable Launch Vehicle (RLV), so that it has the same diameter as the Shuttle tank, but is of a different construction. The Shuttle tank is a 20-year-old design using aluminium; but the new tank would use composite materials for the main structure and the liquid hydrogen tank, and aluminium–lithium to contain the liquid oxygen. The engines and propellant delivery systems would be simpler and cheaper than the Space Shuttle systems; and the favoured engine is the RS-68, as discussed in Chapter 8.

The development items for this launcher are, therefore, the tank and the payload attachment, while the rest is more or less off-the-shelf. The guidance system must, of course, not be forgotten. A vehicle of this size will have dynamics different from those of its smaller cousins, and test flights will be required before it can be qualified for launching the components of the Mars expedition.

Figure 9.2. The RS-68 engine under test. (Courtesy NASA.)

The new launcher cannot be developed until a Mars expedition is approved, unless another customer can be found. At present, 10-tonne spacecraft, launched into geostationary orbit, are commercially important; and the same launcher can put more than 20 tonnes into low Earth orbit. Launchers round the world would not have been up-rated to this level unless, in a very competitive market, there were a strong commercial demand for this size of payload. The Mars launcher would have four times this capacity. It is a matter of *when*, not *if*, the commercial market will require these much bigger spacecraft. Perhaps a combination of commercial and strategic demand, and a new will to carry out the Mars expedition, will merge to tip the balance and trigger the development of the 80-tonne launcher.

9.3   THE AEROSHELL

The aeroshell is required to carry out two manoeuvres: the capture of the Earth return vehicle into Mars orbit; and the atmospheric entry and deceleration of the cargo and crewed vehicles, prior to landing on the surface. Both of these manoeuvres have to be carried out autonomously. The whole science and engineering of atmospheric entry is very well established, having been in use since the earliest days of the space programme, and, indeed, for missiles tool. The thermal protection system and the ablative cooling shell are both well-established technologies. Aerocapture is a relatively new concept, but it has been used successfully to place planetary probes in orbit round Venus and Mars – most recently, for Mars Observer. The real departure for the Mars expedition is the size of the payload to be aerocaptured or landed. The payload, including the aeroshell, weighs about 80 tonnes – slightly less than the Space Shuttle orbiter, which weighs 99 tonnes on re-entry into Earth's atmosphere. The crucial item will be the dynamic stability of the aeroshell during passage through the martian atmosphere; without control surfaces, like those on the Space Shuttle orbiter, there is no possibility of correcting the flight path, particularly during entry. The entire process must therefore rely on the natural stability of the shaped aeroshell, and its presentation at the correct, head-up, angle at the instant of entry. Aerocapture is a similar manoeuvre, requiring precise trajectory control – again without control surfaces. The reaction control system, or RCS, can be used to control attitude, and to adjust velocity, for both aerocapture and entry, but it can only work in a vacuum, before or after, but not during, atmospheric passage.

None of this is new or difficult; but it requires careful modelling and design, followed by wind tunnel tests, and finally by Earth atmospheric tests. Data on the martian atmosphere, returned by the many robotic probes being sent to Mars, will be fed into the modelling and the design. A full-scale test flight, and landing on Mars, would probably not be required, and would instead be replaced by the first real cargo landing.

9.4   THE NUCLEAR PROBLEM

The nuclear rocket stage, and the nuclear electric power generator for propellant production, are crucial to the Mars expedition, and will probably prove the most difficult to develop – not because of their inherent technical difficulties, but because of the political and public health issues surrounding their use in space. Nuclear power plants are ubiquitous on Earth, and most economically developed countries have at least one. Nuclear reactors are used to generate a significant fraction of the electricity used for domestic consumption in several major economies, including France and the United Kingdom. They are part of the background of life; and yet, in these developed countries there is a deeply held public abhorrence for anything 'nuclear'. This is in marked contrast to the situation during the 1950s and 1960s, when nuclear power – as opposed to nuclear weapons – was seen, quite generally, as safe and beneficent.

We cannot go back to the attitudes of the mid-twentieth century, and we have to deal with the present situation and the present public perception, which is unlikely to change. Nuclear power, for electricity generation and for propulsion, is essential for a Mars expedition, and, along with the decision to go to Mars, it must be subject to public acceptance. Public safety will be the primary concern. The issues were discussed in Chapter 7, and here it will be sufficient to reiterate the primary contributors to that public safety.

The nuclear power reactor and the nuclear rocket engine will be launched in an inert state, and will not be radioactive. The active principle, U^{235}, is not significantly radioactive – it has a half-life of more than 100 million years. The fission process is different, and will be disabled on the ground and during launch. The main safety issue will be to ensure that neither of the two fissile cores – in the engine, and in the power reactor – can become critical before or during launch. In practical terms this means that the fuel 'rods' must be kept apart from the moderator and the reflector (see Chapter 6). They could in principle be launched separately, but it may be sufficient to put in place adequate and well-tested fail-safe precautions, to convince the appropriate safety authorities. Public perception can best be dealt with by a programme of transparent honesty and open decision-making. It would be tragic if the exploration of Mars, by representatives of the human race, were to be prevented by the misconceptions of their fellow humans.

The real, as opposed to the perceived, safety issue is the testing of the nuclear rocket engine, which will, of course, never be used other than in space. When nuclear rocket engines were developed, atmospheric testing of nuclear bombs was current, and the releasing of a little hydrogen, contaminated with a few fission fragments, into the air of the Nevada desert, was a drop in the ocean compared with atmospheric contamination from bombs. The release of nuclear rocket exhaust into the atmosphere would not, even today, significantly increase the risk to nearby communities, but there is no possibility of open atmospheric testing of a nuclear rocket engine being allowed. The research that is being carried out today, towards a nuclear engine, uses non-fissile, electrically heated fuel elements to simulate the heat output from the nuclear process. This is perfectly safe. It is also not so distant from the real thing as might be supposed. The real nuclear fuel elements are likely to be the tricarbide type: mixtures of uranium, niobium and silicon carbide. Such tricarbide fuel rods can easily be heated to the same temperature as would arise from the fission reaction, by electrical means, and a full-size working engine, which does everything except nuclear fission, can be made. In this way the whole engine design – including nozzles, propellant distribution, cooling, and so on – can be tested in a normal engine test facility. In particular, the abrasion of the fuel elements can be monitored, and the probable contamination of the exhaust stream by fission products established, without fissile activity actually occurring.

One missing element of this test programme is the control loop that balances the level of fission-produced heating, and the flow of cooling, hydrogen propellant, through the engine. Fission, without the cooling hydrogen, would lead to dangerous overheating of the engine, and the control loop, that shuts down the fission if the hydrogen flow is interrupted, must be tested in a real situation. Note that this overheating would never lead to a nuclear (fisson) explosion; but the engine would suffer major structural damage, and some radioactivity would be released. This overheating was, in fact, tested during the NERVA programme in the 1960s, and so the process is well understood. The other effect that has to be tested, in the real situation, is the heating of the engine structure and the internal shielding by neutrons and gamma-rays produced by the fission process. The shield and structure are provided with cooling ducts to take away this heat, but the heat is not generated unless the fission process is active. This leads to the conclusion that there must be a test of the engine in fully operational mode. The only practical way of doing this is in a closed system, in which the exhaust is not directly released into the atmosphere, but is cooled, recovered, and scrubbed to remove any fragments of radioactive material it might contain. It can then be safely released into the atmosphere. The facility to do this – probably underground – will be expensive to build and maintain, and will no doubt generate significant public concern. But it is a necessary cost for a human expedition to Mars, and it will have to be created and used.

The nuclear rocket engine may be regarded as having already been developed, as described in Chapter 7; and the design of the engine for the Mars expedition can be evolved using testing as described above. Nuclear reactors for use in space are also rather well developed, particularly in the Russian programme. The remaining significant technical issue, for these, is the electrical power conversion machinery. After being landed on the surface, the power plant must operate on Mars for four years; and it has to provide power for the explorers, as well as for the propellant factory. The conversion of heat into electricity really needs to make use of a mechanical system, as opposed to the well-tried thermoelectric systems used with RTGs on many missions today. The thermoelectric system is simply not efficient enough for the Mars expedition. Brayton turbines and Stirling engines have been developed for use in space, but mainly for the reverse operation – electrically powered refrigeration. Nevertheless, space engineers are very wary of moving parts. Traditionally, they are avoided wherever possible, because of the inherent difficulties they present in long-term use. Lubrication failure, abrasion of bearings, metal fatigue, the integrity of seals – all these, and many other aspects, increase the difficulty of designing a reliable mechanical generator. Such systems on Earth undergo regular maintenance, and shut-down periods for refurbishment of bearings, seals, and other components; but none of this is possible on Mars. A significant development programme will therefore be required to produce thermomechanical electrical generators that will work reliably for four years, having gone through the vibration and acceleration of a launch from Earth, eight months in space, atmospheric entry, and a landing on the martian surface. In contrast to the other developments, the mechanical generators will certainly find customers in other fields. The production of sufficient electrical power is a major problem for modern spacecraft, including communication and Earth observation satellites. Solar cells are very inefficient, and a mechanical system, using solar heating, may well be the next major step in improving this situation. The generating machinery would have much in common with the type required for the Mars expedition.

The above are the only real technological challenges that hinder the mounting of a Mars expedition. None of them, however, is in any way comparable to the challenges that had to be met during the Apollo programme. We now need to examine what steps the world's space agencies have made towards taking up the challenge of the Mars expedition.

Figure 9.3. A Stirling engine cooler, developed for use in space. This device uses mechanical energy to transfer heat from a cool body to a hot body. Used in reverse, it can generate mechanical energy by the transfer of heat from a hot body to a cold body. (Courtesy NASA/Ball Brothers.)

Figure 9.4. A Brayton turbine cooler, as used on the Hubble Space Telescope. The physics is the same as for the Stirling engine cooler illustrated in Figure 9.3. (Courtesy NASA.)

9.5   THE POLITICAL CHALLENGE

The United States could begin a human expedition to Mars today. It has all the capabilities, and an agency – NASA – that has the experience and structure to make it happen. ESA could not do it alone, as it has a much smaller programme, and the Ariane launcher would need to undergo very significant development to reach the 80-tonne capability; nuclear rockets would also be difficult to develop and test in a purely European context. Russia could do it, as it has the technical capability, but it is doubtful whether it could do it alone in present economic conditions; but the hardware is there. China has ambitions to engage in a manned space programme, and will no doubt succeed in it; but it will take many years before it has capabilities similar to those of the three major space agencies of the world. Smaller national agencies such as the Japanese NASDA,[1] and the Indian space agency, could make

---

[1]   Now the Japan Aerospace Exploration Agency, JAXA.

useful contributions, but could not lead such a programme. This leaves the United States, or an international consortium led by the United States, with Russia as a major contributor. But how far could we proceed in creating such a partnership?

9.6   THE ISS AS A MODEL FOR THE MARS EXPEDITION

The International Space Station is a prototype for such a partnership. It is led by the United States, but there have been major contributions in hardware from Russia, and significant hardware contributions from ESA, NASDA, and many other smaller national space programmes. In terms of the number of launches necessary to build it, it exceeds the mass into low Earth orbit required by the Mars expedition. Some forty Space Shuttle launches, together with eight Russian Proton launches, were planned for its construction, and ESA is providing a consumables cargo capability with its Ariane V launcher and the Advanced Transportation Vehicle, or space tug. Thus the ISS is a good prototype for the amalgamation of contributions from many agencies to produce a composite whole. In conception and planning it has all the organisational and practical elements required for a Mars expedition; but in

Figure 9.5. The International Space Station. Complete assembly was to be accomplished with forty-one flights of the Space Shuttle and eight flights of the Proton launcher. The US, Russia, ESA, Canada, Japan and Brazil are providing hardware and other contributions. (Courtesy NASA.)

execution it has, as we know, been dogged by difficulties, from the economic traumas of the Russian space programme to the recent tragedy with the Space Shuttle *Columbia*. Nevertheless, it is worth examining its key motivations in order to understand where these can, and cannot, be applied to the Mars expedition.

The driving forces for the ISS are, firstly, the space agencies, which have, as a long-term aim, human exploration of space, even if for the time being it is only near-Earth space. The second major force is world aerospace industry, which has a large capacity and major capabilities. From the commercial point of view, the development of new technology and provision of components for the ISS is seen as highly desirable. This desirability has two elements: large government or agency contracts to provide advanced technology, keep the skilled technical work-force in being and pay its wages; and, more importantly, the development of new technology, to meet some of the challenges of the ISS, can improve the competitive edge of an industry or, indeed, of a nation. This latter motivation is very important, not only for individual companies or nations, but also for society as a whole and for quality of life. Many developments from the space programme find their way into everyday life; and these are not just non-stick saucepans, as denigrators of the Apollo programme used to claim. We shall look further into this aspect a little later in this chapter.

Science, in general, comes a poor third in the list of major motivations. The vast majority of space scientists use only robotic space probes, or observatories, to do their work. This arises naturally, because of the developments in miniaturisation of computers and electronics that make it easy to compress major autonomous capability into quite small packages, which have high reliability. This applies both to space instruments and to the spacecraft that carry them. Thus the natural way to execute space science is to build an unmanned spacecraft that will run for many years in space, receiving its instructions from the ground, and transmitting the results of its observations back to Earth. This has been going on for forty years, with increasing capability. Most space scientists would not want to involve an astronaut in their experiments or observations. Interestingly enough, in the field of astronomy, this approach – the removal of people – has been extended to ground-based astronomical telescopes situated on remote mountain-tops. They are beginning to be operated remotely over the Internet, and the local staff members are being dispensed with. Amongst most space scientists, therefore, there is very little interest in human spaceflight.

On the other hand, in the fields of biological and medical science there is significant interest, mostly with the humans themselves as experimental subjects, although plant and animal life is also studied. The ISS can be seen, for these areas of science, as a biological and medical laboratory. The motivation for this research is partly pure curiosity – as it should be – and partly to form the basis of future human exploration of space. Much work is still needed to understand the adaptation of the human body to zero-g, and to evolve mitigating strategies – not least for the Mars expedition. Another major subject of study is sustainable life support: how to produce food in space, and how to regenerate the atmosphere naturally, as happens on Earth. Thus plants are studied both for use as a food source and as a means of absorbing carbon dioxide and replenishing oxygen.

There is also significant industrial interest in the zero-g environment of the ISS, for many types of developmental work. Typical investigations include combustion processes, in which convection is absent in zero-g, and protein crystal growth, in which the concentration gradients in the mother-liquor, caused by gravity, are absent. It is sometimes questioned whether or not human beings are essential to such experiments, and whether the experiments could quite easily be automated; but with human presence, the ISS crew can operate the experiments, and, very importantly, can react in real time to the results, and change the experimental conditions intelligently, just as in the laboratory. This will probably be more productive, in many experiments, than a preplanned and immutable sequence of operations.

There are thus interesting parallels between the Mars expedition and the ISS. The US-led international consortium is one, and the general apathetic attitude of scientists is another. The main drivers – the agencies and industry – would also stand very much the same. So, the ISS is a possible model for Mars exploration, although it would, of course, require significant improvement. The timetable of a Mars expedition is strictly defined by the celestial mechanics of the planetary orbits, and there is no allowance for late delivery of components or changes in the manifest to be supplied by each agency. There are also fewer opportunities for short-term modifications to the plan. The cycle of flights to and from the ISS is short, and can easily be changed; but such contingencies cannot be applied to the Mars expedition. Once the expedition begins it must be completed in its entirety. Nevertheless, the ISS is in many ways a rehearsal for the expedition to Mars.

There are, of course, some major differences. The ISS uses existing launcher technology, whereas the Mars expedition would require the new 80-tonne launcher. The nuclear elements of the Mars expedition also have no parallel in the ISS. (As an aside, it should be noted that the vast acreage of solar panels required to power the ISS, and which is a major contributor to the atmospheric drag that slows it down, could be replaced with a dustbin-sized nuclear reactor.) The key new technologies would have to be developed under international agreements, before the Mars expedition itself could be mounted. The launcher, and the nuclear reactors for engines and electric power, are ideal candidates for a US–Russian development programme, if the financial and organisational issues can be properly addressed. As mentioned above, the Russian programme has significant experience, in both of these areas, that would be very valuable to the Mars programme.

The probable form of a human expedition to Mars, therefore, would in some ways follow the organisational model of the ISS – but avoiding the pitfalls. In addition, it would require significant collaboration between the US and Russia to develop the necessary new technology for launchers, and nuclear rocket and power systems. The expedition itself would be international, with many different agencies providing sub-systems both large and small, and with an international crew.

9.7   THE ISS AS A TECHNICAL PRECURSOR TO THE MARS EXPEDITION

Apart from its political and organisational aspects, the ISS is also a practical precursor to the Mars expedition. It does not go anywhere, of course, but in several areas it lies in a direct technical developmental path towards the Mars expedition. The main element of this is preparation for long-duration space flight. The conditions in the ISS, for the crew, are similar to those of the Mars expedition members, and six months on board would in many ways be similar to the six-month journey to Mars, at least in physiological terms. Psychologically, it would be very different: the comforting view of the Earth, just a few hundred kilometres away, would be absent. But all the physiological problems in human adaptation can be studied in depth using the ISS. Indeed, this kind of programme is already going on, with wide international participation. There are, of course, many serious developments in life support which could contribute to improving the lot of the expedition members. Amongst these are: refinement of the necessary quantities of life support consumables required under various recycling scenarios; the best way to keep the crew healthy under different gravity and atmospheric conditions; and, in the longer term, the psychological aspects of living in a 'can' for up to four years, with a limited number of companions. All of these contribute to the practical preparations for the human aspects of the Mars mission.

The ISS can also help to refine the nature of the apparatus to be taken for scientific research on Mars. The ISS is in zero-g conditions, and the martian surface will be at 0.38 g; but as a closed system, the ISS is a prototype for the cramped laboratory and living space that will be available in the Mars habitat. Experimental apparatus that is light and compact, and can be adapted for many purposes, will be a major requirement for the Mars surface laboratory; and here, experience of carrying out experiments in the ISS can help. The closed system is a useful discipline, as the windows on the ISS or in the Mars habitat cannot be opened if a noxious substance is produced in an experiment. No spares can be flown up to Mars, and so experimental apparatus will have to be repairable *in situ*. Deciding which spares and tools will be required to keep the laboratory running, and at the same time optimising the mass to be taken to Mars, will also benefit from experience on the ISS.

Another area in which the ISS is leading the way is in the use of robotic arms to manipulate and assemble structures in space. On the Mars expedition, the in-orbit docking of the nuclear rocket stage with the cargo vehicle or crew vehicle will probably not require human or robotic arms, as it is a well-established technology; but robotic arms and manipulators may well be important for exploration and for the assembling of parts of the expedition base camp on the surface. Experience gained on the human–robot interface, during complex building operations on the ISS, will be very useful in planning the best use of robotic systems to support the human exploration of Mars.

The ISS is therefore important to the Mars expedition in political, organisational and practical ways. It is also useful in keeping the manned space programme 'warm', while the will to explore Mars is nurtured and developed. Without the ISS, the manned spaceflight programme would be in the doldrums, with little sense of direction or purpose. The 10–15-year period during which the ISS will be expanded and its use developed, is an important time interval for the Mars plan to be matured, and for the political, social and economic obstacles to the Mars expedition to be overcome.

9.8   ROBOTIC EXPLORATION OF MARS

We are not yet on Mars – and we shall not be there for perhaps twenty years. But the robotic exploration of Mars, begun in the 1960s and then neglected for decades, is now undergoing a major revival. The list of landers and orbiting spacecraft is impressive, and many new results have been accrued over the past few years. In particular, the presence of water – possibly in permafrost, or possibly in combination, over much of the surface of Mars – is now fairly well established. Of course, it is a long way from the remote detection of water, to deciding if it can be used as an indigenous resource to support the expedition.

The robotic exploration of Mars began with intense competition between the United States and the Soviet Union. Many probes were sent, and there were a remarkable number of failures before Mariner 4 accomplished the first successful fly-by and sent back a few pictures in 1964. In 1971, Mariner 9 orbited Mars and sent back several thousand pictures; but despite many attempts by both agencies, it was not until 1975 that Viking 1 sent back the first good pictures from the martian surface. Mariner 2 followed a month later, and together they sent back more than 50,000 pictures of the surface, taken from space and from the surface. Then the impetus was lost, as the impact of Apollo was felt. It was not until 1996 that the next successful mission to Mars, Mars Global Surveyor, arrived in orbit there; and it is still working. But, in keeping with the long record of difficulties, an earlier mission, Mars Observer, had failed on arrival in 1992. So began a series of successes interleaved with failures, in the modern robotic exploration of Mars.

The two modern orbiters – Mars Global Surveyor and Mars Odyssey – have been remarkably successful in mapping the entire surface of the planet. Mars Global Surveyor has provided high-resolution visual images which have not only led to the production of very good topographical maps, but have identified a vast number of particular landforms, the most exciting of which have been those that suggest the past, or even recent, presence of water. Mars Odyssey creates images in other wavebands in order to identify the chemical composition of the surface. It has a thermal imager to create temperature maps of the surface, and a gamma-ray spectrometer that can analyse the surface material to a depth of 1 m below. The radiation coming from space, and hitting the surface, excites its atoms, which release gamma-rays and neutrons that the spectrometer can use remotely to determine the composition. The neutrons are particularly important, because they respond to the presence of hydrogen. There are not many common compounds of hydrogen, other than water, and so the famous map of neutron emission from the martian surface is

Table 9.1. Heritage missions to Mars

Name	Nation	Date	Purpose	Results
Mariner 4	US	1964 Nov 28	First successful Mars flyby, 1965 Jul 14	Returned 21 photographs
Mariner 6	US	1969 Feb 24	Mars flyby, 1969 Jul 31	Returned 75 photographs
Mariner 7	US	1969 Mar 27	Mars flyby, 1969 Aug 5	Returned 126 photographs
Mars 3	USSR	1971 May 28	Mars orbiter/lander, arrived 1971 Dec 3	Some data and few photographs
Mariner 9	US	1971 May 30	Mars orbiter, in orbit 1971 Nov 13-1972 Oct 27	Returned 7,329 photographs
Mars 5	USSR	1973 Jul 25	Mars orbiter, arrived 1974 Feb 12	Lasted a few days
Mars 6	USSR	1973 Aug 5	Mars orbiter/lander, arrived 1974 Mar 12	Little data return
Mars 7	USSR	1973 Aug 9	Mars orbiter/lander, arrived 1974 Mar 9	Little data return
Viking 1	US	1975 Aug 20	Mars orbiter/lander, orbit 1976 Jun 19-1980, lander 1976 Jul 20-1982	Combined, the Viking orbiters and landers returned >50,000 photographs
Viking 2	US	1975 Sep 9	Mars orbiter/lander, orbit 1976 Aug 7-1987, lander 1976 Sep 3-1980	Combined, the Viking orbiters and landers returned >50,000 photographs

Table 9.2. Modern missions to Mars

Name	Agency	Date	Purpose	Results
Mars Observer	NASA	1992 Sep 25	Orbiter	Lost just before Mars arrival, 1993 Aug 21
Mars Global Surveyor	NASA	1996 Nov 7	Orbiter, arrived 1997 Sep 12	Currently conducting prime mission, science and mapping
Mars Pathfinder	NASA	1996 Dec 4	Mars lander and rover, landed 1997 Jul 4	17,000 images; last transmission, 1997 Sep 27
Nozumi (Planet-B)	NASDA	1998 Jul 4	Mars orbiter, currently in orbit around the Sun	Mars arrival delayed to 2003 Dec, due to propulsion problem
Mars Climate Orbiter	NASA	1998 Dec 11	Orbiter	Lost on arrival at Mars. 1999 Sep 23
Mars Polar Lander/Deep Space 2	NASA	1999 Jan 3	Lander/descent probes to explore martian south pole	Lost on arrival 1999 Dec 3
Mars Odyssey	NASA	2001 Jul 4	Orbiter	Currently conducting prime mission, science mapping
Mars Express	ESA	2003 Jun	Orbiter and lander (Beagle 2)	
Mars Exploration Rovers	NASA	2003 May/Jul	Two 180-kg surface rovers; range 100 m per martian day	
Mars Reconnaissance Orbiter	NASA	2005	Orbiter with 25-cm mapping resolution	

Figure 9.6. A Mars Odyssey image obtained with the GRS instrument using epithermal neutrons that respond to buried water up to 1 m below the surface. The dark areas indicate the presence of permafrost bearing at least 2% water within 1 m of the surface; pale areas indicate up to 14% water. (Courtesy NASA/JPL/Los Alamos/University of Arizona.)

taken to indicate the distribution of water. Hydrogen gas would not remain long on Mars, because it is so light. Deposits of petroleum are possible but improbable, and so water is the best guess as to the source of the neutrons. Mars Odyssey also maps the radiation from space – an essential measurement for the safety of humans on Mars.

The most successful of the early missions, the Viking landers, returned much data from the martian surface. Of the modern landers so far, one has failed and one was successful. Mars Polar Lander – intended to investigate the fringes of the ice cap – was lost during the landing operation; but on 4 July 1997, Mars Pathfinder landed safely in Ares Vallis – a northern plain covered with rocks of many different types, carried there by ancient flood-water. Pathfinder was the first probe to use airbags instead of rockets for the final landing. It worked, and the Sojourner rover – a 10-kg solar-powered vehicle – was deployed to investigate rocks over a wide area. This combination operated for twelve weeks, and returned 17,000 images, fifteen chemical analyses, and meteorological data. The general indications from this work are that Mars once had a warm wet climate, with liquid water on the surface.

The outstanding success of Odyssey, Global Surveyor and Pathfinder have contributed towards the regeneration of confidence in our ability to succeed with Mars missions. Two important new missions are to be launched in 2003: Mars Express – the first Mars mission led by ESA – and NASA's two Mars Exploration Rovers. Mars Express carries a small lander, built by the UK, and called Beagle 2. After landing on the surface and deploying the PAW – a robotic arm carrying the instruments – it will perform surface rock analysis, and will also analyse the atmosphere for trace gases that might indicate life. The success of the first rover – Soujourner, deployed by Mars Pathfinder – led to NASA preparing two much more

Figure 9.7. The surface of Mars, imaged by Viking. (Courtesy NASA.)

capable rovers. These weigh about 180 kg and, powered by solar panels, will range the surface of Mars for a much longer period than the first rover. They have a range of 100 m per martian day, or sol – equivalent to about two Earth days. The rovers are landed by a combination of aeroshell, parachute and airbags, as is Beagle 2. This is becoming the standard method of conducting a 'dumb' landing. The parachutes drop the speed to about 60 m/s, and the airbags then deploy to protect the payload from the relatively hard landing on the surface. The bags will bounce several times, and perhaps roll, before coming to rest. They will then deflate, and withdraw into the lander, to expose the payload. In the case of Beagle 2, the clamshell then opens to expose the solar panels, and the PAW deploys to begin rock analysis. The exploration rovers emerge from the deflated and retracted airbags, to begin their independent exploration.

Future plans for robotic exploration include further surveyors. In 2006, Mars Reconnaissance Orbiter will map the surface to a precision of 25 cm; and later, *smart* landers will be sent, which can chose their landing spot in real time, and pave the way for *sample-return* missions. These will extract samples of martian rock and dust, and ship them back to Earth for analysis. This is not a new idea. It was accomplished by the Russians in 1970, when 100 grammes of Moon-rock were drilled out of the surface and shot back to Earth on a ballistic trajectory, by Lunar 16; the 35-kg capsule containing the sample was soft-landed in Russia. To accomplish this from Mars is, of course, much more difficult; and forty years on it is still regarded as sufficiently so that it is not programmed before 2010. In the 1970s, the lunar sample return mission of the Russian programme was based on soft-landing vehicles, originally intended for the aborted Russian manned lunar programme. In a real sense, the sample-return from Mars is a direct precursor to a human mission to Mars. As one space engineer has stated: 'There is no hope of bringing a human back from Mars, if we cannot bring a brick back'. So, those who care about human exploration of Mars will watch the Mars sample-return missions, of any agency, very closely. Apart from the comforting idea that something can be brought back from Mars, the smart, soft landing, carried out completely autonomously, is a precursor to the landing of the Mars expedition cargo vehicle.

The space agencies are certainly doing *something* on Mars – even though it is not human exploration. But does it help? In one sense it is a deviation from the direct course of exploring Mars with human beings; and it also encourages those who believe that human exploration is a waste of resources. But all of this robotic exploration has to be undertaken before humans can safely set foot on Mars; so the robotic programme could be regarded as part of a slow build-up to the human missions. It is certain that, at present, to ask the US Congress to support a human expedition to Mars is to invite a negative response; but on the other hand, lauding the economy of exploring Mars by robotic means gains sympathy for the space programme. We live in hard times for space exploration. Given the difficult situation for a human expedition to Mars, it is well to consider the efforts currently being made by the agencies to explore the planet, as precursors to the inevitable moment when a human being sets foot – or, at least, boot – on the surface.

Figure 9.8. An artist's impression of a sample return mission leaving for Earth. The rovers have collected the samples, and the ascent vehicle assembly has prepared its propellant in the factory to be seen on the front of the launch platform. In this case, the factory is powered by the solar arrays deployed to the right and left of the platform. The ascent vehicle will directly enter the transfer orbit to Earth. (Courtesy NASA/JPL.)

There will eventually be a Mars expedition; and it is worth examining what the latest data sent back by the robotic probes has to say about where it should land, and what the explorers should do. Of course, like all great expeditions it is the going there that matters in the long run; but managers and accountants require to know why we are going and what we will do – and they would also like to know about the benefits to those left behind on Earth: health, wealth and happiness?

9.9   THE LANDING SITE, AND THE TASKS OF THE EXPEDITION

A journey to Mars is certain to be the most tremendous adventure for humanity; but only six people can go, while the rest of humanity will be left on Earth. It is vital for human progress that this first step beyond the Moon is shared by those left behind. It is not enough to simply go and come back with pictures and stories. The first representatives of the human race to journey to Mars have to involve their stay-at-home brethren in the mission, and they have to carry out tasks that are of importance to all humanity. They can do this by gaining as much knowledge as they can on two main questions. The first of these is simply: is there life on Mars; and if not, was there ever any life? We need to know if we are alone in the Universe; and we need to know whether Earth is the only planet ever to produce life. A positive report on the existence of life on Mars, past or present, would change everything for us. The second question is: can human beings eventually live permanently on Mars; can it be colonised? A positive answer to this is the key to expansion of the human race into space – the key to the greatest dream of all.

Mars has a very thin atmosphere – about 1% of the Earth's – and it contains little free oxygen; but at some time the atmosphere must have been much denser. Mars is a very dry planet, with no surface water; but in earlier times water seems to have been present on the surface. Water and oxygen in large quantities are the basic necessities for human life. If water and oxygen could be found, then both martian life and human life on Mars could be sustained. So, the place to explore is the place where there is the greatest probability of finding water. If there *is* any water, in an uncombined form, then there could be martian life, and humans could use the water to sustain *their* life. We could use it to drink, but we could also use it, initially, to produce oxygen to breathe. The atmosphere, with somewhat elevated pressure, is perfectly adequate to support plant-life, provided water is available, and so food could be grown; and the plants would eventually produce oxygen from the atmospheric carbon dioxide.

The 'canals' of Mars are, alas, just a figment of our imagination, but the picture is by no means as bleak as was at one time thought. The worst case assumed by astronomers, before any spacecraft reached Mars, was that the polar caps were pure carbon dioxide, and that there was no water anywhere. But, thanks to the spacecraft, orbiters and landers which have been sent to Mars, we know now that there is water ice at the poles, as well as carbon dioxide. We know also that there is strong evidence for water, not far from the surface, over most of the martian globe. Of course, the water is in a solid form, as ice or permafrost or perhaps as water of crystallisation; but it is there. Furthermore, there is now overwhelming evidence for old watercourses, and even a possible dried-up lake-bed. It will be locations such as these where the Mars expedition will make its base-camp.

One of the most interesting discoveries from the recent detailed mapping and photography of the martian surface is the presence of land-forms indicative of water having flowed on the surface relatively recently. In regions where canyons or craters break through the crust, many gullies are seen (examples are shown in Chapter 1); and they look temptingly like the upper regions of glaciated terrain on Earth where steep slopes have many small gullies generated by seasonal snow and melt-water.

The first theory put forward to explain these gullies – dried-up water courses – was that permafrost, deep beneath the surface, melted at some point in the past, and emerged where the surface was broken by craters or canyons. Having run down the sides, it evaporated, leaving the dry gullies behind. This could have taken place a long time ago. Liquid water cannot exist on the surface today because the low atmospheric pressure lowers its boiling point to within a few degrees of its freezing point, and besides, most of Mars is well below freezing point all the time. It could only be liquid if the temperature were to rise locally above $0°$ C, as is sometimes seen, and it was under moderate pressure. Layers of permafrost could be exposed where deep canyons or craters cut through; and at some time, when the temperature rose, the layer melted, creating the gullies This suggests that there is a large amount of water, permanently frozen, deep under the surface. This is, of course, consistent with the epithermal neutron data (shown in Figure 9.6).

A more recent theory suggests that some of the gullies are formed by present-day liquid water that runs under a crust of melting snow. These gullies are, in fact, mainly seen on the pole-facing sides of crater walls – north in the northern hemisphere, and south in the southern hemisphere – where snow might be expected to accumulate. The Sun then melts the snow as the martian seasons change, and it falls on the previously shadowed areas. The melt-water would evaporate quickly if exposed to the atmosphere, but if it were trapped beneath the snow layer it would remain liquid long enough for the gullies to form. Gullies are often seen adjacent to very smooth areas of the crater walls, which could be interpreted as snow-cover. This implies a very different location and quantity for the water. In permafrost there could be ancient water deposits everywhere under the surface; but if the gullies are created by snow-water run-off, then it is new water, deposited from the atmosphere during the cycle of the seasons. In this case there would be less water locally for a putative martian colony to mine. In either case, the presence of water, if it is confirmed, suggests that life could exist on Mars.

The choice of landing site is therefore still open, because it depends on which of these hypotheses is correct. If snow is the cause of the gullies, then surface water flows regularly down the sides of craters, and life could be found now. Perhaps it would be similar to the red algae found under snow on high mountains, where there is only rock beneath. A landing site would then need to be found near the pole-facing inner slope of a snowy crater wall. If, on the other hand, the permafrost hypothesis holds good, then the watercourses could be ancient, and fossil life would be possible. Different locations could be chosen, perhaps where the gullies end, at the feet of the slopes. In either case, the location has to be both safe to land on, and accessible from the approach orbit. But there is a real difficulty in exploring these gullies, which appear to be on very steep slopes similar to glaciated terrain on Earth. Mountaineering on Mars, in a suit, will be difficult and dangerous, even in 0.38 g; and a fall would be very serious, especially if the suit were to be torn or punctured. This will need to be taken into account in choosing the site, and in providing the correct equipment for exploration. Other types of terrain should, ideally, also be accessible from the base camp – on foot, or using a buggy. However, all should not be staked on a particular theory. These considerations – and more results from the orbiters and landers, to establish the truth about the gullies and the snow – will be taken into account before selecting landing sites; and before the final selection, candidate sites will be explored as far as possible by rovers.

The rapidity with which different theories are proposed, as each new image emerges, suggests that more images and other data will be needed, and that a concerted effort to understand definitively the distribution of water on the surface of Mars must be completed, before the final landing site for the human expedition is chosen. In this sense, the programme of robotic exploration planned for the next ten years is a vital precursor to our Mars expedition.

Whichever site is eventually chosen, the prospects for discovering answers to the

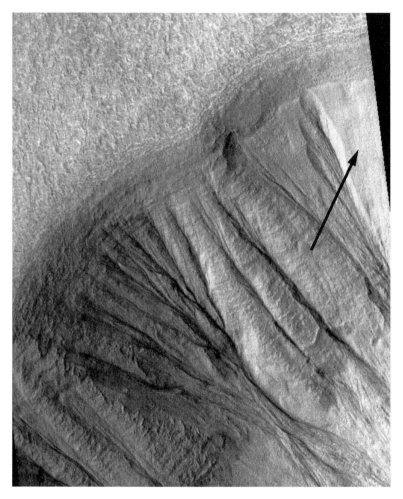

Figure 9.9. Gullies in a pole-facing (winter-shadowed) crater wall. These gullies could have been caused by melt-water from snow cover. The arrow indicates a smooth area that could be snow-covered terrain, with a sharp edge where the snow has all melted and evaporated. Note the apron of water-washed debris at the lower right of the image. (Courtesy NASA/JPL.)

two questions posed above are very good. All the evidence that is being accumulated points towards water being a major presence on Mars, in the past, if not now. Water and sunlight should be all that is necessary to support life of some kind. The Mars expedition has a good chance of finding it – or of not finding it. In either case, the result is of great importance to humanity. If sufficient reserves of water can be found, then a self-sustaining expedition on Mars is at least a possibility, provided that enough electrical power can be provided. The expedition could then be the precursor to a genuine attempt to colonise the planet.

9.10 THE CHALLENGE

We now come to the final challenge. All the technical problems that require to be solved have been examined and found tractable, and the tasks of the expedition have been identified. The preparations for a serious expedition to Mars could start today. It only needs one decision to be made – so let it begin. And yet it is not begun; it is not in any space agency's programme. If this book has been useful at all, it should have demonstrated that the technical problems *can* be overcome, with less effort than was devoted to the preparations for the Apollo programme. Perhaps the spirit of adventure that characterised that period is dead. It is time to examine our motivations, or lack of motivation.

It was mentioned earlier that the motivation for the ISS was driven mainly by the aerospace industry and the space agencies. Considered in this way, similar motivations for a Mars expedition would appear, to some, to be a dubious combination of industry and agencies – to provide interesting jobs for the former, and to put taxpayers money into the pockets of the latter. Today, it is difficult for any major human endeavour to escape such calumny. But there is another way of considering it. If individuals cannot bring themselves to engage in, or support, some large enterprise which is for the general good, it is the duty of government to do so, perhaps against the opposition of many individuals or organisations. Many major scientific activities have been instigated in this way. Much of the current space programme is accomplished like this, as are other major experimental programmes; for example, in physics, or genomics. And it will be similar for the Mars expedition. If it is for the general good of mankind, and can be achieved, then it will eventually *have* to be undertaken, even if many people are against it or are apathetic towards it.

The begged question is, of course, 'the good of mankind' – a major imponderable. Much research that is carried out today would not be well understood, or approved of, by the general public. It is the essence of science that significant training is required before arguments for or against any particular kind of research can be appreciated. But it is not unique to science; the same might be said of economics, law, art, musical composition, or many other areas in which specialist knowledge is required. This does not mean that the new experiment, the new law, the new opera, and so on, should not be formulated, enacted, or composed; rather, that people of sufficient knowledge and judgement should make a decision on behalf of the rest of us. This is the situation for the Mars expedition. The decision to proceed will be taken by an informed group of people. What factors, for or against, will they have to take into account when making the decision?

9.11 REASONS FOR NOT GOING TO MARS

Many reasons have been advanced to argue that we should stay here on Earth and should not explore Mars. The most cogent are: the cost, in human and cash terms; the inefficiency of using human beings to do what robotic probes could do more cheaply; the irrelevance of the knowledge to be gained by exploring Mars; and the much greater need for the same resources to be spent on making human life better here on Earth – these are tangible arguments that can be examined. But there are other arguments that are less easy to counter: human beings should not invade the purity of outer space and pollute another planet; technological progress is a bad thing, and should be slowed down or stopped; space exploration is a waste of money; space exploration is just disguised military activity; space exploration is all a hoax. The list is endless; and these arguments cannot be grappled with, for or against, when there is nothing tangible on which to base the discussion.

Returning to the main arguments against an expedition, we first consider the cost. It is difficult to obtain a firm figure for the cost of the Mars expedition; the Reference Mission was put together with the intent of establishing, among other factors, the cost of the mission. Cost figures are very political for a mission that has not yet been approved. If it appears too high, then it will not be approved; and if it is set too low, then no-one will believe that the programme will be capable of completion. Then there is the question of cost accuracy. It is easy to generate a *precise* cost estimate – anyone can do it using a spreadsheet – but what matters is the *accuracy* of the input figures: the validity of the assumptions on which they are based. It will be appreciated that many elements of the Mars expedition cannot yet be accurately costed, because we have little data on which to base the costs of elements which have not been previously developed or manufactured. This is, of course, a universal problem. What most businesses or agencies do is to take the cost of something previously developed or manufactured, identify the differences between this and the new item, and then make an informed guess as to the impact of those differences on the base cost. The result will not be completely accurate, but the error is only in the differences, not in the base cost. Provided the base item is chosen to be relatively close to the new item, the error will not be large. For some items the calculated cost will be very accurate, particularly when no new development is involved; and in others the error could be large, because intangibles will enter, particularly for radical new developments. The procedure to deal with this is to add a contingency figure to the calculated cost, to take account of the inaccuracy. Where the cost is very uncertain, the contingency will be larger than where the cost is reasonably secure. With most space programmes, contingency is generally set between 10% and 25% overall, with larger figures on the new developments and a smaller figure on the secure items. It is a very complex business, and sophisticated cost models are set up on the computers of the space agencies so that cost estimates can be as accurate as possible.

The first attempt to plan and cost a Mars expedition – the 90-Day Report – produced a figure of $450 billion – an unpleasantly large sum. This was not a figure calculated to warm the heart and loosen the purse strings of Congress; and it did not. President Bush's Mars programme – the Space Exploration Initiative – was never approved. It is easy to realise why such a large figure emerged: everything was treated as a new development; the costs of all new developments were uncertain; and each new development required a large contingency. Adding all these together led to a very large figure indeed. But it was not a stupid thing to do. By arriving at a cost model and a cost figure, it stirred those who wanted to go to Mars into looking more carefully into the cost of their dream. It also made it clear that simply to produce a new technological solution for every element of the Mars mission would make it impossibly costly. To bring it within bounds, therefore, maximum use of existing technology should be enabled. The smaller the number of new developments, the smaller should be not only the cost, but also the uncertainty leading to high contingency figures.

This was the approach used for the NASA Mars Reference Mission (discussed in Chapter 8). The cost estimate is around $55 billion – still a large sum, but one that is much more digestible. Of course, in proposing a cost that is one tenth of the previous estimate, credibility with the planners in government is damaged. They would need to be sure that this was not simply an attempt to persuade them to commit to a programme with which they would then have to continue, even if it turned out to be much more costly. In the Mars Reference Mission document, considerable space is devoted to the cost estimate, and more importantly to the justification of the costs, as outlined above. The costs are based on those of existing technology, where possible, with appropriate factoring for the novel applications of this technology. For the new technological developments, cost models based on past experience have been used, with appropriate contingency factors for the uncertainties involved. The figure of $55 billion may not be the actual cost of the Mars expedition, but it is certainly a figure more firmly based on reality. It is, at present, the best figure available.

It is dangerous to switch off one's critical faculty when confronted with figures like this. In everyday terms it seems an impossibly large number of dollars; it is more than any individual could expect to accumulate. It is dangerous to switch off, because two serious arguments against human exploration of space emerge from the cost figure. 'So large a sum cannot be found, therefore we cannot do it.' 'Such a large sum should not be spent on human exploration, because it would just be wasted, with nothing tangible to show for it.' Neither is true.

NASA's annual budget is about $15 billion, and so it is clearly not impossible to pay the cost of the Mars expedition. Given a twenty-year programme, no more than 18% of the current annual budget, or $2.8 billion, annually, would pay for the Mars expedition. Compare this with the current annual cost of the Space Shuttle programme: $3.9 billion. So, for something between what NASA pays for the Space Shuttle programme, and the slightly lower annual cost of the International Space Station – $1.7 billion – the Mars expedition could be mounted. If major international players entered the arean, then the cost to NASA would be correspondingly lower. This is one way to consider the large figure. Another is to compare it with other major government expenditure: $55 billion is considerably less than the annual expenditure of large government departments. Again, spread over twenty years it is not an outrageous sum to spend on an expedition to Mars.

The other theme of objection – cost – is that money put into space missions is 'wasted'; it all goes up on a rocket and is lost for ever. It is a common misapprehension that the actual hardware sent into space is, of itself, very costly. This, too, is not true. The vast majority of the cost of any space hardware is in the wages of people who manufacturer it and test it; and usually, much more effort goes into testing than into the actual manufacture. The money therefore does not disappear, and is not lost. Just as with any other government project, it appears in the various local communities, as the engineers and technicians spend their wages.

A more challenging argument is that if $55 billion is to be spent, then it could be better spent in improving life here on Earth than in sending people to Mars. This is always a difficult argument to counter, because, of course, it is true. More and better hospitals; the reduction of pollution; aid for the third world; clean water for everyone – the world would be a better place if we had all these. The issue here is twofold. Would it indeed be better to have all these things, and remain ignorant about Mars and the Universe in general? And if this money were available, would it indeed be spent beneficently, and not on smart bombs and Kalashnikov rifles? Alas, recent history shows that the tendency is to take the least beneficial option. Money cannot be found to go to Mars; but money can be found to go to war. In the end, it has to be concluded that *not* to go to Mars would *not* make life better on Earth. There are too many other considerations, inimical to a better life, which would take priority. What 'not going to Mars' will do is to contribute to human ignorance; and the human race cannot afford to lightly remain in a state of ignorance about any aspect of its environment. Ignorance will, in the long run, prove fatal. It has always proved so.

But ignorance is, perhaps, not an inevitable outcome. Perhaps the same knowledge can be obtained by a much cheaper route: the use of robotic probes. Our modern detailed knowledge of Mars has all been obtained with these devices, and the robotic approach is embedded in the mainstream of space science. This is where the true motives of both sides of the argument are revealed. Those who want humans to go to Mars have a complex set of motivations, which inevitably have a very human dimension. But when confronted with a so-called 'rational argument', this human dimension is seen as a weakness. On the other hand, those who espouse the robotic approach will be left on the sidelines if a human expedition goes ahead; and more damagingly, funding for the robotic programme would be diverted to the human programme. It is difficult to achieve a meeting of minds in such a fraught discussion. Setting the human aspects aside for the moment, we can examine the advantages of a purely robotic exploration programme. First, it will be cheaper: less mass will need to be sent to Mars, and so current launchers can be used; and no human lives need to be put at risk. These are the positive aspects. The crux of the argument is, however, the quality and depth of the investigations. Can they be as good as those carried out by the human explorers? Well, if science could be carried out with the exclusive use of robots, scientists would not even be needed on Earth. There therefore has to be a human dimension to Mars exploration, just as in any exploration on Earth.

In the case of robotic exploration of Mars, the humans are back home, monitoring the data, and redirecting the experiments as the results arrive. For human exploration, the link is, of course, much closer. But is this of any consequence? It all a matter of adaptability. The concept of designing a probe, and sending it to Mars, contains a fundamental intellectual flaw: to design the probe implies foreknowledge of the conditions that it is intended to measure. The instruments on board are intended to measure certain properties, decided upon beforehand. Instruments only have a certain dynamic range, so even the magnitude of a parameter has to be estimated beforehand. Space scientists are old hands at this game, and can design instruments that can react rather well to whatever situation they find; but there is always an element of presupposition in any robotic probe design. Humans will be much better at dealing with the unexpected, and at reconfiguring apparatus on the spot, in order to measure new things as they arise. Despite the advances in mobile robots, we still do not have anything that can match a human being in traversing rough ground in an intelligent way. As yet, we do not have effective robotic vacuum cleaners or lawn-mowers, so how can we expect to make robots that will move freely, and safely, on Mars, with the same facility as a human being? It seems that a combination of humans to explore physically the surface of Mars, and remote-controlled rather than autonomous robotic probes, will be most efficient. Remember the 43-minute delay time from Earth to Mars: it is very difficult to control a robotic vehicle with such a long time delay. Anyone who has watched 'Robot Wars' on television will have seen how easy it is to achieve this if the link is more or less instant. In general terms, therefore, it appears that the arguments for and against humans on Mars, as experimenters, are about even. It will be much cheaper and safer not to use them, but the quality of the science will be lower – and it will be far less fun.

9.12 REASONS FOR HUMANS TO GO TO MARS

The primary reason for humans to go to Mars is that it is there – the famous remark about Mount Everest remains true for all 'firsts' in exploration. The urge to explore is a fundamental part of human nature. It may be genetically implanted: the whole of human history is punctuated by migrations – presumably preceded by explorers. Some people believe that migration into space is the destiny of the human race. Apollo was the first step in this direction, and the expedition to Mars will be the second step. This aspect cannot be replaced by a robotic probe programme; human beings will have to go there for it to be a true step on the way to the stars. Ultimately, it is this aspect that drives the urge for a human expedition to Mars; and we have to accept it for what it is, and value it in human terms.

But there are other reasons for human exploration of Mars. It has been argued above that chemical, physical and biological experiments can be much more productive, if monitored, on the spot, by people. Geological and other forms of exploration can be carried out much more effectively on foot than with robots. The key factors here are the ability of human beings to adapt themselves and their activities to their immediate environment, the superb locomotive ability of the human body, and the ability of our hands and minds to carry out the most complex of tasks. The Mars expedition will explore Mars, and answer the big questions about Mars that we identified above, much more comprehensively than could any robotic programme. If we really want to find evidence for life on Mars, and to find out if we could live there independently, then we had better send people there. There is no other way.

For those of us left here on Earth there are also benefits. Most people are thrilled by achievement in any field, where fellow human beings pit themselves against odds, and succeed; we see examples every day. The Mars expedition will be the greatest achievement, against the toughest odds. But in practical terms it will also provide impetus to many aspects of our civilisation. In the crudest terms we may expect significant advances in technology to arise from such a technically advanced programme. Scientists and engineers, both in industry and in the academic world, will be challenged by the many problems, small and great, that will have to be solved for the expedition. It is difficult to imagine what these will be; but they will range from the minutiae of life – such as the best way to prepare and package food that has to keep for four years and remain appetising, or devising recreation for the expedition rest hours – to major technical advances. The technologies required for Mars will have unique properties, even if versions already exist for Earth use. The compact nuclear power plant is one example. Reliable long-term operation of complex systems; the electricity-generating machinery; the high-voltage transmission line; switches that will not spark in the low-pressure atmosphere – the list is endless. Many of the solutions to these and the myriad other martian problems can be applied back here on Earth. Anyone who has wrestled with the power cord of an electric lawn-mower or vacuum cleaner would welcome technology that will prevent kinks and snags, whatever the terrain the cable might be dragged over. If this can be done for a kilometre-long, high-voltage cable, on the rough terrain of Mars, then the same technology might just possibly solve the lawn-mower cable problem. This is a trivial example, and the real technical advances that will result from the Mars expedition will not be known until they actually happen; but we may expect that at least some developments from the Mars expedition will eventually make a fundamental difference to our lives.

We should consider the Apollo programme for some examples. In technical terms it was not the space pen or the non-stick saucepan that mattered; it was the requirement to develop small and lightweight computers with low-power consumption, that led to the crucial development of large-scale integration – the silicon chip, and the microcomputer. There are many trivial and annoying examples of domestic technology that use these, but the improvements in industrial automation, navigation, space technology and, of course, communication and the dissemination of knowledge via the Internet – to mention but a few – have transformed everyday life in a way that could not have been conceived in the 1960s, when the crucial developments were being made. In the late 1960s, the computer I used for my PhD work had 0.0008% of the memory of my present laptop computer. It filled a large room, required a sub-station to provide the power, and took all night to complete a relatively simple numerical integration. This book was written on my laptop computer. The processor card contains half a gigabyte of RAM; it can complete 500 million operations per second; it could fit comfortably into a cigarette packet; and with a small battery it will operate for four hours. Apollo was the seed that sparked that revolution in technology. It would be foolish not to believe that similar advances will certainly come from the Mars expedition; but we shall have to wait and see what they are.

The impact of a Mars expedition on culture and human identity will be dramatic.

If Apollo, for the first time, showed us the Earth as a closed, fragile ecosphere – with all the impact of this idea on our changed relationship to the planet and its resources – what can be expected from the Mars expedition? Neil Armstrong's 'small step' shook our human self-perception; the first footsteps on another world broke a taboo that had held from the dawn of human existence; the heavens were no longer inviolate. The shock to the human psyche has died down, more than thirty years have passed, and generations have grown up with the history but without their own corresponding achievement. Now the old memory will be reawakened; humans will walk again on another world; and this time, human presence may become permanent. The impact on our culture will be even greater than that of Apollo. Moreover, if life, or evidence of life, is found on Mars, then our perceived place in the Universe will change fundamentally; and the same will happen if it is *not* found. If Mars proves potentially habitable, even in the distant future, then again our self-perception will be changed fundamentally. The future of humanity will be changed.

There is one final remark to be made in support of a Mars expedition: its effect on the young. It is a fundamental tenet of our society that the young need inspiration, and inspiring deeds to motivate them. The rebelliousness that accompanies a settled, complacent lifestyle is often all too obvious. As the greatest human adventure, the Mars expedition can hardly fail to inspire. Educationalists list space – second to dinosaurs – as the major scientific interest of youngsters under the age of twelve. Alas, interest in science and technology declines rapidly through the teenage years, with the result that we have too few scientists and engineers in training. While the usual adolescent preoccupations are no doubt responsible for some of this decline, the lack of inspiring scientific achievements, of real public impact, has to be a major factor. In my generation, nuclear power and space travel provided the major inspiration for many physics students; in biology, DNA followed closely; and a general belief in science *per se*, was helpful in other disciplines. But that confidence and inspiration is no longer there, with the inevitable decline in new science and engineering students. The Apollo programme caused a major surge in the number of science and engineering graduates in the United States; and it has been cogently argued that these graduates – reaching maturity and entrepreneurial freedom in the 1980s – created Silicon Valley and the technological surge of the 1990s. Who knows what could be expected from a similar effect following the expedition to Mars?

9.13 THE CHOICE TO BE MADE BY HUMANITY

The arguments, for and against, have been set out. The technical readiness has been reviewed. The Mars expedition is both feasible, and, in the long run, beneficial to humanity. The immediate post-Apollo generations shirked the decision, and it is now the task of the present generation to decide. Will the next generation set foot on Mars? Or will the human race remain trapped here on Earth, until some time in the distant future? The time for the decision is now; it cannot wait much longer. The technological impetus is there, the problems have been identified, and solutions have been proposed that are achievable today. It is no use calling for more information;

that information has been provided; the only way to discover if Mars can be explored, now, is to attempt it.

To decide 'no' is to firmly fix human development exclusively on this planet, for generations to come. It has the validity of a firm decision, and no doubt a good attempt can be made to improve life for everyone; but human nature and recent history indicate that universal good is an improbable result. To decide 'yes' is to open a new world for exploration, to answer scientific questions of interest to us all, and to focus our energy and might on what is surely a noble endeavour, far removed from the petty, self-seeking anxieties that have led to bloody wars and economic conflicts, even in the new millennium. The choice, then, is clear: to open humanity's route to the stars, or to close that path and attempt to make the best of things here on Earth. The next generation will decide.

Further reading

Werner von Braun's book *The Mars Project* is a classic, and contains much of the physics of Mars expeditions, with technological solutions, appropriate to the 1950s but surprisingly durable. (University of Illinois Press, 1953). A reprint is available (ISBN 0-252-06-227-2).

Robert Zubrin and Richard Wagner's book *The Case for Mars* remains an inspiration to all interested in Mars exploration. (Touchstone, 1997, ISBN 0-684-82757-3.)

Race to Mars, edited by Frank Miles and Nicholas Booth, contains much useful information on 1980s Russian plans for human missions to Mars, but is now out of print. (Macmillan, 1988, ISBN 0-333-46177-0.)

The most comprehensive document is the *Reference Mission of the Mars Exploration Study Team* (NASA, SP6107). This is available direct from NASA, but can be downloaded (280 pages) from the NASA Johnson Spaceflight Center web site. The 'Addendum', *Mars Reference Mission 3.0*, is also available from the same site.

Nuclear propulsion and electric propulsion are dealt with in later editions of George P. Sutton's classic *Rocket Propulsion Elements* (John Wiley & Sons), and in the second edition of my own book *Rocket and Spacecraft Propulsion* (SpringerPraxis, 2004). More information on the 1960s NERVA programme and its derivatives can be found on NASA web sites. Papers by Dr Stanley K. Borowski, of the NASA Glen Research Center, describe theoretical schemes for nuclear missions to Mars, although these are accessible only through the NASA GRC web site.

The web site of the Mars Society contains much useful information on human exploration of Mars, as does the 'marsscheme' site of the students of Caltech. The Jet Propulsion Laboratory (JPL) and Johnson Spaceflight Center (JSC) sites reveal useful information and images when searched for under 'human spaceflight' or 'human exploration'.

Index